Time series techniques for economists

Time series techniques for economists

TERENCE C. MILLS
Professor of Economics, University of Hull

Published by the Press Syndicate of the University of Cambridge
The Pitt Building, Trumpington Street, Cambridge CB2 1RP
40 West 20th Street, New York, NY 10011–4211, USA
10 Stamford Road, Oakleigh, Melbourne 3166, Australia

First published 1990
First paperback edition 1991, 1992, 1994, 1996

Printed in Great Britain at the University Press, Cambridge

British Library cataloguing in publication data

Mills, Terence C.
Time series techniques for economists
1. Economics. Time series. Analysis.
Econometric models
I. Title
330′.01′51955

Library of Congress cataloguing in publication data

Mills, Terence C.
Time series techniques for economists/Terence C. Mills.
p. cm.
Bibliography.
1. Economics, Mathematical. 2. Time-series analysis. I. Title.
HB135.M54 1990
330.028–dc20 89–7187 CIP

ISBN 0 521 34339 9 hardback
ISBN 0 521 40574 2 paperback

UP

To Thea and Alexa

Contents

Preface

Although I specialised in econometrics as a student at the universities of Essex and Warwick in the early 1970s, and was appointed as a lecturer in econometrics at the University of Leeds in 1976, I first became acquainted with the theory and techniques of time series analysis while completing my doctoral thesis in the late 1970s. I suspect that this relatively late exposure to such an important area of statistics was a common occurrence amongst many econometricians of my generation.

My time series education, both in terms of theory and practice, was significantly accelerated by the time I spent in the Economics Division of the Bank of England during the period 1980–2, where I was able to benefit from the expert guidance and tuition of Peter Burman.

Convinced of the importance of time series techniques in applied econometric analysis, I returned to Leeds to introduce a significant time series component into the Applied Econometrics course in the B.A. degree in Economics and Econometrics then available but, alas, no longer offered. At the same time, a course in Time Series Analysis was also developed for the M.Sc. in Economic Forecasting, which has also fallen by the wayside in recent years. My thanks to the students of those courses, and also those of the M.Sc. in Statistics, who, unwittingly, suffered early drafts of what has turned out to be the basic material of many of the chapters of this book.

Notwithstanding these teaching ventures, and the vast intellectual debt I obviously owe to such pioneers as George Box, Gwilym Jenkins, Clive Granger and George Tiao, the major impetus to my interest and enthusiasm for time series techniques in economics has been the collaboration over the years with a number of colleagues whom I regard not just as co-research workers but as close friends: Forrest Capie, Nick Crafts, Steve Leybourne, Mike Stephenson, and Geoffrey Wood. My thanks go to all of them for making joint research such an enjoyable and fruitful experience.

At this point I must also pay tribute to the various members over the years of the 'Fenton Research Centre'; in addition to Nick, Steve and Mike mentioned above, these must include Garry Phillips, Brendan McCabe, Andy Tremayne, Karen Bevins, John and Jane Binner, Mike Hudson, and Louise Webster. Although they are now widely scattered, my thanks go to them all for helping to provide an island of sanity in an environment that was often hostile to serious economic research.

More specific thanks go to Pat Hatton for her expert typing and to Patrick McCartan, Susan Beer and Carmen Mongillo at Cambridge University Press for their input into this finished product. Finally, but of course they should be at the head of any list of acknowledgements, my thanks and love go to my family, Thea and Alexa, to whom this book is dedicated.

1 Introduction

The empirical analysis of time series has a long tradition in economics, the historical development of which has been surveyed by both Spanos (1986, chapter 1) and Nerlove et al. (1979, chapter 1). While the basis of time series analysis was laid in the inter-war period by Yule (1927), Slutsky (1937) and Wold (1938), the main thrust of econometric research was aimed at integrating the 'Fisher paradigm' – the amalgamation of descriptive statistics and the calculus of probability within the framework of the linear regression model – into econometrics (Koopmans (1937), Haavelmo (1944)). This was taken an important stage further by Haavelmo's (1943) formulation of the simultaneous equation system, the statistical analysis of which provided the research agenda for the Cowles Foundation Group. This distinguished collection of statisticians and econometricians introduced and developed new techniques of estimation and hypothesis testing, based on the concept of maximum likelihood, within the framework of the simultaneous equation model. Their results, collected in Koopmans (1950) and Hood and Koopmans (1953), provided the main research program in econometrics for the next 25 years, which was thus dominated by the linear regression model and its associated misspecification analysis, and the simultaneous equations model and its identification and estimation.

Research in time series analysis was primarily concerned, on the other hand, with developing techniques for modelling a single series. Economists were particularly concerned with the decomposition of a series into a set of unobserved components, traditionally taken to be the trend, cycle, seasonal and irregular, these being associated with the ideas of secular evolution (or long swings), the concept of the business cycle, seasonal variation, and transitory influences, respectively. Operations researchers, on the other hand, were primarily concerned with forecasting, developing techniques which were computationally feasible and based upon local trends and levels (Holt et al. (1960), Winters (1960), Brown (1963)). These 'exponential smoothing' techniques were later shown to have unobserved component representations (see, for example, Harrison (1967)), hence providing a more unified framework for analysis.

Thus, by the late 1960s the research of econometricians and time series analysts seemed to be so far apart that no synthesis of the two frameworks would have appeared possible. This dichotomy was, though, to change dramatically in the years after the publication in 1970 of the original edition of Box and Jenkins' (1976) famous book in which a feasible model building procedure for the general class of autoregressive-integrated-moving average processes was developed.

The practical manifestation of the Cowles Foundation's research program was the development of large-scale macroeconometric models designed for forecasting and policy simulation; see, for example, Klein and Burmeister (1976) for a survey and evaluations of a number of such models. The appropriateness of this form of model began to be questioned when they were found to be unable to compete with Box–Jenkins models in terms of forecasting comparisons; see Cooper (1972), Naylor et al. (1972) and Granger and Newbold (1975). Macroeconometric model builders responded by arguing that univariate time series models were irrelevant for policy considerations: that macroeconometric models could also be was forcefully argued by Lucas (1976) and Sims (1980).

The importance of a synthesis between traditional econometric and time series techniques was brought into even sharper focus by Granger and Newbold's (1974) illuminating analysis of the spurious regression problem, in which it was shown that regressing independently generated random walks would lead to a very high probability of rejecting the correct hypothesis of no relationship between the two series. The groundwork for such a synthesis was laid by Zellner and Palm (1974) and further developed in Wallis (1977) and Nerlove et al. (1979). A great deal of research in the last decade has thus been directed at producing a methodology for the econometric modelling of economic time series. One strand of this literature is probably best encapsulated in the papers by Hendry and Richard (1982, 1983), with a complete textbook treatment of this methodology being provided by Spanos (1986). Other viewpoints are provided by Sims (1980) and Leamer (1983), and current accounts of these competing views are to be found in Hendry (1987), Leamer (1987) and Sims (1987), with a review being provided by Pagan (1987). Although many arguments remain to be resolved, this general methodology does provide an important synthesis of traditional econometric and time series techniques.

Nevertheless, even given the undoubted success of this new synthesis, economists face many problems in the analysis of time series which do not fall comfortably, if at all, into the (single or simultaneous equation) regression framework, and this necessarily leads to the examination of a much wider class of models and techniques. Serious difficulties may be encountered in this enterprise for, since time series occur naturally in a wide range of disciplines, so the literature on time series techniques appears in a wide range of journals. Many of these may be inaccessible to economists, both physically and in terms of the notation and terminology used (this is particularly so for the time series literature in the electrical engineering field, as any economist who has attempted to read articles in control theory journals will surely testify!). Furthermore, the increase in the sheer volume of literature over the last decade has been quite phenomenal; as an indication of this, of the more than 400 citations referenced in this book, 62% have been published since 1980. Many economists may therefore be simply unaware of some of the new techniques and insights that may be of potential importance to the analysis of economic time series. The primary aim of this book is, therefore, to bring together those techniques that we feel are of importance to analysts of economic time series, to indicate by discussion and examples why they are of importance, and to provide a consistent notation and terminology by which they may be developed.

It may be important to state at this juncture what the book is *not* about. It is not an econometrics text, and therefore it does not cover regression analysis and its associated simultaneous extensions in any formal way. For such treatments see, for example, Judge et al. (1985) and Spanos (1986). Neither is it a textbook on forecasting techniques, nor does it discuss in great detail the 'Box–Jenkins' approach to analysing time series: for treatments of the former, see Abraham and Ledolter (1983) and Makridakis et al. (1983), for the latter, see Vandaele (1983) and Pankratz (1983). It is unashamedly non-rigorous; formal treatments of the theory of time series are provided by Anderson (1971) and Fuller (1976). Finally, it does not discuss spectral analysis. Our experience suggests that the frequency domain is unfamiliar territory for the majority of economists and analysis is therefore carried out entirely in the more natural time domain.

The book is structured in the following way. After this introductory chapter, the rest of the material is divided into four parts. The first part, comprising three chapters, is devoted to an area of analysis covered by few textbooks, that of the exploratory analysis of time series, a topic inspired by the work of Tukey (1977) (sometimes also referred to as initial data analysis, see Chatfield (1985)). Part II is devoted to the analysis of univariate time series and, as befits the most completely developed area of the subject, is the longest part of the book, containing seven chapters.

Multivariate time series are analysed in Part III, in which there are three chapters, while the final part, just two chapters, discusses non-linear time series models. Each part begins with an introductory discussion setting out the subsequent material.

Throughout the chapters, numerous examples are provided to illustrate the techniques discussed. Certain time series are used in a sequence of examples, thus showing how applied analysis progresses and evolves as the theoretical base of material becomes richer. Other examples draw heavily on the author's own research, both published and unpublished.

One final note of caution. Anyone familiar with the time series literature will be aware of, and perhaps irritated by, the vast number of acronyms this literature has spawned and, indeed, they have inspired a journal article (Granger (1982)). Irritating or not, such acronyms are invaluable to the textbook writer and we make no apologies for making liberal use of them throughout the book!

Part I

Exploratory analysis of economic time series

As a consequence of the highly influential book by Tukey (1977), exploratory data analysis, often referred to by its acronym EDA, has gained enormously in importance in all areas of applied statistical research over the last decade. Time series analysis is no exception and, indeed, time series data lend themselves extremely well to such exploratory analysis. It appears to be the case, however, that texts on time series and forecasting have not in the past dwelt overmuch on this aspect of data analysis, preferring to emphasise the more formal aspects of modelling techniques, although a notable exception to this is Levenbach and Cleary (1981).

This part of the book is explicitly concerned with three areas of exploratory time series analysis: (i) graphical display, both of one variable and scatter plots of the relationships between two or more variables (Chapter 2), (ii) summarising time series in terms of frequency distributions and measures of location and dispersion (Chapter 3), and (iii) the transformation and smoothing of time series to provide easier forms of analysis (Chapter 4).

Although introduced as exploratory data analysis, many of the techniques discussed here will also be used, and may recur in different guises, throughout the remaining parts of the book dealing with more formal modelling techniques. Indeed, it cannot be stressed too highly the importance of good preliminary data analysis: it will lead to better models, more efficient computing and a greater understanding of the relationships between the data at hand, the underlying economic theory, and the modelling techniques employed.

2 The graphical display of time series

2.1 Introduction

It is often the case that graphical data displays are easier to interpret than tabular representations of the same data. It is also the case that such graphical displays can be unhelpful, exaggerated and even deliberately deceiving. As Tufte (1983) says, in a highly entertaining and provocative book on the visual display of data,

> Much of...statistical graphics has been preoccupied with the question of how some amateurish chart might fool a naive viewer...At the core of the preoccupation with deceptive graphics was the assumption that data graphics were mainly devices for showing the obvious to the ignorant. It is hard to imagine any doctrine more likely to stifle intellectual progress in a field.
>
> (Tufte (1983, page 53)).

The revolution in statistical graphics came in the late 1960s, primarily through the work of John Tukey (see Tukey (1972) and Tukey and Wilk (1970)) and has since led to such masterpieces of graphical excellence as those in Tukey (1977), Chambers et al. (1983), and Cleveland (1985). Cleveland (1987) provides a survey of recent research in statistical graphics.

Graphical displays of time series data are the most frequently used form of graphic design and, as such, are subject to misuse and abuse just as any other form of visual display. They can also be one of the most powerful techniques of exploratory analysis available and, in the hands of a master such as Tukey, become a means of conveying complex and changing relationships in a simple yet flexible manner. A great deal of research has been carried out recently on the theory of graphical perception, leading to a much deeper understanding of the scientific foundations underlying graph construction: see, for example, the survey by Cleveland (1985, chapter 4) and the recent research by Cleveland and McGill (1987).

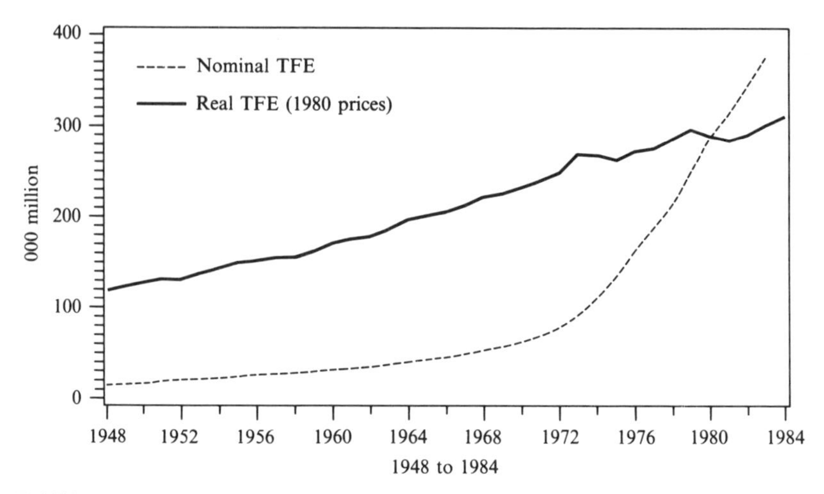

Exhibit 2.1 UK total final expenditure

Hopefully, such research findings will quickly find their way into the important area of statistical, and hence time series, graphics.

2.2 Time series plots

A time series plot is simply a graph in which the data values are arranged sequentially in time. Usually, time series are generated by taking observations at equally spaced time intervals, and hence these values must be plotted at equally spaced intervals along the time axis, conventionally taken to be the horizontal. Although the construction of a time series plot would appear to be a trivial exercise, certain commonsense guidelines concerning labelling of axes and variables, choice of scale, use of grid lines, and so on, should be adhered to. Schmid (1983, pages 17–19) and Cleveland (1985, chapter 2, particularly pages 100–1) provide convenient lists of 'good practices' to follow when designing time series plots.

Time series of economic data display many different and important characteristics and a plot is an effective way of quickly perceiving the evolution of a single, or a group of, time series: for an interesting case study, see Chatfield and Schimek (1987). Exhibit 2.1, for example, plots nominal and real UK Total Final Expenditure (TFE) annually for the period 1948 to 1984. It is immediately apparent that nominal TFE displays a very smooth upward trend which has steepened considerably during the observation period, a consequence, of course, of it being denominated in current prices. The real TFE series, deflated using 1980

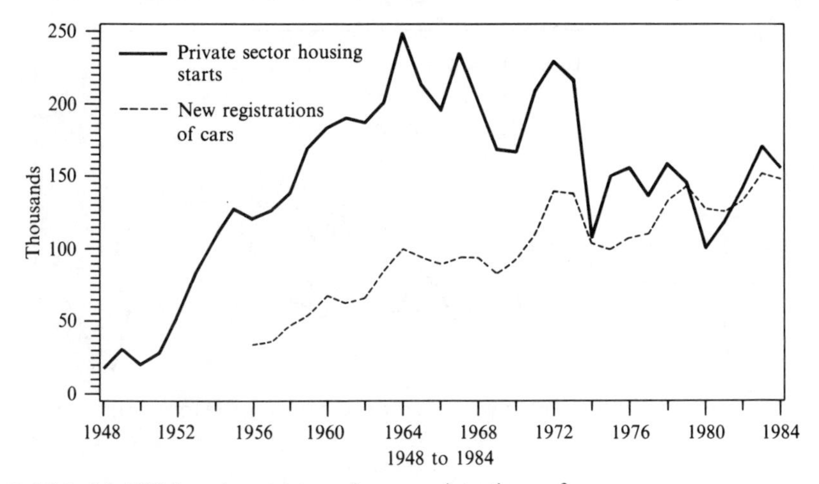

Exhibit 2.2 UK housing starts and new registrations of cars

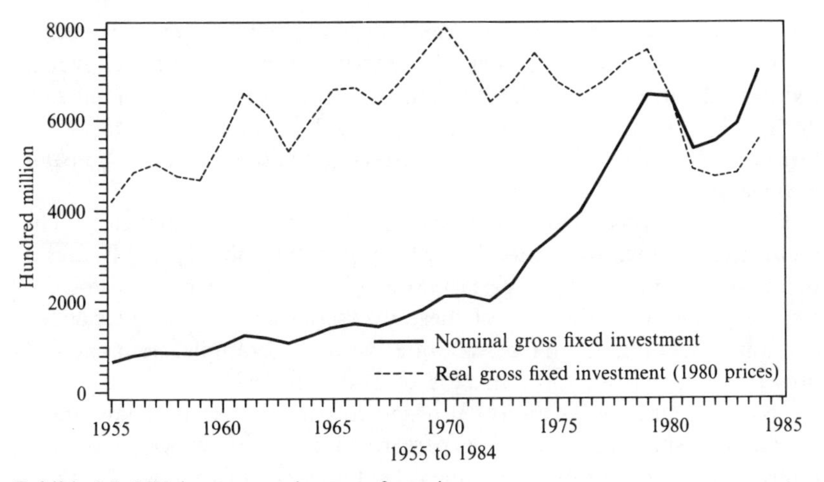

Exhibit 2.3 UK investment in manufacturing

prices, displays a less pronounced trend, and is also rather less smooth, showing a number of cyclical movements.

Not all annual economic time series evolve so smoothly. The new registration of cars in the UK, shown in Exhibit 2.2, has a general upward trend but is dominated by short-term erratic fluctuations around this underlying movement. The UK private sector housing starts series, on the other hand, which is also shown in Exhibit 2.2, displays no general trend

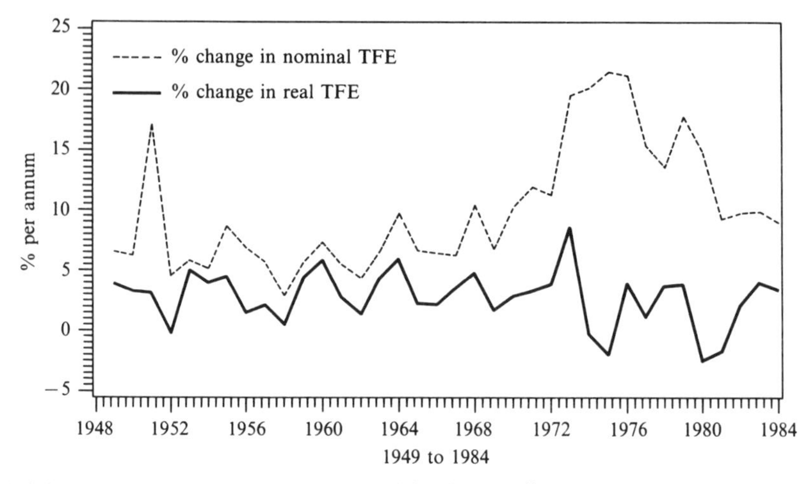

Exhibit 2.4 Annual growth in UK total final expenditure

but fluctuates widely around a mean level that itself changes fairly often. Nominal gross fixed investment in manufacturing industry, shown in Exhibit 2.3, has both a cyclical pattern and a trend that alters dramatically during the 1970s. Also shown in Exhibit 2.3 is the real version of this series, which may be best characterised as fluctuating around a constant mean level.

Transformations of time series can alter their plots dramatically. This is effectively shown by Exhibit 2.4, which plots the annual growth rates of the nominal and real TFE series whose levels were shown in Exhibit 2.1. The very smooth evolutions of these two series are, in fact, revealed to contain widely fluctuating growth rates with, for example, nominal TFE growth ranging from 2.8% in 1958 to 21.3% in 1975.

Economic time series are often observed over shorter intervals than a year and most applied research is performed using data collected at either a quarterly or a monthly time interval. Time series of this type display a further characteristic, other than trend and cyclical movements, which is not present in annual data, that of seasonality. Exhibit 2.5 displays quarterly time series from 1955 to 1984 of both the value and volume of UK retail sales. Both series are dominated by seasonal effects, a consequence of the annual Christmas 'spending spree', but whereas the volume series displays a fairly stable seasonal pattern superimposed on a slowly increasing trend, the value series has a seasonal pattern whose amplitude increases substantially as the trend steepens. When monthly time series are considered, it is also possible to observe 'double-seasonal' patterns. The bank lending to the UK private sector series shown in

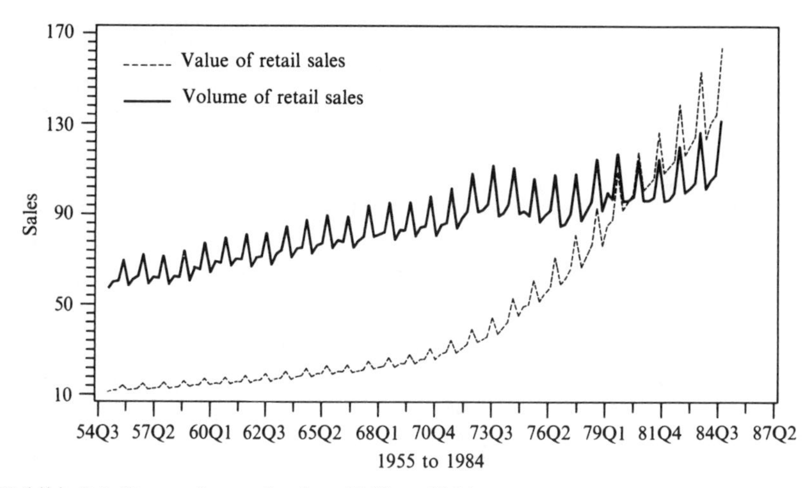

Exhibit 2.5 Quarterly retail sales: 1955 to 1984

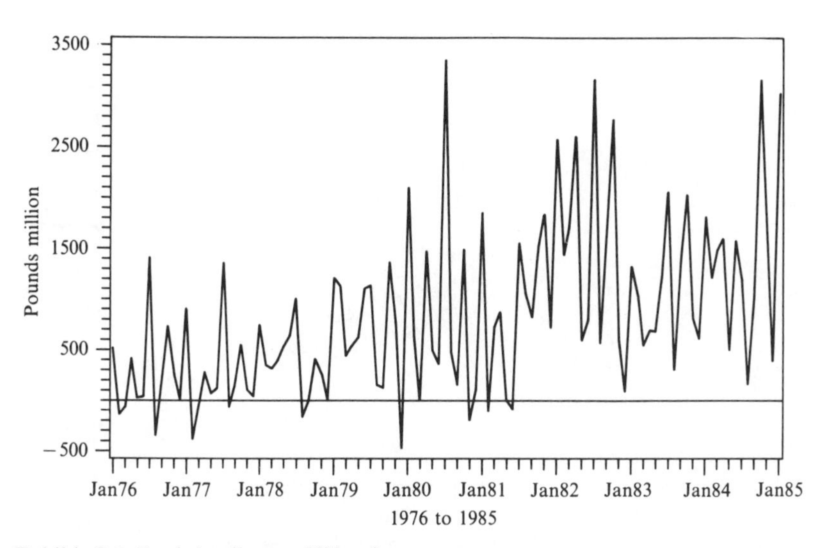

Exhibit 2.6 Bank lending to UK private sector

Exhibit 2.6 demonstrates this, with a quarterly (three-month) seasonal pattern being superimposed upon a conventional annual (twelve-month) pattern.

Not all series observed at intervals shorter than one year will display a seasonal pattern. Interest rates have typically shown little or no seasonal fluctuation, as is demonstrated by the plot of the monthly UK Treasury

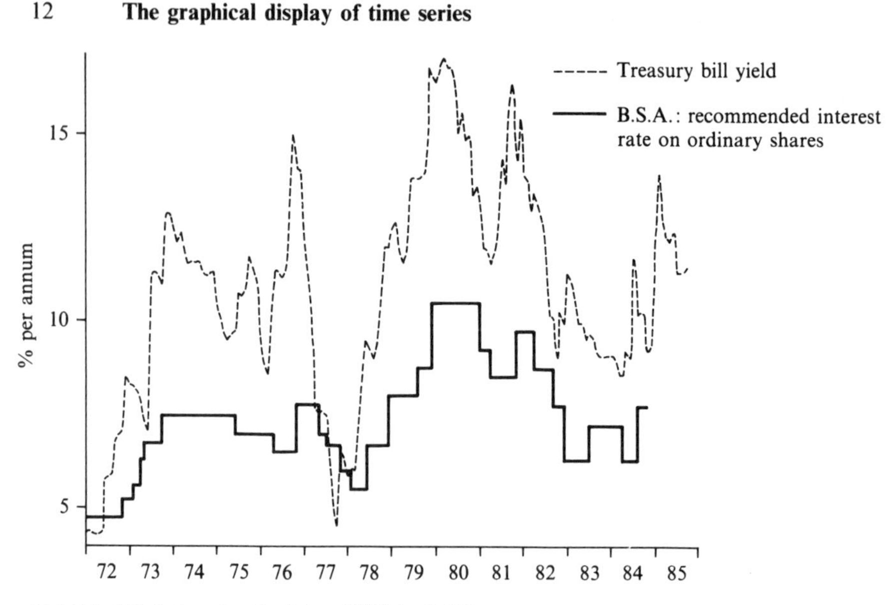

Exhibit 2.7 Interest rate data: 1972 to 1985

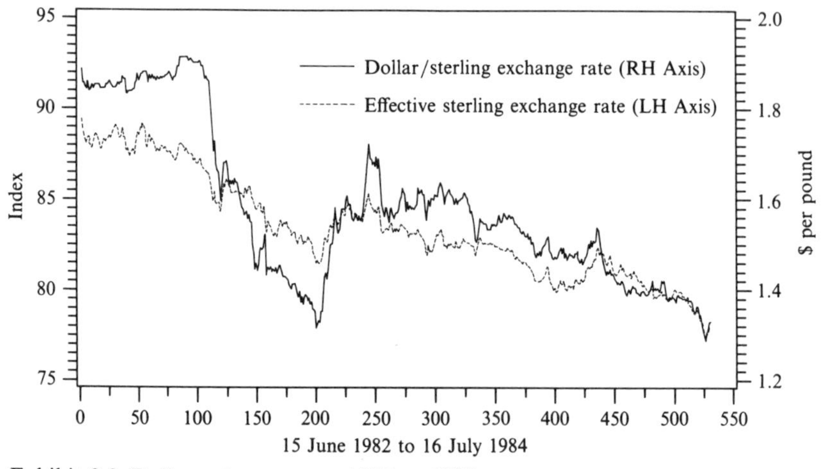

Exhibit 2.8 Daily exchange rates 1982 to 1984

bill yield from 1972 to 1985 given in Exhibit 2.7. This figure also shows another type of time series, one in which institutional arrangements only allow the series to move in discontinuous jumps. The series used to illustrate this feature is the Building Societies Association recommended interest rate on ordinary shares, which ceased to exist in November 1984 when the traditional building society cartel broke up.

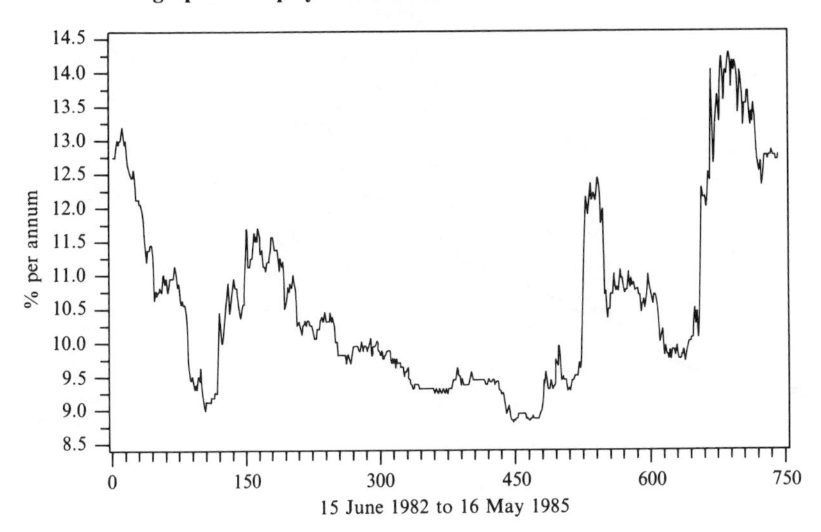

Exhibit 2.9 3-month interbank rate

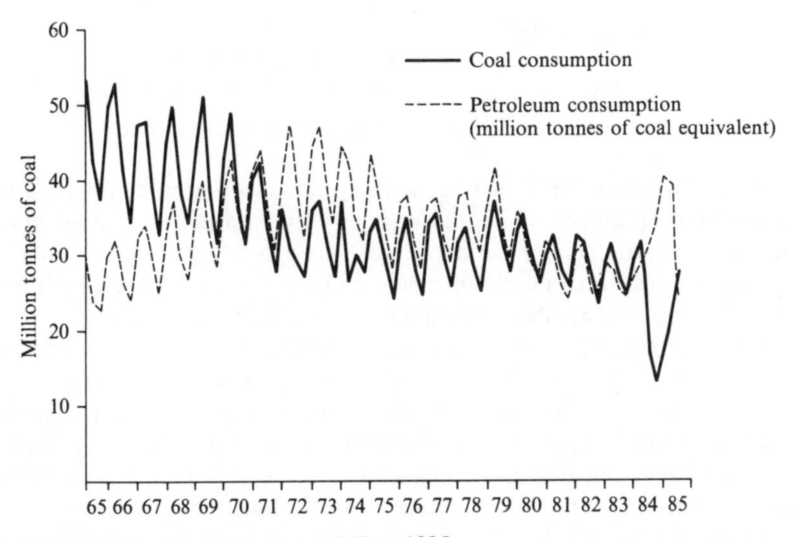

Exhibit 2.10 Fuel consumption: 1965 to 1985

Periodic patterns may occur even in daily data, although from the time series plots of the $/£ and effective £ exchange rates shown in Exhibit 2.8, any such patterns, say of a weekly periodicity, are very difficult to discern even if they actually exist. Both these time series seem to show downward trends, but from our knowledge of the foreign exchange market it is highly

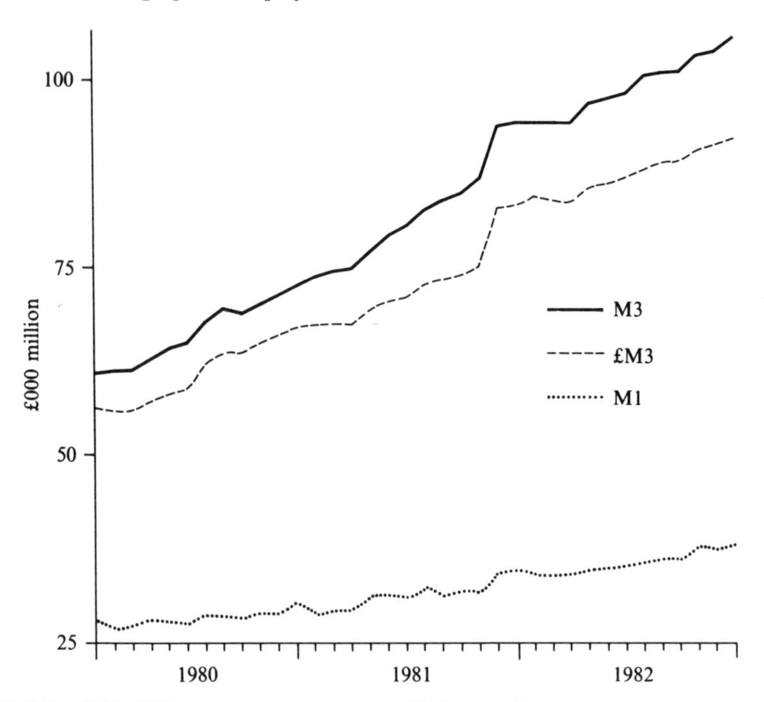

Exhibit 2.11 UK monetary aggregates: 1980 to 1982

unlikely that such trends can be anything but short term, as opposed to those found in other annual and quarterly time series and, indeed, they may even be artefacts of the data itself. A further illustration of a daily time series is the three-month interbank rate shown in Exhibit 2.9, and this series is used in a number of examples throughout the book.

Plotting time series quickly reveals the effects of external influences on data. Exhibit 2.10 shows time series for quarterly UK coal and petroleum consumption. The effect of the miners' strike in 1984 in both decreasing coal and increasing petroleum consumption is quite dramatic, but careful examination of this plot also reveals the effect of an earlier miners' strike on coal consumption in 1972. Alternatively, a time series may undergo a once-and-for-all shift in level as a result, for example, of a definitional change. Exhibit 2.11 shows the levels of three UK monetary aggregates, M1, £M3 and M3, from 1980 to 1982. As a consequence of the widening of the Banking Sector in November 1981, the levels of all three series, to differing extents, were increased by the inclusion of more reporting institutions.

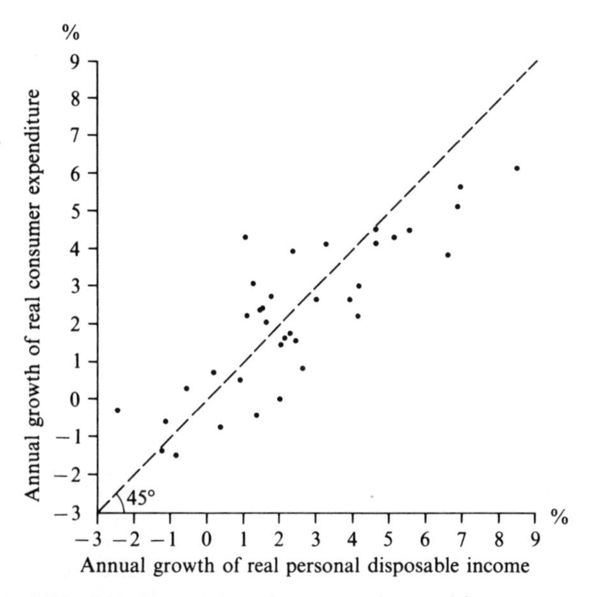

Exhibit 2.12 Scatterplot of consumption and income growth

2.3 Scatterplots

Bivariate scatterplots are probably the simplest, most widely used and easily understood graphical method for displaying the relationship between two variables and, of course, they can be used to study the association between two time series. A typical scatterplot is shown in Exhibit 2.12, depicting the relationship between annual growth rates of real UK consumer expenditure and real UK personal disposable income for the period 1949 to 1984. A 45° line is superimposed to highlight the result that fitting a linear relationship to the data would yield a slope flatter than this, thus suggesting that the elasticity of consumption with respect to income is less than unity.

It is perhaps difficult to see how the traditional scatterplot may be improved upon, but consider Exhibit 2.13, which reproduces the consumption-income data in the form of a *dot-dash-plot* as developed by Tufte (1983, pages 130–5). In this graphical display, two extensions to the frame of the scatterplot have been made. The tails of the frame have been erased so that the axes only extend to the maximum and minimum of the two plotted variables and thus also show their ranges. 'Dashes' have been introduced onto the axes to show the marginal distributions of the two variables. The dot-dash-plot now shows both the bivariate and marginal distributions and thus conveys much more information than the

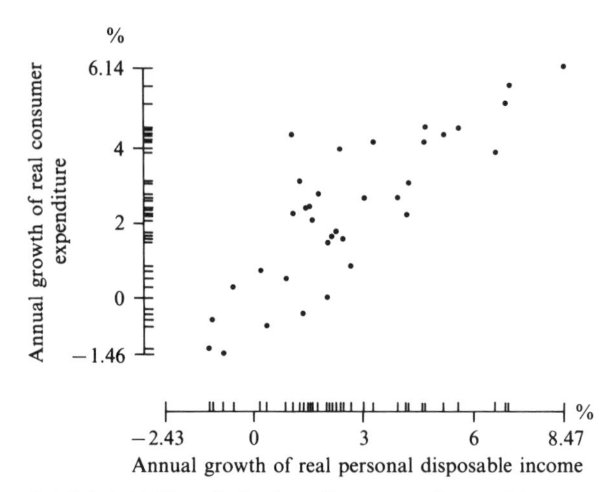

Exhibit 2.13 Dot-dash-plot of consumption and income growth

traditional scatterplot. An alternative to using dashes on the axes to represent the marginal distributions is to display these distributions schematically by drawing 'boxplots' in the scatterplot margins. Boxplots are introduced in Section 3.4 and Chambers et al. (1983, chapter 4) present examples of this technique.

When dealing with economic time series, we often wish to consider the relationships between a number of variables and much progress has been made recently in graphically displaying data in three or more dimensions (see, in particular, Tukey and Tukey (1981)). A useful device when dealing with three or more time series is to construct a *scatterplot matrix*, which is an array of pairwise scatterplots (see Cleveland (1985, pages 210–13)). A useful feature of such a display is that the scatterplots are arranged so that adjacent plots share a common axis, thus eliminating some scale and legend clutter and enabling the viewer to scan across or down the display, matching up points that correspond to the same observation in different plots. Exhibit 2.14 shows a five-dimensional scatterplot matrix using the following important UK macroeconomic time series for the years 1970 to 1984: real consumption growth (C), real income growth (Y), the rate of inflation (P), the unemployment rate (U) and the savings ratio (S). As well as the positive consumption-income relationship already observed, Exhibit 2.14 reveals a negative relationship between inflation and each of income growth, consumption growth and unemployment. There would appear to be a positive relationship between inflation and the savings ratio, but only weak, if any, associations between all other pairs of series.

Many other techniques are being developed for graphically displaying

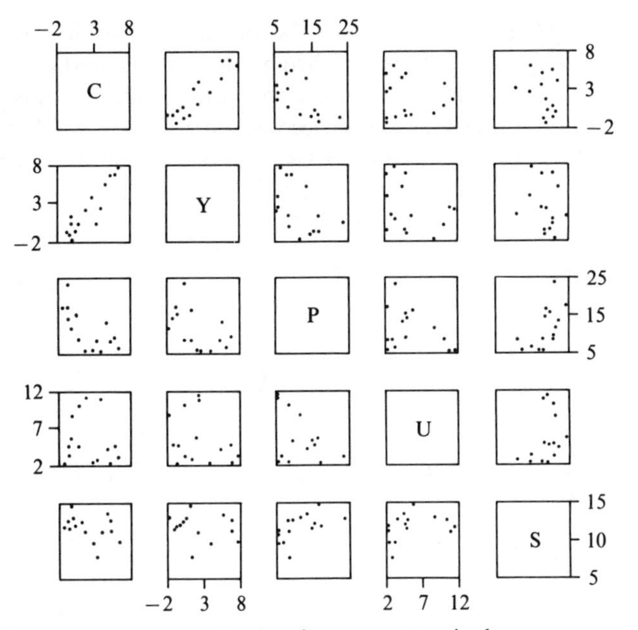

Exhibit 2.14 Scatterplot matrix of macroeconomic data

multivariate data and the interested reader is urged to consult, for example, Chambers et al. (1983, especially chapters 4 and 5) and Cleveland (1985, chapter 3) for excellent discussions of such methods. We should also note that Tufte (1983, page 135) introduces a display similar to, but less systematic than, the scatterplot matrix, called a *rugplot*.

2.4 Computing software

Most statistical packages include routines for plotting a series against time, or for plotting one series against another. For example, SAS has routines PROC TIMEPLOT and PROC PLOT (see SAS (1985a, chapters 52 and 42 respectively)), SPSS^X has the SCATTERGRAM routine (SPSS^X (1983, chapter 30)), while MINITAB has its sequence of PLOT and TSPLOT commands (Ryan et al. (1985, chapter 3)). A drawback with most software packages is that lineprinter output is restricted to plotting points which are fixed by choosing a particular character position to represent the horizontal coordinate and choosing a print line on the page to represent the vertical coordinate. This makes many of the plots produced by the packages rather 'granular' in form, and it is often better to produce such plots by hand using ruled graph paper, perhaps tracing the plots when completed onto transparencies. (Tukey (1977, pages 42–3), discusses basic ideas of plotting which the present author has found invaluable.) For a really professional finished product, specialised graphics packages are now available in conjunction with standard statistical packages: for example, the GPLOT procedure on SAS/GRAPH (SAS (1985d, chapter 15)), and the LINECHART routine on SPSS^X (SPSS^X (1983, chapter 24)).

3 Summarising time series

3.1 Introduction

The previous chapter has emphasised that graphical displays of time series data can reveal a great deal about the patterns existing both within a single series and between a set of series. This chapter continues the theme of exploratory analysis by concentrating attention on the distributional aspects of time series data and on methods of summarising and displaying such features.

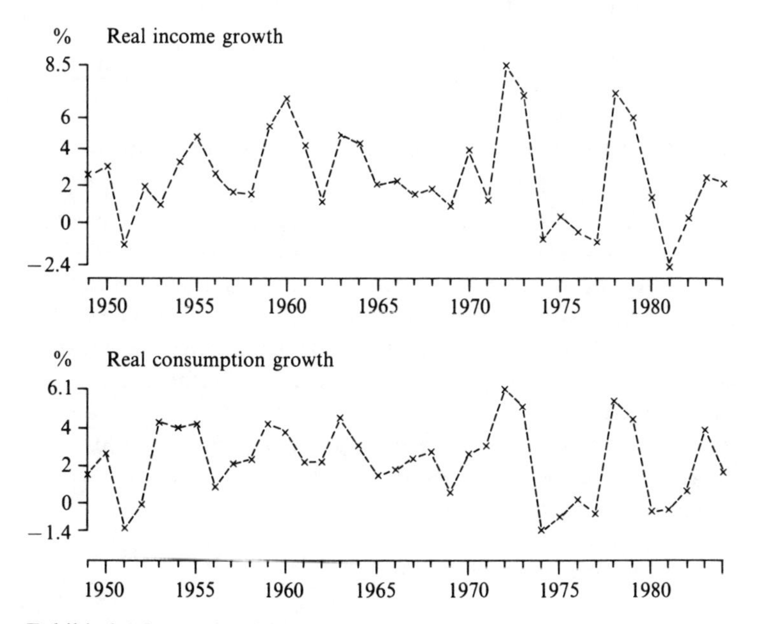

Exhibit 3.1 Income and consumption growth

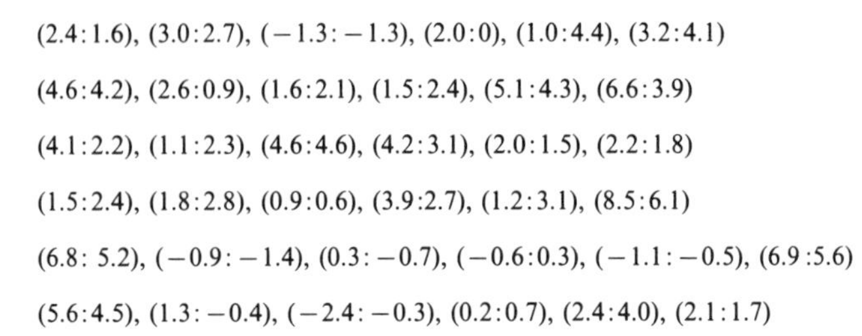

Raw data: (Income: consumption)

(2.4:1.6), (3.0:2.7), (−1.3: −1.3), (2.0:0), (1.0:4.4), (3.2:4.1)

(4.6:4.2), (2.6:0.9), (1.6:2.1), (1.5:2.4), (5.1:4.3), (6.6:3.9)

(4.1:2.2), (1.1:2.3), (4.6:4.6), (4.2:3.1), (2.0:1.5), (2.2:1.8)

(1.5:2.4), (1.8:2.8), (0.9:0.6), (3.9:2.7), (1.2:3.1), (8.5:6.1)

(6.8: 5.2), (−0.9: −1.4), (0.3: −0.7), (−0.6:0.3), (−1.1: −0.5), (6.9:5.6)

(5.6:4.5), (1.3: −0.4), (−2.4: −0.3), (0.2:0.7), (2.4:4.0), (2.1:1.7)

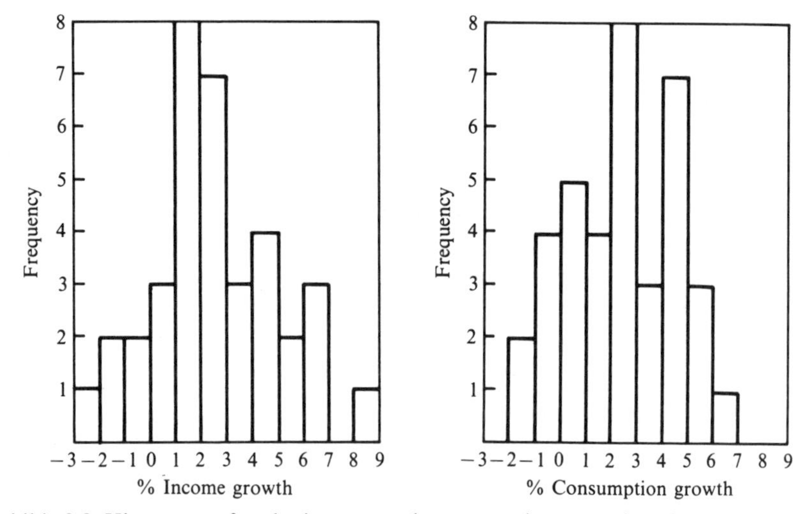

Exhibit 3.2 Histograms for the income and consumption growth series

When details of the distributional shape of a time series are needed, such as when stock market price changes are being analysed, a plot of the data will typically fail to convey the required information. Exhibit 3.1 shows plots of real UK income and consumption growth from 1948 to 1984 and differs from the plots of Chapter 2 by amending the conventional plot-frame to show the ranges of the series and by altering the plotting convention to focus more attention on the individual observations. Nonetheless, the plot still fails to provide sufficient detail to allow any serious judgements to be made about distributional shape and hence alternative approaches must be considered.

3.2 Stem-and-leaf displays

The usual 'descriptive statistics' approach to assessing distributional shape is to represent the empirical frequency distribution in the form of a histogram or polygon. Exhibit 3.2 shows histograms, constructed on standard principles, for the income and consumption growth series introduced earlier. Although histograms are simple to construct and allow the analyst to obtain an overall feel for the shape of the frequency distribution, they do have one major disadvantage: by moving to grouped data, which is necessary to observe general shape, the specific information contained in the original observations is lost.

Stem-and-leaf displays have been developed to allow the analyst to ascertain general shape without losing the original data. Many modifications of stem-and-leaf displays have been developed to handle different data sets and the reader is encouraged to refer to Tukey (1977, chapter 1) and Velleman and Hoaglin (1981, chapter 1) for many examples and extensions of the basic displays used in Exhibit 3.3(a) to illustrate these ideas for the consumption and income growth data. One such extension, useful when comparing two distributions, is the *back-to-back* stem-and-leaf shown in Exhibit 3.3(b). The 'stem' in such displays is analogous to the class interval in a classical histogram, whilst the 'leaves' allow the analyst to obtain the same distributional shape as the histogram but to retain the original data in an easy to interpret format. (For example, the first row in the display for income growth shows the value −2.4 %.) Further information can be conveyed by, for example, 'flagging' outlying data points.

It can be seen from Exhibit 3.3 that income growth displays greater variability than consumption growth and that there is a tendency for both series to be skewed, income to the right and consumption to the left. More detailed summarisation of the distributions, though, requires the computation and interpretation of more informative statistics measuring locational and dispersional characteristics and it is to these that we now turn.

3.3 Letter-value displays

When summarising frequency distributions it has become standard practice to calculate measures based on the moments of the data. Typically, the first and second moments are used to calculate the mean and variance, but sometimes higher moments are employed to obtain measures such as skewness and kurtosis. While these measures are often informative, particularly for 'well behaved' data sets, it is well known that

Exhibit 3.3 *Stem-and-leaf displays for the income and consumption growth series*

(a) '*Side-by-side*' *display*

Income (units: % point)			(#)	Consumption (units: % point)			(#)
−2	4	(1981)	1	−2			0
−1	31		2	−1	43	(1974, 1951)	2
−0	96		2	−0	7543		4
0	239		3	0	03679		5
1	01235568		8	1	5678		4
2	0012446		7	2	12344778		8
3	029		3	3	119		3
4	1266		4	4	0123456		7
5	16		2	5	26		2
6	689		3	6	1	(1972)	1
7			0	7			0
8	5	(1972)	1	8			0
			Total: 36				Total: 36

(b) '*Back-to-back*' *display*

(#)	Income (units: % point)			Consumption (units: % point)		(#)
1	(1981)	4	−2			0
2		13	−1	43	(1974, 1951)	2
2		69	−0	7543		4
3		932	0	03679		5
8		86553210	1	5678		4
7		6442100	2	12344778		8
3		920	3	119		3
4		6621	4	0123456		7
2		61	5	26		2
3		986	6	1		1
0			7			0
1	(1972)	5	8			0
Total: 36						Total: 36

since all are functions of the sample mean, the presence of extreme observations, or 'outliers', can have an important influence on the calculated values of these statistics. As an example, the sample mean and variance of the income growth data used in Exhibits 3.1 to 3.3 are 2.47 and 6.31 respectively. When the outlying growth rate of 8.5% for 1972 is omitted from the calculations, the two values are reduced to 2.30 and 5.39. Economic time series often have outlying observations and thus it is important to consider alternative measures of location and dispersion that are more 'robust' to the presence of extreme values.

Such measures can easily be obtained from a stem-and-leaf display. A simple yet robust measure of location is the *median*. To define formally this measure, the concept of the 'depth' of a data value is required. This is just the value's position in an enumeration of values that starts at the nearer end of the ordered set (or batch) of data. Thus, each extreme value has a depth of 1; the second largest and second smallest values have depths of 2; and so on. In a batch of size n, the ith and $(n+1-i)$th values both have depth i.

The median is then defined as the 'deepest' data value. If n is odd, this occurs when $i = n+1-i$ and thus the depth of the median is $(n+1)/2$, which may be abbreviated to

$$d(M) = (n+1)/2.$$

If n is even, there will be two deepest data values, each having depth $(n+1)/2-\frac{1}{2}$. It is then conventional to define the median as the average of these two values. Hence the median, labelled by the letter M, is that value which splits an ordered batch in half.

It would then seem natural to ask about the middle of each of these halves. These mid values are called *hinges*, labelled by the letter H, and are given by the values at depth

$$d(H) = ([d(m)]+1)/2 = \left(\left[\frac{n+1}{2}\right]+1\right)\Big/2,$$

where the [...] symbols are to be read as 'integer part of' and indicate the operation of omitting the fraction. The division by 2 indicates that the two data values surrounding that depth are to be averaged. The hinges are similar to the more familiar *quartiles*, which are defined to be at a depth of $[n/4]$. Hence the hinges will be slightly closer to the median than will be the quartiles. This process of successively halving the intervals between the most recently calculated *letter value* and the extremes can be continued until the depth of the extremes, 1, is reached, obtaining in succession *eighths*, labelled E, *sixteenths*, labelled D, and so on. It is rare, though, for analysis to go further than eighths.

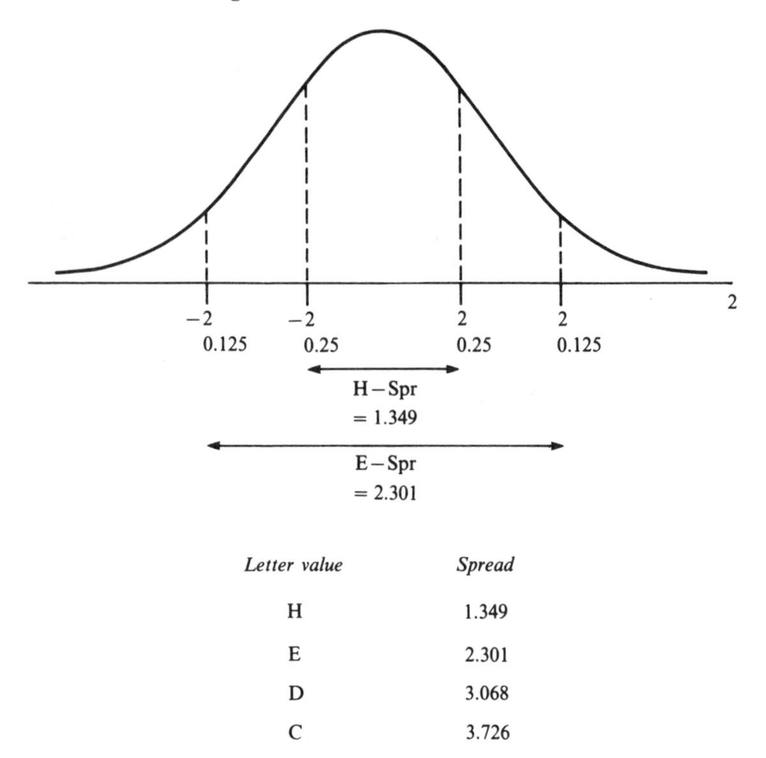

Letter value	*Spread*
H	1.349
E	2.301
D	3.068
C	3.726

Exhibit 3.4 Spreads for the standard normal distribution

3.3.1 *Midsummaries and spreads*

As well as providing positional measures for the data set, the sequence of letter values can also be used to generate further measures that will provide information on symmetry and dispersion. Since letter values come in pairs symmetrically placed at the same depth, it is useful to consider whether their values are also symmetric. The average of each pair of letter values is termed a *midsummary* and hence the average of the two hinges is called the *midhinge* (midH), with further midsummaries being the *mideighth* (midE), the midD, and so on. The *midextreme* is also called the *midrange* and the median, by being in the middle of the batch, is itself already a midsummary.

If all the midsummaries are approximately equal, then the values of the hinges, eighths, and so on, are nearly symmetric about the median. If the midsummaries become progressively larger, the batch is skewed towards high values, i.e. it is skewed to the right, whereas if they decrease steadily the batch is skewed to the left.

Exhibit 3.5 *Letter-value displays for the income and consumption data shown in Exhibit* 3.2

		Lower	Upper	Mid	Spread
(a) *Income*: $n = 36$					
M	18.5	2.05		2.05	
H	9.5	1.05	4.15	2.6	3.1
E	5	−0.6	5.6	2.5	6.2
D	3	−1.1	6.8	2.85	7.9
C	2	−1.3	6.9	2.8	8.2
	1	−2.4	8.5	3.05	10.9
(b) *Consumption*: $n = 36$					
M	18.5	2.35		2.35	
H	9.5	0.65	4.05	2.35	3.4
E	5	−0.4	4.5	2.05	4.9
D	3	−0.7	5.2	2.25	5.9
C	2	−1.3	5.6	2.15	6.9
	1	−1.4	6.1	2.35	7.5

The sequence of letter value pairs can also be used to define *spreads*, the difference between the pair of letter values. Thus the *H-spread* (H-spr for short) is the difference between the hinges and shows the range covered by the middle half of the batch, while the *E-spread* gives the range of the middle three-quarters of the data. The difference between the extremes is simply called the *range*. The greater the variability in the data, the larger will the spreads be. However, if the batch is reasonably symmetric, as revealed by an examination of the sequence of midsummaries, the spreads can be used to compare the shape of the data with a common standard, that of the normal (or Gaussian) distribution. Since the normal distribution is symmetric, a severely skewed batch should not be compared to this standard although, as will be shown in the next chapter, certain 're-expressions' of the data can transform a batch of data to approximate symmetry.

The spreads for the standard normal distribution are easily obtained and are shown in Exhibit 3.4. To obtain spreads for a normal distribution with standard deviation σ, the values in Exhibit 3.4 are just multiplied by σ. A simple way to compare the shape of a batch to the normal distribution is to divide the spread values of the data by the normal spread values. If the data resemble a sample from a normal distribution, then all quotients will be roughly equal, whereas if there is a clear trend in the quotients an indication is obtained of how the data depart from the normal shape. If the quotients increase, the tails of the batch are heavier than normal tails, while if the quotients decrease, the tails of the batch are lighter.

The sequence of letter values, midsummaries and spreads can be brought together and presented as a *letter-value display*. Such displays for the income and consumption data are shown as Exhibit 3.5. The first two columns of the display show the letter value labels and associated depths. The next two columns, labelled 'lower' and 'upper', give the two letter values for each label and the 'mid' and 'spread' columns contain the midsummaries and spreads. Note that the median has only one value, which straddles the 'lower' and 'upper' columns, and therefore no spread.

It is readily seen that median income growth is 2.05% and median consumption growth is 2.35%, that income growth ranges from −2.4% to 8.5% and consumption growth ranges from −1.4% to 6.1%, and that the middle halves of the 36 years of growth rates run from 1.05% to 4.15% and 0.65% to 4.05% respectively. The sequence of midsummaries for income growth shows a pronounced increase, from which it is concluded that this series is positively skewed, whereas the midsummaries for consumption growth remain fairly stable, suggesting that the series is roughly symmetric, with perhaps a slight skew to the left. On comparing the consumption growth spreads to the normal spreads given in Exhibit 3.4, the sequence of quotients, calculated as 2.52, 2.13, 1.92 and 1.85, is seen to decrease, thus suggesting that the tails of the consumption growth distribution are lighter than those of a normal distribution.

3.4 Boxplots

It is often desirable to present a letter-value display graphically. The most useful letter values are usually the median, the hinges and the extremes. This set is often called the *5-number-summary* and provides the basis for a graphical display called a *boxplot*. In constructing a boxplot, attention must be paid to outlying values. Although the letter-value display provides the two extreme values, it does not explicitly differentiate between extremes that are, nevertheless, close to the main body of the data and extreme values, and perhaps values next to them, that are so low or so high that they seem to stand apart from the rest of the batch. It is very important to identify such outliers; they may have been caused by some kind of reporting or measuring error, in which case they require correcting or, if this is not possible, excluding from further analysis, or they may genuinely reflect unusual circumstances and can therefore convey valuable information.

Rules of thumb have been developed for dealing with outliers that only

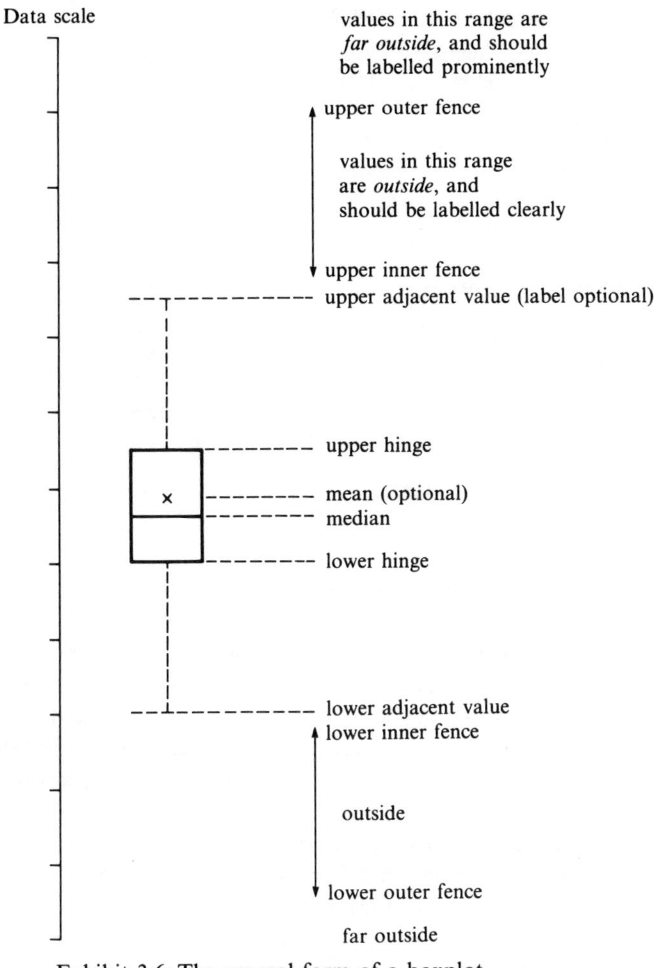

Exhibit 3.6 The general form of a boxplot

require knowledge of the hinges and the associated H-spread. *Inner fences* are defined as

$$\text{lower } H-(1.5\times H\text{-spread})$$

$$\text{upper } H+(1.5\times H\text{-spread}),$$

while *outer fences* are defined as

$$\text{lower } H-(3\times H\text{-spread})$$

$$\text{upper } H+(3\times H\text{-spread}).$$

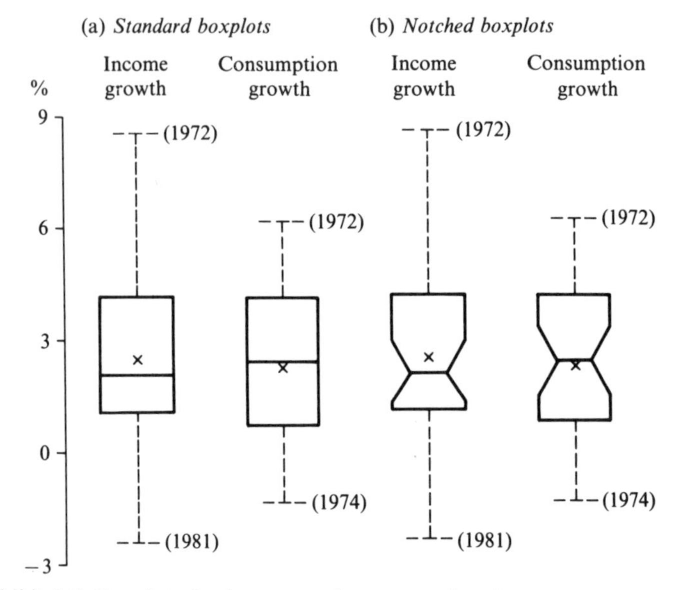

Exhibit 3.7 Boxplots for income and consumption data

Tukey (1977) names the expression (1.5 × *H*-spread) as the *step*. Inner fences therefore lie one step outside the hinges, and outer fences lie two steps outside. Any data value beyond either inner fence is termed *outside* and any data value beyond the outer fences is called *far outside*. The data values that are nearest to, but not outside, the inner fences are known as *adjacent* values.

With these rules for identifying outliers, a boxplot can easily be constructed. The general form of a boxplot is shown in Exhibit 3.6 and boxplots for the income and consumption growth series are presented as part (a) of Exhibit 3.7. Since the steps for the two series are 4.65 and 5.1 respectively, no values lie outside and hence the extremes are also the adjacent values. These extreme values have therefore been labelled on the boxplot. The positive skewness of the income growth series is immediately apparent and the inclusion of the sample mean on the boxplot only serves to emphasise this. The consumption growth series is confirmed to be almost symmetric.

3.4.1 *Notched boxplots*

When boxplots of different batches are placed side-by-side, as in Exhibit 3.7, they invite comparison, and it is then tempting to note batches that are 'significantly' different from each other. A natural approach to assessing significance is to look for non-overlapping central boxes, but unfortunately the hinges, which determine the extent of the box, are inappropriate guides to significance. An alternative is to use regions of overlap of special intervals around each median of a boxplot. The intervals used are symmetric and may be calculated by using

$$M \pm 1.58 \times (H\text{-spr})/\sqrt{n}.$$

The details underlying this formula may be found in McGill et al. (1978) and Velleman and Hoaglin (1981, pages 79–81), and two sets of data whose intervals do not overlap can be said to be significantly different at roughly the 5% level. To aid in the assessment of whether such intervals overlap or not, *notches* are put into the side of the central box at those values marking the end of the interval, thus producing a *notched boxplot*. Part (b) of Exhibit 3.7 incorporates the calculated notches and, not surprisingly in view of the similarity of the two central boxes, the large overlap of the notched intervals confirms that the median growth rates of consumption and income are not significantly different.

3.5 Goodness-of-fit tests

Having assessed the shape of the batch of observations available on a time series, it is often the case that this shape is compared with some well known 'standard' distribution. In the previous section, we introduced one method of comparison: that of comparing observed spreads with normal spreads. A more rigorous approach is to calculate a 'goodness-of-fit' test based on the χ^2 distribution (see Cochrane (1952) for a detailed review of this test and Mendenhall et al. (1986), for example, for a good textbook exposition). In this test the observed numbers of observations falling into a set of class intervals, often referred to as the set of *counts* in *bins*, is compared to the number of observations expected to fall into the bins if the hypothesised 'standard' distribution actually did generate the data. If n_i is the observed count for the ith of a set of K bins, and $\hat{n}_i$ is the corresponding expected (or *fitted*) count, then the statistic

$$X^2 = \sum_{i=1}^{K} (n_i - \hat{n}_i)^2 / \hat{n}_i$$

is approximately distributed as χ^2 with $(k-1-m)$ degrees of freedom, where m is the number of parameters in the hypothesised distribution that have to be estimated from the data. If the null hypothesis is true, the difference between n_i and $\hat{n}_i$ should be attributable solely to sampling fluctuations, and the null will therefore only be rejected for large values of X^2. Note that as only the counts for a chosen set of bins are required, the information on individual observations provided by a stem-and-leaf display is unnecessary and a conventional histogram display will suffice. A detailed example of this test is to be found below as Example 3.2.

3.5.1 *Q–Q plots*

A second approach to examining goodness-of-fit is a graphical display known as the *quantile–quantile* (*or Q–Q*) *plot* (see Wilk and Gnanadesikan (1968) and Chambers et al. (1983, chapter 6) for expository discussion of this technique and Poskitt and Tremayne (1986) for some interesting applications to time series data). This display plots the observed quantiles of a set of data against the quantiles of the hypothesised distribution. The two sets of quantiles are defined in the following way. Suppose that $x_1, x_2, \ldots, x_n$ are the n observations (ordered by time) on the series under consideration and that $x_{(1)}, x_{(2)}, \ldots, x_{(n)}$ are the same observations sorted from smallest to largest. The tth *order statistic*, $x_{(t)}$, is then said to be the p_t observed quantile, where $p_t = (t-0.5)/n$, for $t = 1, 2, \ldots, n$. In general, we will denote the p_t observed quantile as $Q_0(p_t)$. Similarly, if $F(x)$ is the cumulative distribution function of the hypothesised distribution, then the p quantile of F, where $0 < p < 1$, is a number, which we will denote $Q_h(p)$, which satisfies

$$F(Q_h(p)) = p$$

or

$$Q_h(p) = F^{-1}(p).$$

In other words, a fraction p of the probability of the hypothesised distribution occurs for values of x less than or equal to $Q_h(p)$, just as a fraction p (approximately) of the observed data are less than or equal to $Q_0(p)$. The Q–Q plot is then constructed by plotting $Q_0(p_t)$ against $Q_h(p_t)$, for $t = 1, 2, \ldots, n$, where, as above, $p_t = (t-0.5)/n$. When the hypothesised distribution is the normal, the result is commonly called a *normal probability plot*.

3.5.2 *Properties of Q–Q plots*

If the two distributions are identical, the plot will be a straight line configuration pointed towards the origin, i.e. $Q_h = Q_0$. Of course, random fluctuations in any particular data set will cause the points to drift away from this line, but if the hypothesised distribution is 'correct', the points should remain reasonably close to the line. If any large or systematic departures from the line occur, then these should be taken as an indication of the lack of fit of this distribution to the data.

What kinds of important departures from the line $Q_h = Q_0$ might we expect? One possibility is that the points may fall near some straight line other than this reference line. If the observed plot follows a line parallel to $Q_h = Q_0$, then an appropriate constant could be added to all data points to shift the configuration onto $Q_h = Q_0$: the observed distribution differs from the hypothesised distribution only in terms of location (as measured by means or medians). Alternatively, the observed plot may have a straight line configuration that passes through the origin but is not parallel to $Q_h = Q_0$. Since it is always possible to find a single positive constant by which we may multiply all observations and so, in effect, compress or expand the plot vertically to make it follow $Q_h = Q_0$, in this case the two distributions match except for a difference in dispersion (as measured, for example, by the standard deviation or the *H*-spread). If we also need to add a constant to map the observed plot onto the reference line, then the observed and hypothesised distributions are of the same 'family', but differ in both location and dispersion. Thus it is the straightness of the *Q–Q* plot that is used to judge whether the data and the hypothesised distribution are compatible, with shifts and tilts away from the reference line $Q_h = Q_0$ indicating differences in location and dispersion, respectively.

It therefore follows that large or systematic departures from straightness, or linearity, in a *Q–Q* plot provide an indication of a mismatch between the observed and hypothesised distributions and, moreover, the actual form of the departure may suggest the nature of the divergence. To judge linearity it is useful to fit a straight line either by numerical estimation or, more often in exploratory analysis, by eye, to either the whole data set or to a portion that seems reasonably straight.

When there are departures from linearity in a *Q–Q* plot they often match one of four descriptions: (i) outliers at either end, (ii) curvature at both ends, indicating long or short tails, (iii) convex or concave curvature, suggesting asymmetry, and (iv) horizontal segments, plateaus, or gaps.

Outliers have already been discussed in Section 3.4, where rules of thumb were suggested for their identification. The question here is whether the most extreme observations in a set of data are larger than could reasonably be expected for samples of this size from the hypothesised distribution. Again we emphasise that if outliers are encountered, it is always wise to go back to the original data source to verify their values. If the values have been incorrectly reported or measured and need to be omitted from further analysis, it is seldom necessary to redraw the *Q–Q* plot unless the outliers were so extreme as to have severely compressed the rest of the plot. If the values are correct, however, they may often be the most important observations in the data set and should be retained for further analysis.

A common departure from linearity is curvature at the ends of the *Q–Q* plot, most commonly an upward curve at the right hand end and a downward curve at the left. In this case the observations in the tails are further away from the centre than they ought to be for a sample from the hypothesised distribution: we say that the observed distribution is long-tailed. Occasionally, the curvature at the ends of the plot is reversed, and in such a case the observed distribution is said to be short-tailed relative to the hypothesised distribution.

Asymmetry in the observed data when the hypothesised distribution is symmetric will show itself in a *Q–Q* plot displaying curvature along its entire length, the typical case of right skewness found in many economic time series results in the slope of the plot increasing from left to right.

Horizontal segments, plateaus and gaps typically occur in *Q–Q* plots as a result of data measurement and rounding, and with time series data these processes will often lead to plots being rather 'granular', having a number of horizontal segments.

Although *Q–Q* plots are a powerful exploratory tool, they must still be used with care. In particular, the natural variability of the data will often generate departures from linearity, even if the hypothesised distribution is valid, and this is especially prevalent with short data series. Furthermore, the plots should not be used in isolation but should be supplemented by other displays and analyses to get a full picture of the behaviour of the data, with attention being particularly focused on other variables which may potentially be able to explain some of the variation in the series being analysed. Indeed, Chapter 12 presents formal modelling techniques for detecting the presence, and examining the influence, of outliers in time series data.

Example 3.1: A *Q–Q* plot of daily interest rate changes

With these strictures borne in mind, Exhibit 3.8 plots the daily changes of the interbank rate series, whose levels were graphed in Exhibit 2.9. (The reason for concentrating attention on the daily changes, rather than the daily levels, of the interbank rate will be discussed in later chapters.) Exhibit 3.9 shows, in one display, a histogram of the data, a box plot, and a *Q–Q* plot of the observed distribution against the standard normal. The histogram is constructed using the commonly employed 10/15/20 rule for selecting the width of bins (see Levenbach and Cleary (1981, pages 75–6)). The range of the data is 2.74 percentage points, and as the number of observations is 740, this rule suggests using a bin width resulting from dividing the range by a value between 15 and 20. The convenient width of 0.15 was thus selected. Both the histogram and the boxplot show that although the data is extremely tightly packed around the centre of the distribution, with both the mean and median being zero

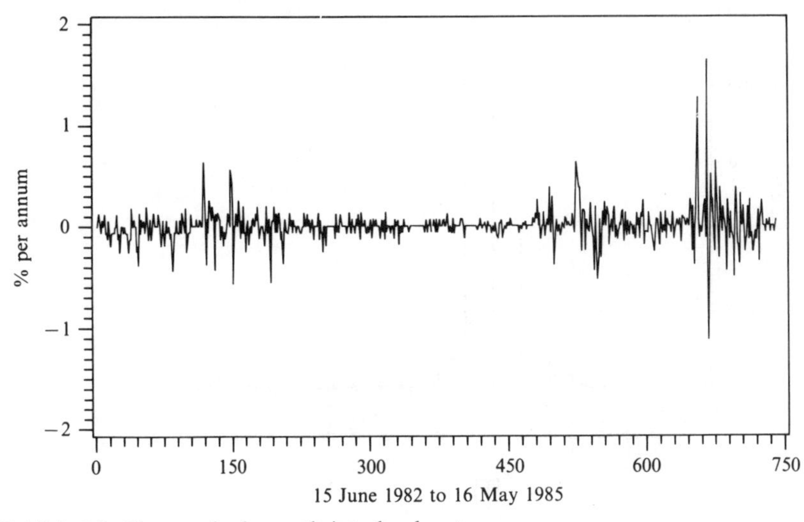

Exhibit 3.8 Changes in 3-month interbank rate

and the H-spread being 0.12 percentage points, the tails of the distribution are very straggling with 19 data values lying far outside and a further 31 lying outside. As interest rates are only recorded to two decimal places, the actual Q–Q plot is very granular, with numerous horizontal segments, particularly in the central part of the distribution. To retain resolution in the plot shown in Exhibit 3.9, only one point from each horizontal segment has been plotted. The Q–Q plot confirms that far too many observations lie in the tails of the distribution than can be explained by a normal distribution and also shows that the observed distribution is 'too peaked' in comparison to the normal, since the plot has a distinct sigmoid shape. This suggests that the distribution of interest rate changes is, in fact, 'stable Paretian', as has been found for many financial price change series: see, for example, Mandelbrot (1963) and Fama (1965). Formal modelling of this characteristic of the interest rate series is presented in Chapter 15.

Example 3.2: A χ^2 goodness-of-fit test of daily interest rate changes

The above example can also be used to illustrate the calculation of a χ^2 goodness-of-fit test. A standard approach to the calculation of this test is to take the bins used in constructing the histogram, perhaps combining adjacent bins to ensure that fitted counts exceed 5, and to use the sample mean and standard deviation (found to be 0 and 0.160

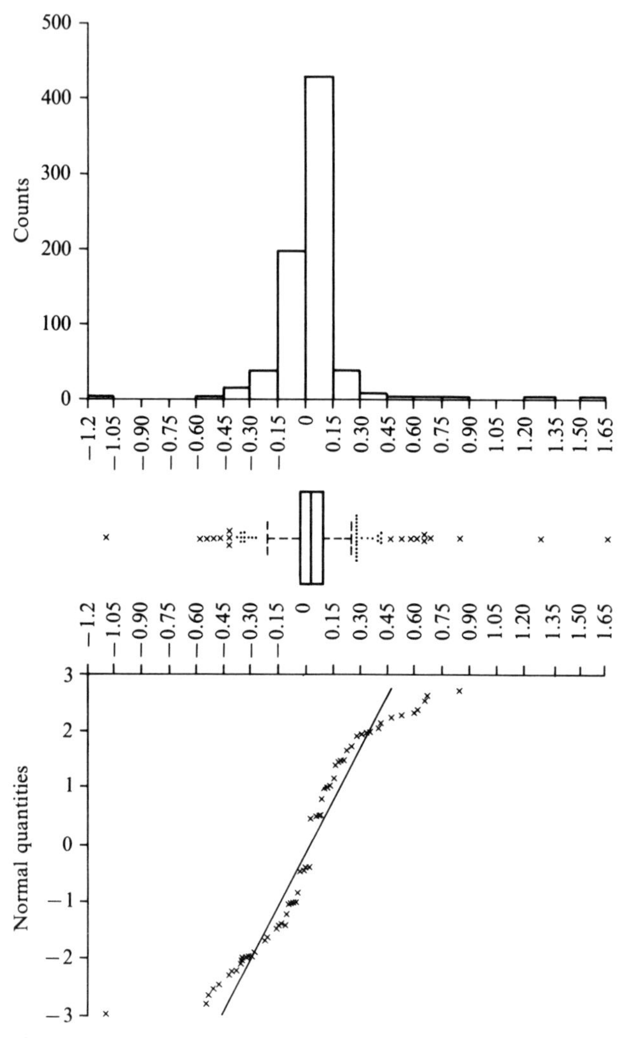

Exhibit 3.9 Histogram, boxplot and Q–Q plot for interest rate data of Exhibit 3.8

respectively in this example) as estimates of the normal population parameters μ and σ. The computations involved in the calculation are set out in Exhibit 3.10 and show, not surprisingly in view of the information revealed in the graphical displays, that the hypothesis that the data was generated by a $N(0, 0.160^2)$ distribution is conclusively rejected. Note, however, that with the given set of bin widths, rejection is brought about by too many observations lying in the centre of the distribution and too

Exhibit 3.10 X^2 *goodness-of-fit test: standard approach*

Change in rate	Count		Fitted	X^2
−1.20 to −1.06	1			
−1.05 to −0.91	0			
−0.90 to −0.76	0	21	23	0.2
−0.75 to −0.61	0			
−0.60 to −0.46	4			
−0.45 to −0.31	16			
−0.30 to −0.16	37		106	44.9
−0.15 to −0.01	197		241	8.0
0 to 0.14	429		241	146.7
0.15 to 0.29	38		106	43.6
0.30 to 0.44	9			
0.45 to 0.59	3			
0.60 to 0.74	3			
0.75 to 0.89	1			
0.90 to 1.04	0	18	23	1.1
1.05 to 1.19	0			
1.20 to 1.34	1			
1.35 to 1.49	0			
1.50 to 1.64	1			
	740		740	$244.5 \sim \chi^2(6-1-2=3)$

few in the 'inflection' regions, rather than there being too many in the tails.

A number of further difficulties arise with the procedure just outlined: (i) as we have already discussed, the sample standard deviation can be badly affected by the presence of outliers and the interest rate data is heavily contaminated by such extreme values. An alternative estimate of σ is, from Exhibit 3.4, given by dividing the H-spread by 1.349, obtaining in this case 0.089, some 45% smaller than the estimate used in the construction of the X^2 test statistic in Exhibit 3.10, (ii) the standard practice of combining bins to ensure that fitted counts exceed 5 has little basis in statistical theory. Cochrane (1952, page 329) suggests that for tests of bell-shaped curves such as the normal distribution, fitted counts in the tails can be allowed to go as low as 1 without any detrimental consequences for inference, (iii) the bin widths used in constructing the histogram will almost certainly not be optimal in terms of maximising the power of the test. Cochrane (1952, pages 332–4) suggests using enough bins to keep the fitted counts down to around 30 for the sample size being analysed here.

Exhibit 3.11 X^2 *goodness-of-fit test: amended approach*

Change in rate	Count	Fitted	X^2
less than −0.20	36	9	81.0
−0.20 to −0.16	22	25	0.4
−0.15 to −0.13	23	30	1.6
−0.12 and −0.11	27	31	0.5
−0.10 and −0.09	8	40	25.6
−0.08 and −0.07	35	50	4.5
−0.06 and −0.05	89	56	19.4
−0.04 and −0.03	14	63	38.1
−0.02 and −0.01	1	66	64.0
0 and 0.01	254	66	535.5
0.02 and 0.03	14	63	38.1
0.04 and 0.05	4	56	48.3
0.06 and 0.07	101	50	52.0
0.08 and 0.09	2	40	36.1
0.10 and 0.11	1	31	29.0
0.12 to 0.14	53	30	17.6
0.15 to 0.19	25	25	0
0.20 and above	34	9	69.4
	740	740	$1061.1 \sim \chi^2(18-1-2=15)$

Although the clear rejection of the null hypothesis in Exhibit 3.10 makes it unlikely that incorporating these suggestions will alter the conclusion concerning distributional shape, Exhibit 3.11 repeats the computations using 0.089 as the estimated value of σ, and with bin widths chosen so that fitted counts are kept roughly equal, although the restriction that the data is recorded at intervals of 0.01 percentage points necessarily raises this number on many occasions to above the recommended size. This amended procedure increases the degrees of freedom to 15 and, of course, still provides a clear rejection of the null hypothesis of $N(0, 0.089^2)$. From an examination of Exhibit 3.11, we see that the observed distribution is indeed thick tailed, but it is also the case that the small bins, comprising only two observable values in most cases, induce considerable variability in the observed counts which may well overstate the discrepancy with the fitted counts.

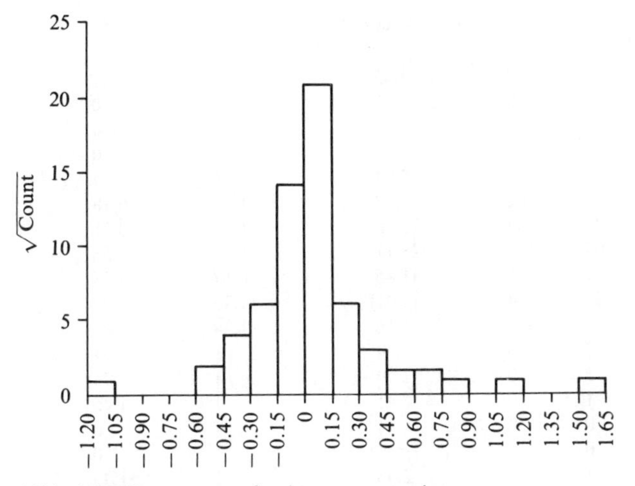

Exhibit 3.12 Rootogram for interest rate data

3.5.3 *Rootograms*

This problem is associated with the general tendency for the variability of bin counts not to be constant but to be proportional to the counts themselves. Tukey (1972) suggests that a suitable 're-expression' or transformation of the counts that will tend to stabilise this variability is the square root. A histogram based on square root counts rather than raw counts is termed a *rootogram*, and Exhibit 3.12 shows such a display for the interest rate data. It is seen that the pattern in the central portion of the distribution is more regular than in the conventional histogram display of Exhibit 3.9 and the straggling tails are given more emphasis. Note, however, that the area of each bin is no longer proportional to the count, reflecting the fact that for data analysis it is more important to stabilise the variability of fluctuations than to be able to picture the raw counts directly in terms of area.

As well as stabilising variability, the square root transformation of the counts has important implications for goodness-of-fit tests. Rather than work directly with the transformed observed and fitted counts $n_i^{\frac{1}{2}}$ and $\hat{n}_i^{\frac{1}{2}}$, some difficulties with small counts are avoided by transforming the observed counts to

$$\begin{array}{ll} (2+4n_i)^{\frac{1}{2}} & \text{if} \quad n_i > 0 \\ 1 & \text{if} \quad n_i = 0, \end{array}$$

and transforming the fitted counts to

$$(1+4\hat{n}_i)^{\frac{1}{2}}.$$

Change in rate	Count	Fitted	$\sqrt{2+4\,(\text{count})}$	$\sqrt{1+4\,(\text{fitted})}$	DRR
−1.20 to −1.06	1	0	2.45	1	1.45
−1.05 to −0.91	0	0	1	1	0
−0.90 to −0.76	0	0	1	1	0
−0.75 to −0.61	0	0	1	1	0
−0.60 to −0.46	4	0	4.24	1	3.24
−0.45 to −0.31	16	0.2	8.12	1.34	6.78
−0.30 to −0.16	37	34	12.25	11.70	0.55
−0.15 to −0.01	197	336	28.11	36.67	−8.56
0 to 0.14	429	336	41.45	36.67	4.75
0.15 to 0.29	38	34	12.41	11.70	0.71
0.30 to 0.44	9	0.2	6.16	1.34	4.82
0.45 to 0.59	3	0	3.74	1	2.74
0.60 to 0.74	3	0	3.74	1	2.74
0.75 to 0.89	1	0	2.45	1	1.45
0.90 to 1.04	0	0	1	1	0
1.05 to 1.19	0	0	1	1	0
1.20 to 1.34	1	0	2.45	1	1.45
1.35 to 1.49	0	0	1	1	0
1.50 to 1.64	1	0	2.45	1	1.45

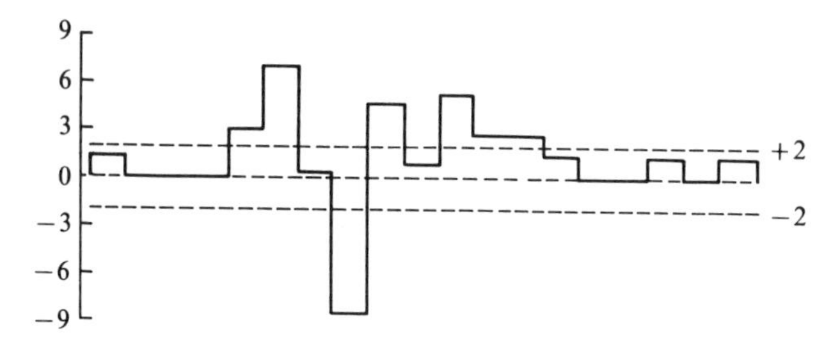

Exhibit 3.13 Double-root residual computations and suspended rootogram display

Velleman and Hoaglin (1981, chapter 9) provide an extended discussion of this particular transformation, using which we may then define the *double-root residual* for the *i*th bin as

$$\text{DRR}_i = \begin{cases} (2+4n_i)^{\frac{1}{2}} - (1+4\hat{n}_i)^{\frac{1}{2}} & \text{if} \quad n_i > 0 \\ 1-(1+4\hat{n}_1)^{\frac{1}{2}} & \text{if} \quad n_i = 0. \end{cases}$$

The computation of the double-root residuals from the rootogram of Exhibit 3.12 is shown in Exhibit 3.13. The usefulness of double-root residuals for assessing goodness-of-fit tests is twofold. Based on the results of Freeman and Tukey (1950), it can be shown that if the model used to compute the fitted counts fits the data well, then an individual double-root residual will behave approximately like an observation from the standard normal distribution. Thus a 'large' DRR is one that is greater than ± 2, although if the fitted count is less than 1 the standard normal approximation may be less satisfactory and it may be wise to set 'significance' limits at ± 1.5. The set of double-root residuals may be displayed graphically through a *suspended rootogram*, and this is also shown in Exhibit 3.13 (see Velleman and Hoaglin (1981, chapter 9) for the methodological arguments underlying this display). Seven of the nineteen bins yield DRRs in excess of ± 2, with some of the more central bins exceeding this value by a large margin, obviously confirming the inadequacy of the normal assumption for this series. Since the set of double-root residuals are individually approximately standard normal, the statistic

$$\sum_{i=1}^{k} \mathrm{DRR}_i^2$$

will follow roughly a χ^2 distribution with the usual degrees of freedom, $k-1-m$. Computation of this statistic for the set shown in Exhibit 3.13 yields a value of 207.9, which should be referred to the $\chi^2(16)$ distribution. This relationship with the usual X^2 test illustrates two advantages of using double-root residuals: individual DRRs can be examined to reveal any bin where the fit is poor, with the $\sum \mathrm{DRR}^2$ statistic then providing an overall measure which will be particularly useful when the fit is generally poor but not unusually bad in any one bin, and no loss of degrees of freedom through the combining of adjacent bins appears to be required.

3.6 Computing software

These techniques are now becoming fairly widely available on the more comprehensive statistical software packages. Histograms, stem-and-leaf displays, boxplots and normal probability plots are available on SAS in the PROC UNIVARIATE routine (SAS, 1985a, chapter 54), and can also be found in SPSS^{X} in the MANOVA routine (SPSS^{X}, 1983, chapter 28). The MINITAB package has a particularly comprehensive set of exploratory data analysis routines, including histograms and rootograms, stem-and-leaf displays and boxplots (see Ryan et al. (1985)).

4 Transforming and smoothing time series

4.1 Transformations

In statistical analysis, it is often easier to work with transformed data rather than the observed, or raw, values and time series analysis is no exception. Indeed, three transformations have already been met in Chapter 3: analysing growth rates of income and consumption, working with changes (or, more formally, first differences) of interest rates, and using square roots of counts to construct a rootogram. We will distinguish between transforming a single time series for the purpose of conducting a univariate analysis, and transforming two, or more, time series for undertaking a multivariate analysis: the importance of making such a distinction will become apparent in later sections of the chapter.

4.2 Transforming an individual series

The usual reasons for transforming an individual series are twofold: to induce symmetry, and possibly normality, to the distributional shape, and to stabilise the location and variability and reduce the covariability of the series through time. This second motive is known as a transformation to *stationarity*, and this concept will be defined formally and discussed in considerable detail in Part II of the book.

4.2.1 *Transformations to symmetry*

Asymmetric distributions often arise when a time series can take only positive (or at least non-negative) values, in which case the distributions will usually be skewed to the right because, although there is a lower bound on the data, often zero, no upper bound exists and the values are allowed to stretch out. Frequently, the asymmetry steadily increases in going from the centre of the distribution to the tails, and when this happens the following class of power transformations for positive data will help to reduce such asymmetry (a similar class of transformations have been introduced by Box and Cox (1964): see Chapters 6 and 15):

$$\begin{aligned} y_t^\lambda \quad &\text{if} \quad \lambda > 0 \\ \log y_t \quad &\text{if} \quad \lambda = 0 \\ -y_t^\lambda \quad &\text{if} \quad \lambda < 0. \end{aligned}$$

The reason for changing the sign when λ is negative is to ensure that the transformed values have the same relative ordering as the original values.

When the data is skewed to the right, values of $\lambda < 1$ will decrease the asymmetry and it is often the case that the logarithmic transformation ($\lambda = 0$) does an excellent job in inducing symmetry in economic time series. As we have seen in Chapter 3, the square root transformation ($\lambda = 1/2$) is commonly used for data in the form of counts, hence its use in transforming histograms into rootograms. The obvious question is how should a value of λ be chosen for a given time series? A useful technique is to consider the sequence of midsummaries in a letter-value display of various transformations of the data. The transformation $\lambda = 1$ is, of course, the 'identity' transformation: the set of observed data. As mentioned above, moving to values of λ less than 1 will pull in a stretched-out upper tail while stretching out a bunched-in lower tail. Increasing the value of λ above 1 will have the reverse effect: skewness to the left will be alleviated. Thus, if in the raw data an upward trend in the midsummaries is observed, transforming the data using values of λ less than 1 should be tried, and the new midsummaries analysed, stopping when the midsummaries are stable. Note that since the power transformation preserves order, only the letter values of the data need to be transformed each time, but the midsummaries of the transformed data are *not* the transformations of the raw midsummaries, they are the averages of the letter values of the transformed data.

Exhibit 4.1 illustrates this technique using the real and nominal total final expenditure (TFE) series graphed originally in Exhibit 2.1. For the real TFE series a slight upward trend is observed in the raw midsummaries, while for the square root and logarithmic transformations a slight decline is found. It is only for the reciprocal square root transformation that a pronounced decline is observed and thus, given that the length of the series is only moderate (37 observations), this would suggest that any transformation between zero and unity is acceptable, with perhaps $\lambda = 1$ being preferred on grounds of simplicity. With nominal TFE a different pattern in the midsummaries emerges. Both $\lambda = 1$ and $\lambda = 1/2$ have pronounced upward trends in their midsummaries and even for $\lambda = 0$ an

Exhibit 4.1 *Midsummaries for various power transformations of TFE*

Letter	Depth	$\lambda = 1$ $(\times 10^5)$	$\lambda = \frac{1}{2}$ $(\times 10^2)$	$\lambda = 0$ $(\times 10)$	$\lambda = -\frac{1}{2}$ $(\times 10^{-3})$
(a) *Real TFE*					
M	19	1.134	3.368	1.164	−2.969
H	10	1.131	3.333	1.151	−2.646
E	5.5	1.148	3.328	1.158	−2.520
D	3	1.152	3.323	1.157	−2.491
C	2	1.147	3.313	1.156	−2.490
	1	1.156	3.319	1.156	−2.463
(b) *Nominal TFE*					
M	19	0.455	2.134	1.073	−4.687
H	10	0.810	2.657	1.101	−4.410
E	5.5	1.449	3.308	1.121	−4.471
D	3	1.802	3.569	1.122	−4.789
C	2	1.965	3.689	1.124	−4.870
	1	2.127	3.803	1.125	−4.967

upward trend is still visible. It is only for the reciprocal square root transformation that the trend in the midsummaries is eradicated, there now being a slight tendency for the mids to decrease. Thus $\lambda = -1/2$ would seem to be the preferred transformation for nominal TFE. These findings are, in fact, consistent with the graphs of the two series shown in Exhibit 2.1. Nominal TFE grows almost thirtyfold during the observation period and is typical of a highly skewed (to the right) economic time series measured in current prices, whereas real TFE only grows by $2\frac{1}{2}$ times and its distribution is thus much less asymmetric.

A graphical display is also available to assess symmetry, and although this requires more computational effort than calculating the midsummaries of various transformations, it may yield more precise information about the form of departure from symmetry, particularly for small or moderate length time series. As in Chapter 3, let us define $x_{(t)}$ to be the tth *ordered* observation (from smallest to largest) of a time series of length n. Note that $x_{(t)}$ will only be the same observation as x_t if the series is monotonically increasing. The time series will therefore be symmetric if

$$\text{median} - x_{(t)} = x_{(n+1-t)} - \text{median}.$$

Symmetry can then be checked by plotting $u_t = (x_{(n+1-t)} - \text{median})$ against $v_t = (\text{median} - x_{(t)})$ for $t = 1, 2, \ldots, n/2$ (or $(n+1)/2$ if n is odd). If the points lie close to the line $u = v$, the distribution is nearly symmetric. As

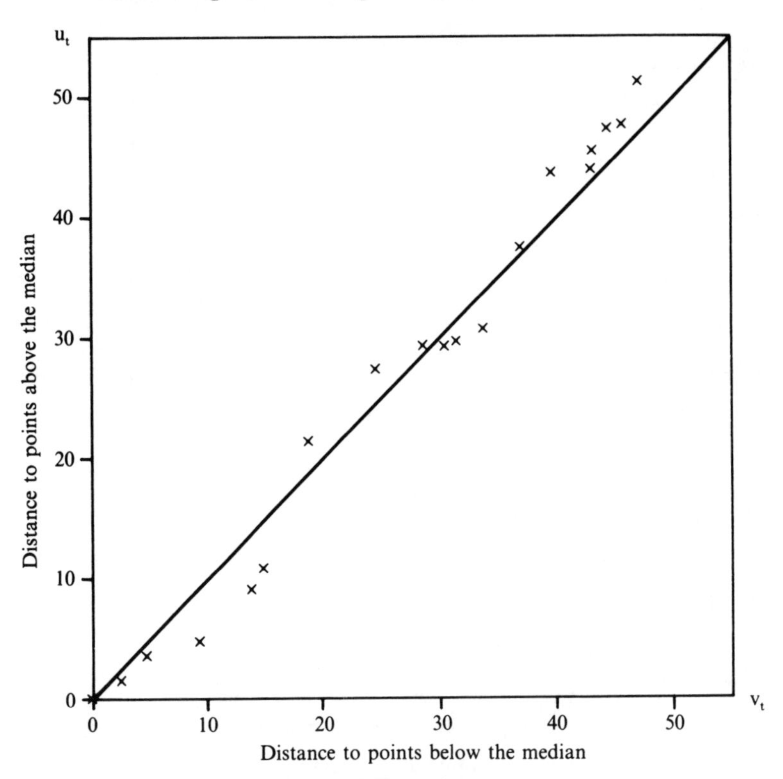

Exhibit 4.2 Symmetry plot for real TFE series

Chambers et al. (1983, pages 29–32) discuss, as t increases, neither v_t nor u_t will decrease, since $x_{(t)}$ and $x_{(n+1-t)}$ both move further from the median, and thus, going from left to right, the ordinates of the points will be nondecreasing.

Exhibit 4.2 displays a symmetry plot for the real TFE series, with the $u = v$ line superimposed. Moving from left to right, the fractions for the quantiles used on the horizontal axis go from 0 to 0.5, while those on the vertical axis go from 0.5 to 1. Thus the points in the lower left corner of the plot correspond to the quantiles close to the median and hence to observations in the centre of the distribution, while the points in the upper right corner correspond to quantiles far from the median and hence to tail observations. We can see that there is no tendency for the points to diverge from the $u = v$ line and thus the distribution appears to be reasonably symmetric, confirming the conclusions drawn from the midsummary analysis.

The symmetry plot of Exhibit 4.3 is for the observed nominal TFE

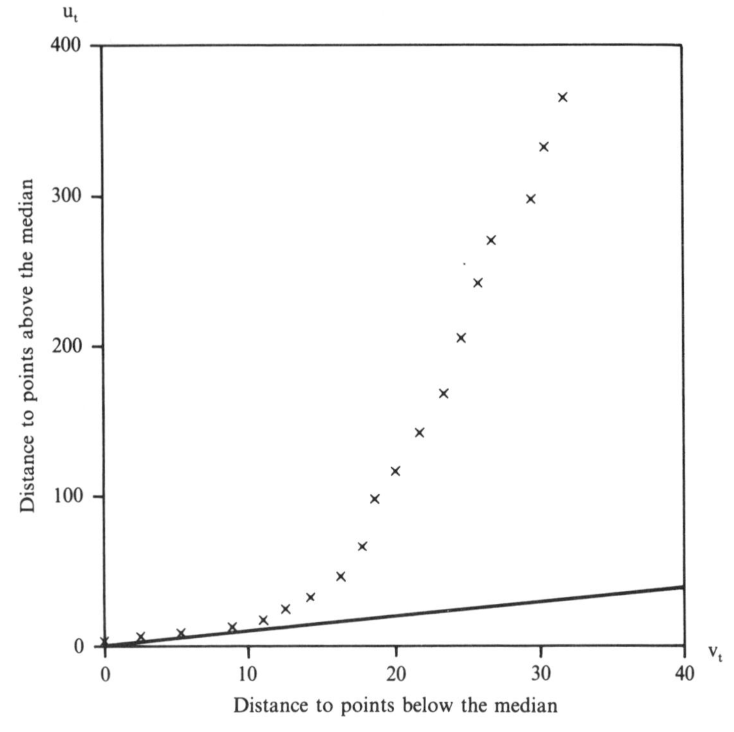

Exhibit 4.3 Symmetry plot for nominal TFE series

series, which confirms the high rightward skewness known to exist in this time series. Exhibit 4.4 shows two further symmetry plots for the logarithmic and reciprocal square root transformations of the series. Now it is apparent that $\lambda = 0$ does not eradicate the asymmetry in the data, and that a negative value of λ is required. The symmetry plot for $\lambda = -1/2$ shows that asymmetry has been virtually eradicated, although there is some suspicion that the transformation is too powerful, and perhaps $\lambda = -1/3$ might be preferred.

Since the assumption of normality is a very convenient one to make in statistical analysis, it is often useful to consider transforming the data to normality rather than just to symmetry. This can be done by using *Q–Q* plots rather than symmetry plots. Exhibit 4.5 shows the *Q–Q* plot for nominal TFE for $\lambda = 1$ and $\lambda = -1/2$, and this latter transformation is seen to do a reasonable job in inducing normality as well as symmetry to the series, although the curvature of the tails of the plot suggest that the distribution has somewhat shorter tails than a normal distribution. (It should be noted here that if a series needs to be power transformed to

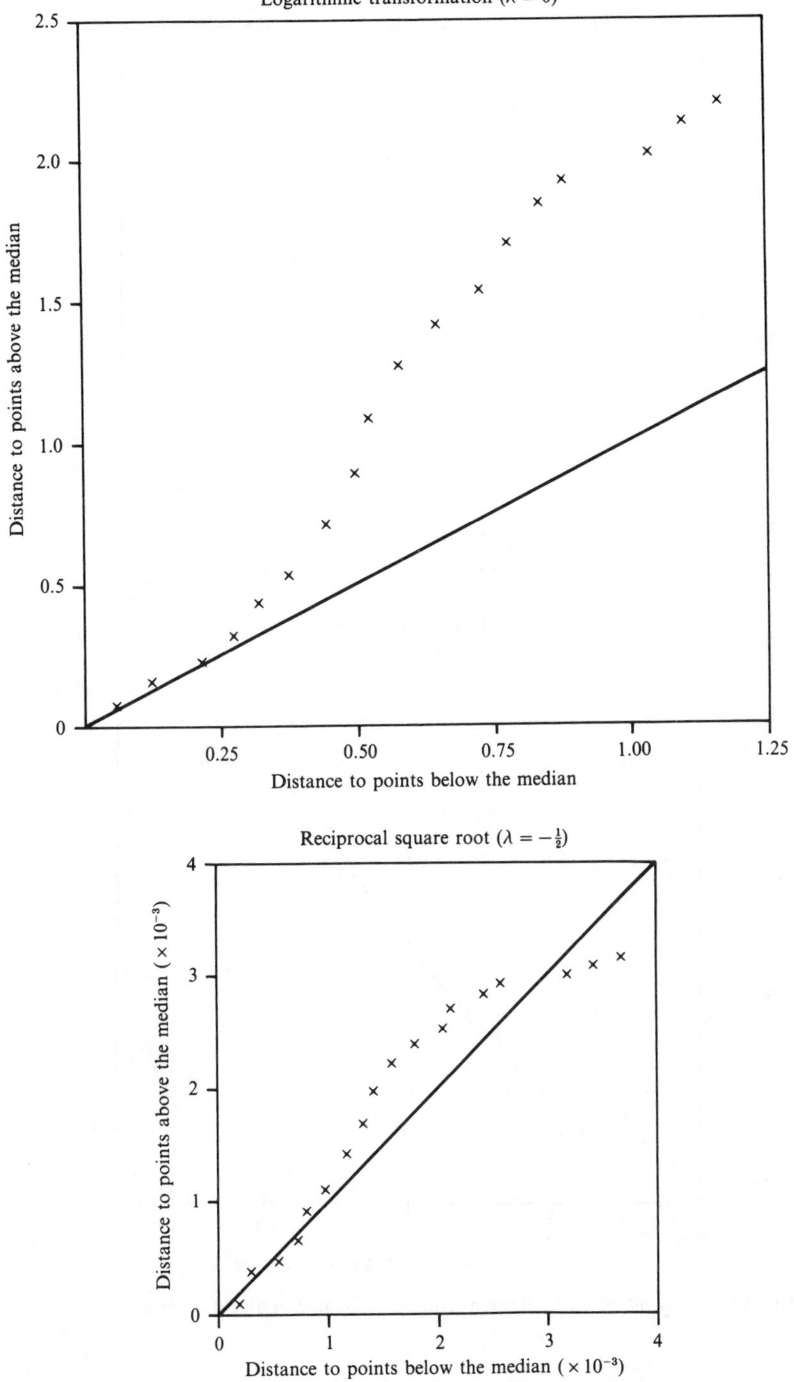

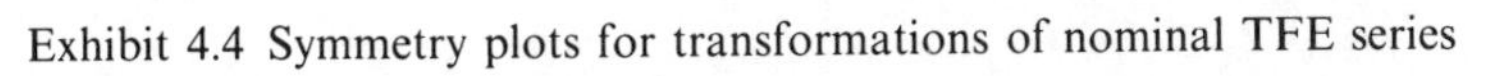
Exhibit 4.4 Symmetry plots for transformations of nominal TFE series

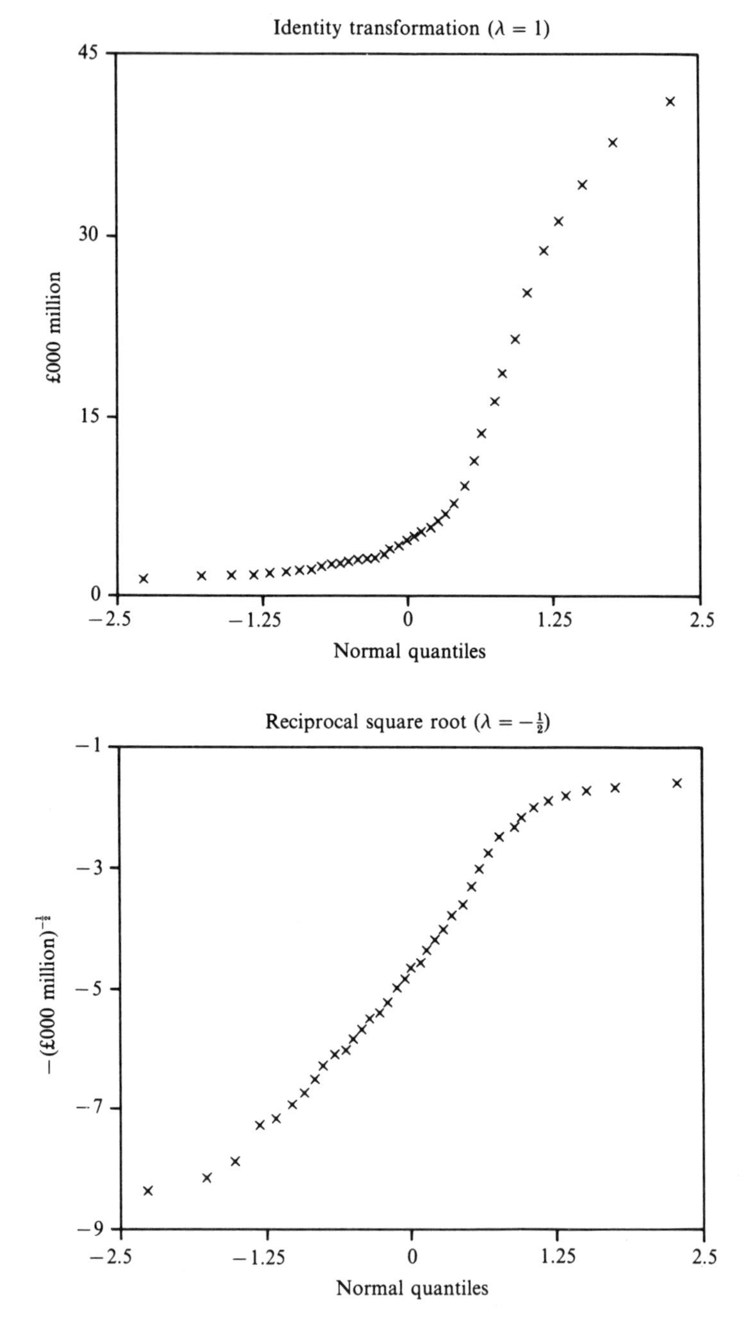

Exhibit 4.5 *Q–Q* plots for transformations of nominal TFE series

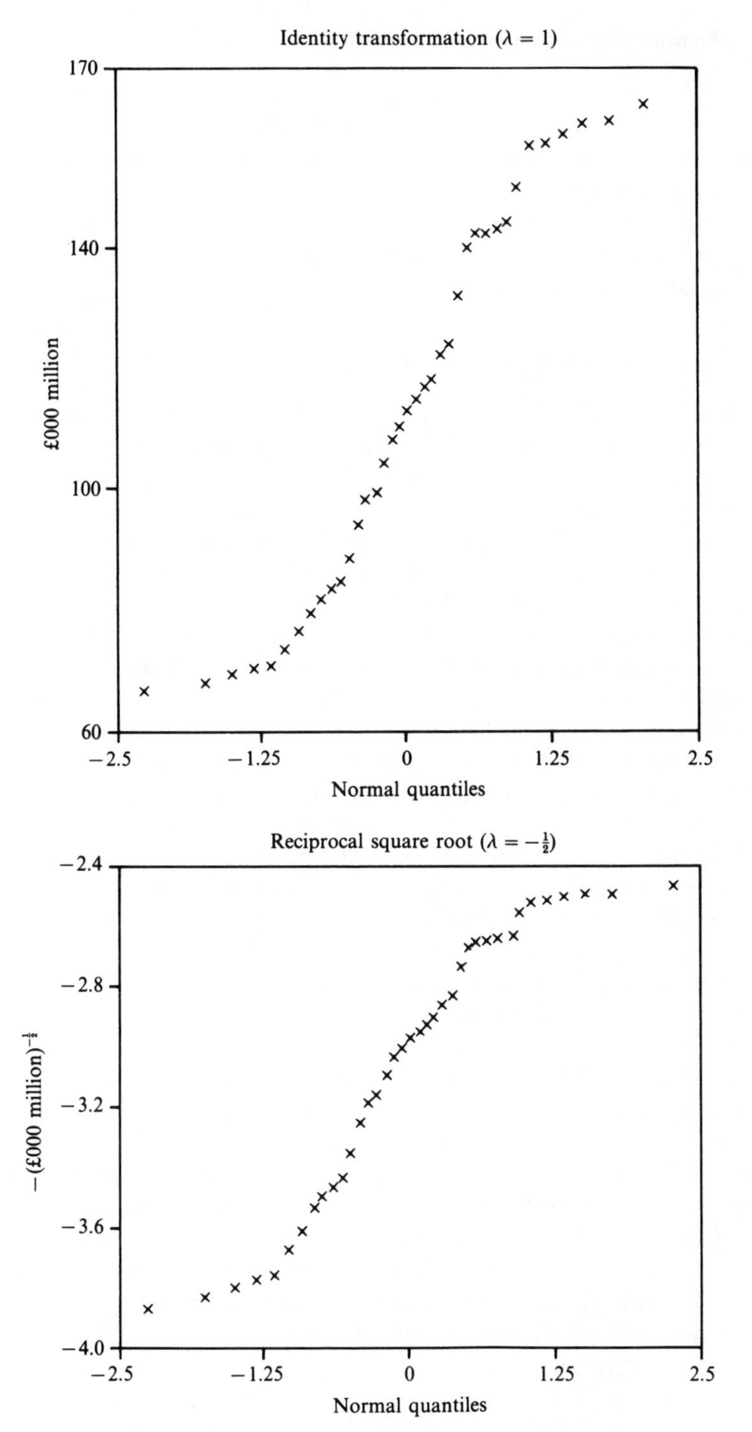

Exhibit 4.6 *Q–Q* plots for transformations of real TFE series

achieve normality, it is said to have a power normal distribution with *shape parameter* λ.) A similar conclusion is arrived at from the *Q–Q* plots for real TFE shown in Exhibit 4.6. The series is shown to be fairly robust to power transformation, with the plots for $\lambda = 1$ and $\lambda = -1/2$ being quite similar. Again, though, the curvature of the tails, being more pronounced than those for nominal TFE, highlight the short-tailness of the distribution.

4.2.2 *Transformations to stationarity*

While symmetry and *Q–Q* plots may be used for any batch of observations, time series data do have one special feature: the time ordering of the data is important, and this allows the moments of the distribution to change through time. We will focus attention here on methods by which the mean and variance can be stabilised, and for the latter the power transformation introduced above can be used in the form of a *range-mean*, or *range-median* plot (see Jenkins (1979)). For this plot, the time series is sliced into sections, with usually between 4 and 12 observations in each, and for each section the range and median are calculated. The sequence of paired range-median values are then plotted and the relationship between them examined. The first plot in Exhibit 4.7 shows how departures from a horizontal relationship (constant variance through time) can be related to a particular power transformation, which can then be used to transform the series to stabilise the variance. The median and range are used rather than the mean and variance for each slice to ensure robustness. The second plot shows the range-median relationship for the nominal TFE series, using slices of 5 years (the final slice containing 7 observations). The strong positive relationship between the range and median shows that, once again, a logarithmic ($\lambda = 0$) or reciprocal square root ($\lambda = -1/2$) transformation is needed for this series.

Since power transformations preserve order, they cannot by themselves stabilise a time varying mean. Certain types of time series, known as nonstationary series, are typically characterised as having increasing (or occasionally decreasing) mean levels and being very smooth. The nominal and real TFE series, for example, are typical of such nonstationary behaviour. If short-term changes in the series are of interest, these may not be apparent from the raw data until the 'trend' has been removed. A number of methods of trend estimation, and hence removal, will be considered in subsequent chapters, but one simple and commonly used transformation is very effective at removing trends. This is the *difference transformation*, defined as

$$\nabla x_t = x_t - x_{t-1},$$

(a) Range-median relationships and implied λ values

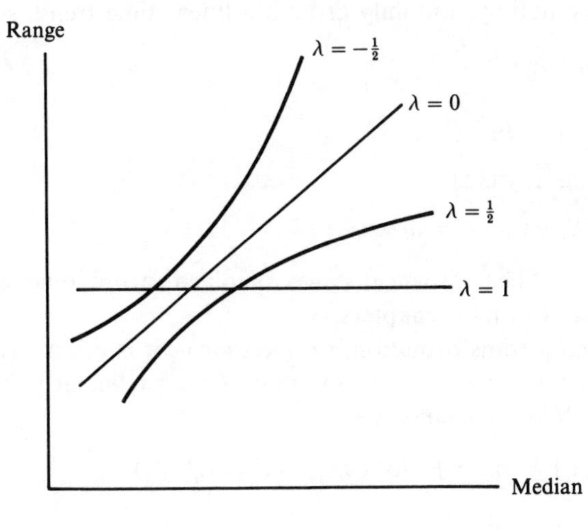

(b) Range-median plot for nominal TFE

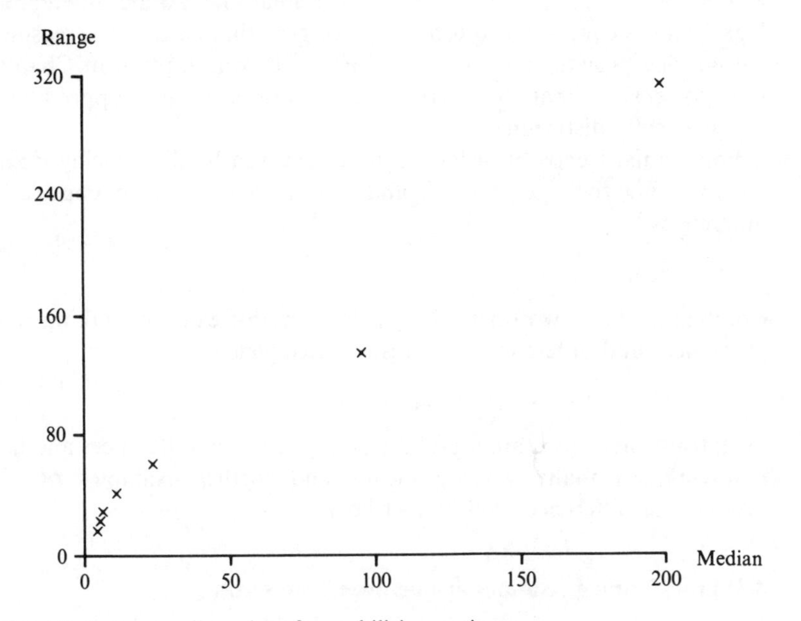

Exhibit 4.7 Range-median plots for stabilising variance

which thus takes successive changes of the series. If the series can be regarded as fluctuating randomly around a linear time trend, so that

$$x_t = \beta_0 + \beta_1 t + u_t$$

and

$$E(x_t) = \beta_0 + \beta_1 t,$$

then differencing x_t yields

$$\nabla x_t = \beta_1 + v_t, \quad v_t = u_t - u_{t-1},$$

with $E(\nabla x_t) = \beta_1$. (The statistical issues involved in differencing will be discussed in detail in later chapters.)

The differencing transformation, or *operator* as it is often termed, can be combined with power transformations. One particularly important combination is $\nabla \log(x_t)$, since

$$\nabla \log(x_t) = \log(x_t) - \log(x_{t-1}) = \log(x_t / x_{t-1})$$

$$\simeq \frac{x_t}{x_{t-1}} - 1 = \frac{x_t - x_{t-1}}{x_{t-1}}$$

as long as the ratio x_t/x_{t-1} is reasonably small, i.e. taking differences of logarithms is equivalent to using rates of growth, and thus the reason for considering growth rates of real income and consumption in Chapter 3 was to ensure that the series were stationary and approximately symmetrically distributed.

Polynomial trends of order d can be removed by differencing d times, i.e. by using the operator ∇^d, and it is also possible to consider sth differences ∇_s,

$$\nabla_s x_t = x_t - x_{t-s}.$$

Combining these two operators yields $\nabla_s^d x_t$; for example, taking fourth differences and differencing the result twice yields

$$\nabla_4^2 x_t = (x_t - x_{t-4}) - (x_{t-4} - x_{t-8}).$$

Such transformations are useful for taking seasonal differences and hence removing seasonally varying means and further examples of using products of differences will be met later.

4.3 Transforming relationships between time series

When considering the relationship between two, or more, variables, it is usually the case that analysis and interpretation is made easier if the form of the relationship is linear. Linearity in the original data, however, may not exist and we must then look for transformations of the data that will induce linearity. If just two variables are being analysed then an obvious starting point is to consider a scatterplot of the data. If the plot shows a bend and a generally consistent trend either up or down, rather than a cup shape, then we may be able to straighten (linearise) the relationship by re-expressing one or both variables by using the family of power transformations.

We can get an idea of how straight the relationship between the two variables, x and y, is by using three *summary points*. These summary points are obtained by dividing the scatterplot into three regions: points with low x-values, points with middling x-values, and points with high x-values. Roughly a third of the points are allocated to each region, judgement being used to obtain a reasonably equal allocation. In each region, the medians of the x and y values are then found, realising, of course, that these median pairs need not be original data points. If the regions are labelled as left (L), middle (M), and right (R), these three pairs of summary points can be denoted as

$$(x_L, \quad y_L)$$

$$(x_M, \quad y_M)$$

$$(x_R, \quad y_R).$$

Part (i) of Exhibit 4.8 illustrates the positioning of these three summary points for an artificial data set. An idea of how straight the relationship is between x and y can be obtained by using the three summary points to approximate the slope in each half of the data by computing left and right *half-slopes*,

$$b_L = \frac{y_M - y_L}{x_M - x_L} \quad \text{and} \quad b_R = \frac{y_R - y_M}{x_R - x_M},$$

and finding the *half-slope ratio*, b_R/b_L. If the half-slopes are equal, the half-slope ratio is 1 and the x–y relationship is straight. If the half-slope ratio is not close to 1 then re-expressing x or y or both may help, although if the ratio is negative, the half-slopes will have different signs and re-expression will not help.

If the half-slopes are not equal, the plotted lines will meet at an angle, as shown in part (ii) of Exhibit 4.8. Following Velleman and Hoaglin (1981, chapter 5), we may think of this angle as forming an arrowhead that points towards re-expressions, using the family of power transformations, that might make the relationship straighter. If we denote by λ_x and λ_y the power transformation applied to x and y respectively, then if the arrowhead points upward, increasing λ_y is suggested, while if the arrowhead points downward, λ_y should be decreased. Changes in the value of λ_x depend on whether the arrowhead points to the left or to the

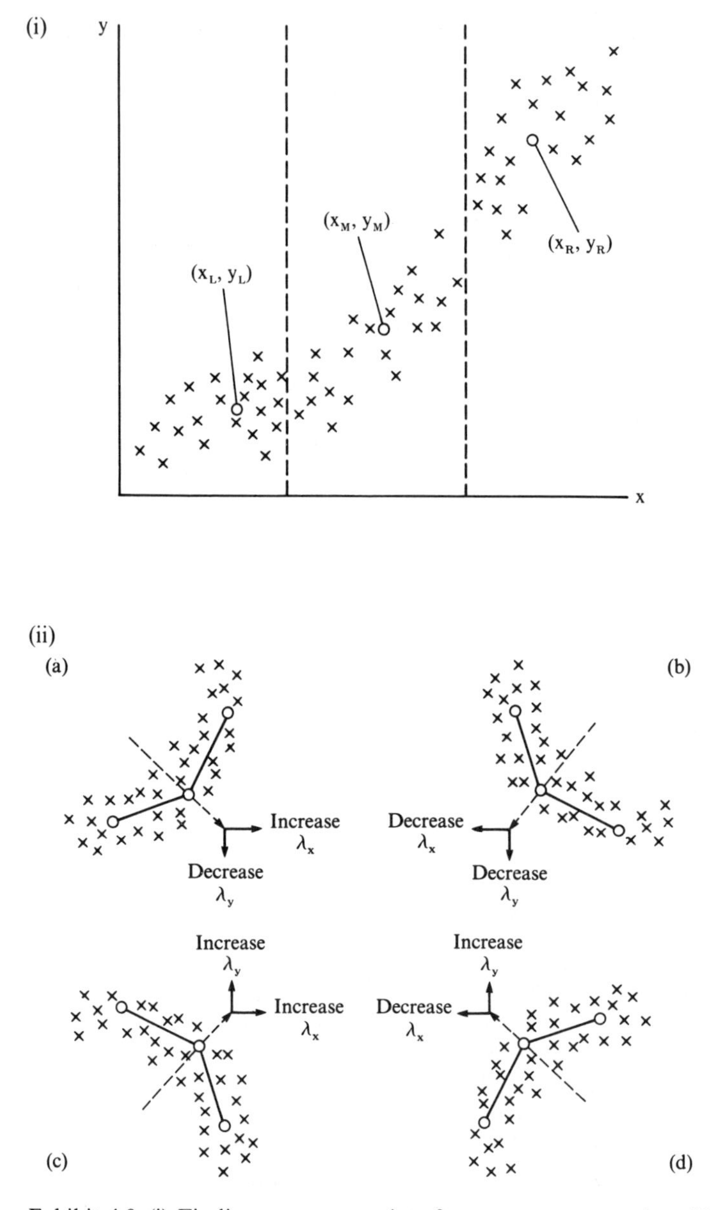

Exhibit 4.8 (i) Finding summary points from an x–y scatterplot. (ii) Choosing power transformations for x and y from half-slopes

right, if the former then λ_x should be decreased, if the latter it should be increased. Part (ii) of Exhibit 4.8 shows the four possible patterns that may occur.

The pair of half-slope lines meeting at the middle summary point will thus suggest re-expressing y or x or both. Often the nature of the data will suggest a preference for re-expressing one or the other, and on occasions there may well be some underlying theory that suggests a preferred 'functional form'. A quick way of examining the effects of alternative re-expressions is to note that, since the family of power transformations preserve order, the transformations of the raw summary points will be the summary points of the transformed series. Hence the half-slopes, and thus the half-slope ratio, can be computed for the transformed data just by re-expressing the summary points.

Example 4.1: Exploring the relationship between the money-income ratio and interest rates

Exhibit 4.9 presents a scatterplot of the M3-nominal income ratio (x) and the consol yield (y) for the period 1920 to 1982, an example inspired by Artis and Lewis (1984). Since there are 63 pairs of observations, the three regions contain 21 points each. From the summary points shown in Exhibit 4.9, the half-slopes are $b_L = -25.73$ and $b_R = -12.86$, yielding a half-slope ratio of 2.00. We obviously have an example of case (b) of Exhibit 4.8, and so decreasing λ_x or λ_y (or both) from unity is suggested.

Since the half-slope ratio is well above 1, and the bend in the plot is very pronounced, a major re-expression seems called for. Transforming the money-income ratio to logarithms ($\lambda_x = 0$) yields half-slopes $b_L = -12.14$ and $b_R = -7.91$, and hence a half-slope ratio of 1.54, which only marginally helps to straighten the plot. Transforming the consol yield to logarithms ($\lambda_y = 0$), on the other hand, yields half-slopes $b_L = -3.89$ and $b_R = -3.41$, with ratio 1.14, a much better re-expression. Artis and Lewis (1984), in fact, develop a model in which both variables are transformed logarithmically ($\lambda_x = \lambda_y = 0$) and such a re-expression obtains $b_L = -1.84$, $b_R = -2.09$, with half-slope ratio 0.88. This 'double log' transformation is more powerful than the previous two and is seen to go a little too far in straightening the relationship. Nevertheless, it would appear to be a useful re-expression and, of course, is a functional form commonly used for analysing the demand for money, since the slope coefficient will be the inverse of the interest elasticity of demand.

It is interesting to note that Artis and Lewis omitted the three outlying points that have been identified by their years in Exhibit 4.9 when

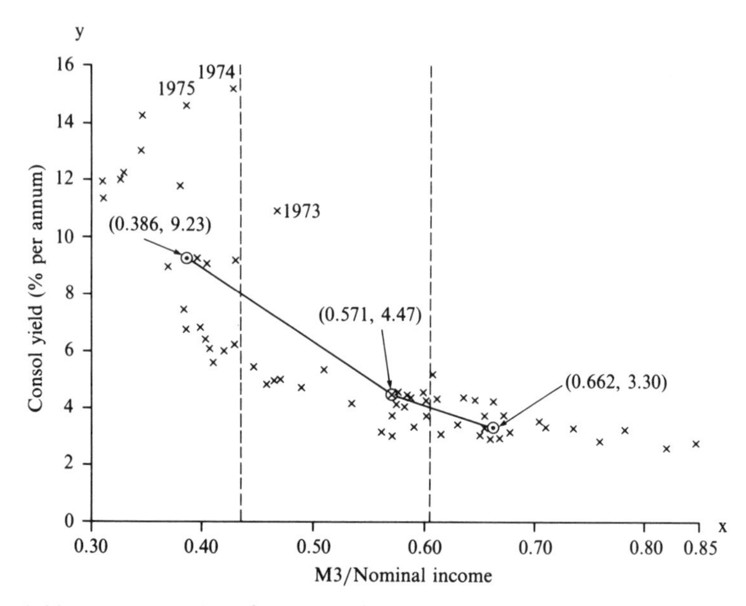

Exhibit 4.9 Scatterplot of (M3/nominal income) and consol yield, 1920 to 1982

computing their regression line. The use of summary points based on medians is fairly robust to outliers, however, and, in fact, omission of these points does little to the calculated half-slopes, the recalculated half-slope ratio for the untransformed data being 2.02. Indeed, computing the slope of the logarithmically transformed data yields a value of 1.91, the inverse of which, 0.52, bears a good correspondence with Artis and Lewis' estimate (0.59) of the interest elasticity obtained by using least squares regression on a truncated data set.

4.4 Smoothing time series

When we analyse time series in an exploratory fashion, we often wish to concentrate attention on 'large-scale' or 'long run' behaviour. To do this requires separating the data values using the following decomposition, borrowing the terminology of Tukey (1977),

$$\text{data} = \text{fit} + \text{residual}.$$

All that is required of the 'fit' at this stage of the analysis is that it be a simple description of the data that, ideally, captures the underlying pattern of the series. Because such a description is more than likely going to lead to a *smooth* series, the above separation is often written as

$$\text{data} = \text{smooth} + \text{rough},$$

where, by contrast, the residuals are termed the *rough*.

There are many methods for separating, or 'decomposing' a time series, and a number of these will be developed in later chapters. Most of these methods, however, require specific assumptions about the model generating the time series and/or specialised computations that can only readily be handled by special computer software. Such requirements are too stringent for exploratory analysis, where the techniques must be both relatively simple and robust. Tukey (1977, chapter 7) proposes using *moving* or *running medians* for smoothing time series and we follow this proposal here. Attention will be concentrated on non-seasonal time series: seasonality can be tackled using median smoothing techniques, but at the cost of much greater complexity (see, for example, Cleveland et al. (1978)).

4.4.1 *Running medians*

The simplest running median, that of order 3, replaces each value of a time series by the median of itself, its predecessor, and its successor. A value that is out of step with its neighbours will then be replaced by one or the other of them, whichever is closer. Medians of three, however, cannot correct for two outliers in a row, and so higher order running medians may be used, with five being a favourite. On occasions, running medians of four are used, calculated by taking the average of the two middle-sized values in each segment of four (middle sized by value, of course, not by position). It is usual to take a second running median, of order 2, to 'centre' an even-length running median.

4.4.2 *Hanning*

An alternative to running medians are *running weighted averages* (often termed moving averages), which replace each data value with the average of the data values around it. Tukey uses a particular form of running weighted average as a gentle smoother after outliers have been removed by a running-median smoother. This replaces x_t by

$$z_t = \tfrac{1}{4}x_{t-1} + \tfrac{1}{2}x_t + \tfrac{1}{4}x_{t+1}$$

and it is referred to by Tukey as *hanning*, after Julius von Hann, a nineteenth-century Austrian meteorologist who advocated its use.

4.4.3 *Resmoothing and reroughing*

The simple running medians introduced above may produce only crude smoothed sequences and may be improved upon by combining smoothing procedures. Applying one smoother to the results of a previous smoother is known as *resmoothing*. A useful resmoothing procedure for exploratory analysis is to use running medians of 3 and resmooth repeatedly until further resmoothing yields no further changes. This repeated combination is denoted '3R', and such repeated smoothers are called *compound smoothers*.

It is possible for running-median smoothers to smooth a time series too much and thus remove interesting patterns. To guard against this, patterns can be recovered by smoothing the rough obtained by residual from the resmoothing process and adding the result to the smooth sequence. By analogy with resmoothing, this is known as *reroughing*. Often the same smoother is used in both smoothing and reroughing, and when this occurs we say the smoother has been used 'twice'. Hence, for example, we could use '3R, twice'.

4.4.4 *Smoothing endpoints*

If we are using the smoother '3', say, then each value x_t is being replaced by

$$z_t = \text{median}\{x_{t-1}, x_t, x_{t+1}\}.$$

However, the end values z_1 and z_n cannot be obtained by this approach since z_0 and z_{n+1} are unknown. Often we can just 'copy-on', using x_1 and x_n as end smoothed values. However, smoothed end values can be obtained by extrapolating from the smoothed values near the ends of the series. From Velleman and Hoaglin (1981, chapter 6) this leads to

$$z_1 = \text{median}\{\hat{x}_0, x_1, z_2\}$$

$$z_n = \text{median}\{\hat{x}_{n+1}, x_n, z_{n-1}\},$$

where $\hat{x}_0 = 3z_2 - 2z_3$

and $\hat{x}_{n+1} = 3z_{n-1} - 2z_{n-2}.$

If we are using compound smoothers, '3R' say, then we will not make this adjustment every time, but it should be used at least once, usually at a late step in the smoothing process.

4.4.5 *Splitting*

The compound smoother '3R' is a very attractive one to use, as it is easily implemented by hand. Unfortunately, '3R' has a tendency to chop off peaks and valleys, leaving flat 'mesas' and 'dales' two points wide. These flat segments can be smoothed by an operation known as *splitting*, in which the data is split into three pieces: the two-point flat segment, the smooth data sequence to the left of the two points, and the smooth sequence to their right. Each point in the flat segment is then replaced by the median of itself, its successor (predecessor) and its extrapolated predecessor (successor), i.e. if, in the smooth of '3R', the sequences

$$\ldots, x_{f-4}, x_{f-3}, x_{f-2}$$

and

$$x_{f+1}, x_{f+2}, x_{f+3}, \ldots$$

are to the left and right of the two-point flat segment

$$x_{f-1}, x_f,$$

then we replace x_{f-1} and x_f by

$$z_{f-1} = \text{median}\{x_{f-2}, x_{f-1}, 3x_{f-2} - 2x_{f-3}\}$$

$$z_f = \text{median}\{3x_{f+1} - 2x_{f+2}, x_f, x_{f+1}\}.$$

After splitting each two-point segment, the entire sequence is resmoothed automatically by '3R'.

This has led to the following compound smoother that is feasible to do by hand and has been found to work reasonably well (see Velleman (1980) and Gebski and McNeil (1984) for detailed discussion of this and other nonlinear compound smoothers). It is termed '3RS(3R) S(3R) H, twice', so that after an initial '3R', splitting, denoted by 'S', is performed twice (each time automatically followed by '3R'), before the smoothing is finally completed by hanning, 'H'. The entire operation is then repeated to rerough the residual series, adding the result to the calculated smooth.

Example 4.2: Smoothing consumption growth

Exhibit 4.10 shows the steps of '3RS(3R) S(3R) H, twice' applied to the consumption growth series for the years 1970 to 1984, and Exhibit 4.11 provides plots of the smooths and roughs so obtained. The observed series is very erratic but the smoothed series obtained from the first application of the smoother eradicates most of the 'transitory' fluctuations, leaving a series that shows a pronounced fall in consumption growth in the mid 1970s followed by an almost linear increase to the end of the period. On reroughing a somewhat different view of the latter part

Exhibit 4.10 *Smoothing consumption growth by '3RS(3R) S(3R) H, twice'*

Consumption	3R[a]	S	(3R)	S[c]	(3R)	H_1	Rough	3R[a]	S	(3R)	S	(3R)	H_2	Smooth $= H_1 + H_2$
2.7	2.7[b]	2.7[b]	2.7[b]	3.1	3.1	3.1	−0.4	−0.4[b]	−0.4[b]	−0.4[b]	0	0	0	3.1
3.1	3.1	3.1	3.1	3.1	3.1	3.1	0	0	0	0]	0	0	0.15	3.25
6.1	5.2]	3.9	3.1	3.1	3.1	3.1	3.0	3.0]	0.8	0]	0.6	0.6	0.45	3.55
5.2	5.2]	−0.7	3.1	3.1	3.1	2.2	3.0	3.0]	−0.2	0.6	0.6	0.6	0.6	2.8
−1.4	−0.7]	5.2	−0.5	−0.5	−0.5	0.4	−1.8	−0.2]	3.0	0.6	0.6	0.6	0.6	1.0
−0.7	−0.7]	−0.7	−0.5	−0.5	−0.5	−0.5	−0.2	−0.2]	0.6	0.6	0.6	0.6	0.6	0.1
0.3	−0.5	−0.5	−0.5	−0.5	−0.5	−0.3	0.6	0.6]	−0.2	0.6	0.6	0.6	0.6	0.3
−0.5	0.3	0.3	0.3	0.3	0.3	0.1	0.6	0.6]	4.1	0.6	0.6	0.6	0.6	0.7
5.6	4.5]	1.9	0.3	0.3	0.3	0.3	5.3	4.1]	0.6	0.6	0.6	0.6	0.6	0.9
4.5	4.5]	−0.3	0.3	0.3	0.3	0.4	4.1	4.1]	−1.0	0.6	0.6	0.6	0.39	0.8
−0.4	−0.3]	4.5	0.7	0.7	0.7	0.6	−1.0	−1.0]	4.1	−0.25	−0.25	−0.25	−0.04	0.56
−0.3	−0.3]	−0.3	0.7	0.7	0.7	0.7	−1.0	−1.0]	−0.75	−0.25	−0.25	−0.25	−0.25	0.45
0.7	0.7	0.7	0.7	0.7	0.7	0.95	−0.25	−0.25	−0.25	−0.25	−0.25	−0.25	−0.19	0.76
4.0	1.7	1.7	1.7	1.7	1.7	1.45	2.55	0	0	0	0	0	−0.06	1.39
1.7	1.7[b]	1.7[b]	1.7[b]	1.7	1.7	1.7	0	0	0	0	0	0	0	1.7

[a] only one pass required
[b] copying-on end values
[c] only end point extrapolation required

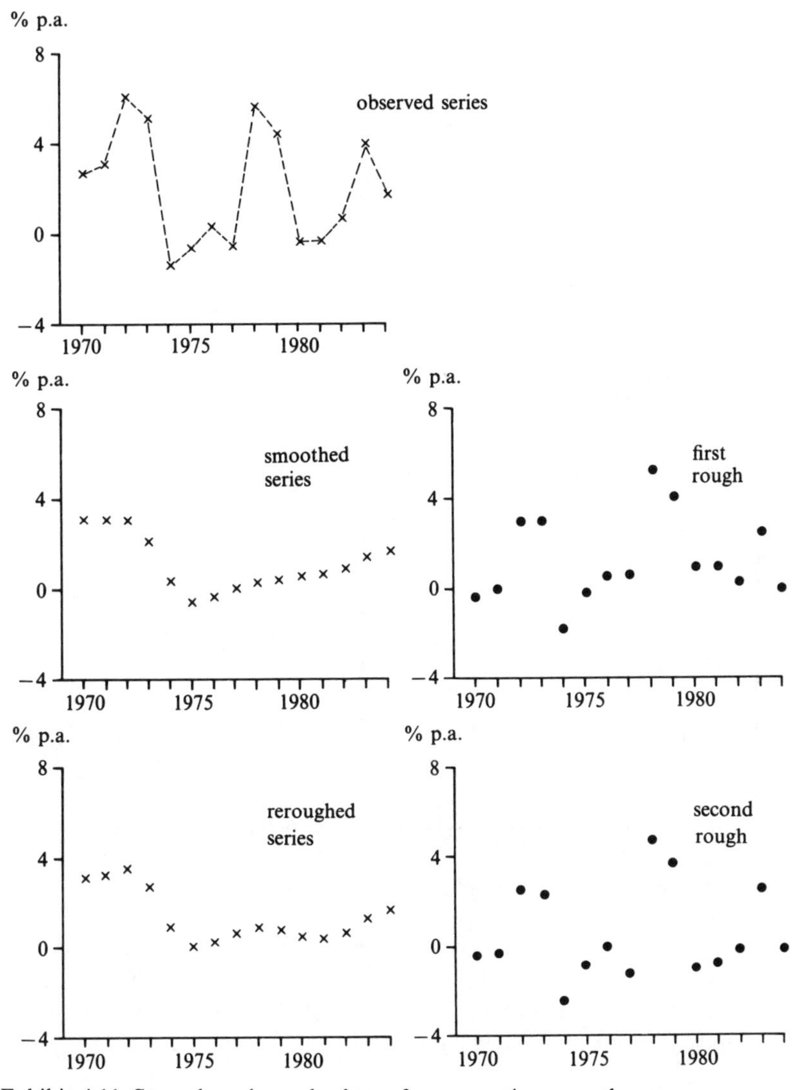

Exhibit 4.11 Smooth and rough plots of consumption growth

of the period emerges. Rather than a linear increase, there is now a second decline between 1979 and 1981, before the series recovers in the last three years of the period, and it may be argued that this provides a better representation of the behaviour of consumption growth during these years. The plots of the rough show that it is the years of high consumption growth that provide the 'outlying' observations.

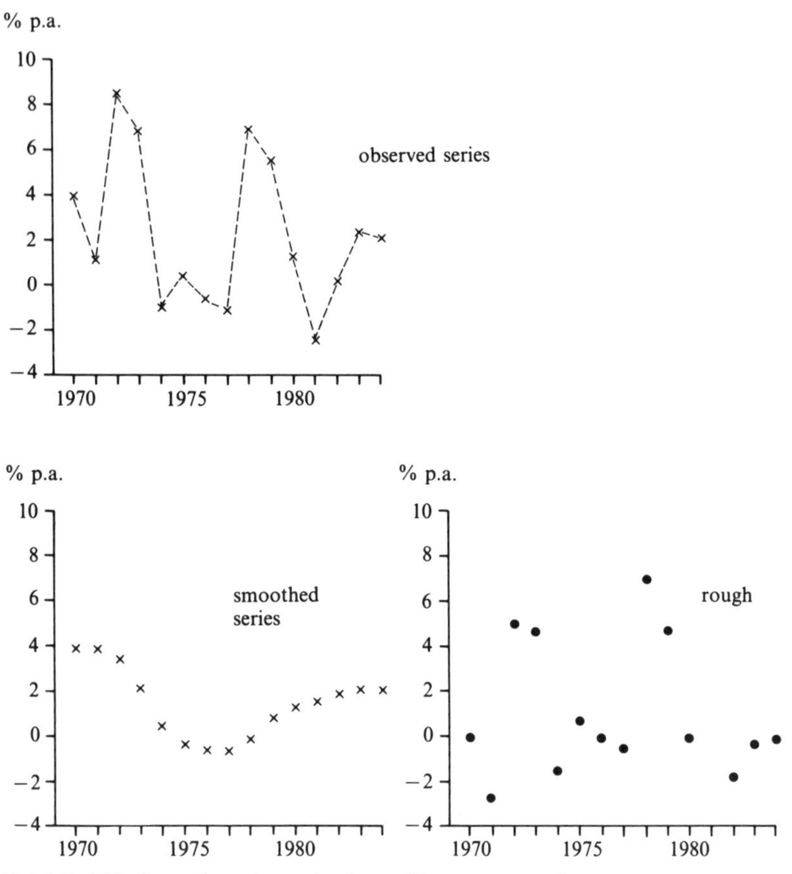

Exhibit 4.12 Smooth and rough plots of income growth

Exhibit 4.12 provides smooth and rough plots for income growth over the same fifteen-year period, with a fairly similar pattern to that of consumption growth emerging. It is interesting to note that for this series, reroughing did not alter the smooth and rough sequences.

4.5 Computing software

Although all the techniques discussed in this chapter are able to be performed by hand using just a calculator, power transformations and differencing operators are available as mathematical functions on all statistical packages. The compound smoother introduced above is available as part of the RSMOOTH command on MINITAB (Ryan et al. 1985).

Part II

The modelling of univariate economic time series

Having introduced various techniques for undertaking exploratory analysis of economic time series, we now move on to developing formal statistical models for individual series analysed in isolation. A great body of theoretical work exists on the analysis of time series that can be regarded as both stochastic and stationary; see, for example, Anderson (1971), Fuller (1976) and Wold (1938). We begin this part of the book by briefly reviewing, in a relatively informal fashion, the salient features of this theory and, in particular, developing the analysis of autoregressive-moving average models, often referred to as ARMA models (Chapter 5). Since relatively few economic time series are actually stationary, Chapter 6 extends this analysis to nonstationary time series, introducing the class of integrated processes, associated with Box and Jenkins (1976), to model a particular form of nonstationarity, that of homogeneous nonstationarity. Although forecasting, per se, is not the primary aim of this book, certain aspects of the theory of forecasting using this extended class of autoregressive-integrated-moving average (ARIMA) processes are extremely important for understanding the full implications and properties of fitting such models to economic time series, and this is developed in Chapter 7.

The actual methodology of fitting ARIMA models to observed time series is the subject of Chapter 8, with particular attention being paid to recent developments in the identification (specification) of models and in the diagnostic checking of fitted models. While ARIMA models have become widespread in economic applications, it is important to emphasise that an alternative methodology, based upon local polynomial trends fitted by the technique of exponential smoothing, has been in common use in forecasting applications for many years. Chapter 9 discusses these techniques, and places strong emphasis on the 'structural' models that underlie them. Recent research has uncovered the links between such general exponential smoothing models and ARIMA processes, and this chapter discusses these connections and the implications they have for the interpretation of both families of models.

As we have already seen, a distinguishing feature of many economic time series is the presence of seasonal fluctuations, and any class of models that has pretensions of being useful to economists must be capable of adequately dealing with seasonality. Chapter 10 extends the analysis of the previous chapters in this direction and also examines the process of seasonal adjustment, intimately linked with the modelling of seasonality, in the light of recent developments in the theory of ARIMA models. The final chapter in this part of the book deals with some recent developments in time series analysis that are of particular interest to economists, examining such diverse topics as state space models and the Kalman filter, trend stationary and difference stationary processes, trend-cycle decompositions and signal extraction, aggregation of ARIMA models, fractional differencing and long memory models, and R^2 measures for time series. This chapter concludes with a brief discussion of the software available for the statistical analysis of univariate time series.

5 Stationary stochastic time series models

5.1 Stochastic processes, ergodicity and stationarity

When we come to modelling time series formally, it is useful to regard an observed series, $(x_1, x_2, \ldots, x_n)$, as a particular *realisation* of a stochastic process. In general, stochastic processes can be described by an n-dimensional probability distribution $p(x_1, x_2, \ldots, x_n)$, so that the relationship between a realisation and a stochastic process is analogous to that between the sample and population in classical statistics. Specifying the complete form of the probability distribution will generally be too ambitious and we usually content ourselves with concentrating attention on the first and second moments: the n means

$$E(x_1), E(x_2), \ldots, E(x_n);$$

n variances

$$V(x_1), V(x_2), \ldots, V(x_n);$$

and $n(n-1)/2$ covariances

$$\text{Cov}(x_i, x_j), i < j.$$

If we could assume joint normality of the distribution, this set of expectations would then completely characterise the properties of the stochastic process. If normality could not be assumed, but the process was taken to be linear, in the sense that the present value of the process is generated by a linear combination of previous values of the process itself and present and past values of any other processes, then again this set of expectations would capture its major properties. In either case, however, it will be impossible to infer all the values of the first and second moments from just one realisation of the process, since there are only n observations but $n+n(n+1)/2$ unknown parameters. Hence further simplifying assumptions must be made to reduce the number of unknown parameters to more manageable proportions. It should also be noted that the procedure of using a single realisation to infer the unknown parameters of the joint probability distribution is only strictly valid if the process is *ergodic*, which roughly means that the sample moments for finite stretches of the realisation approach their population moments as the length of the realisation becomes infinite. For more on ergodicity, see Granger and Newbold (1977, chapter 1) and Nerlove et al. (1979, chapter 2), and since it is impossible to test for ergodicity using just (part of) a single realisation, it will be assumed from now on that all time series have this property.

5.1.1 *Stationarity*

One important simplifying assumption, already informally introduced in Chapter 4, is that of *stationarity*, which requires the process to be in a particular state of 'statistical equilibrium' (Box and Jenkins (1976, page 26)). A stochastic process is said to be *strictly stationary* if its properties are unaffected by a change of time origin; in other words, the joint probability distribution at *any* set of times $t_1, t_2, \ldots, t_m$ must be the same as the joint probability distribution at times $t_1+k, t_2+k, \ldots, t_m+k$, where k is an arbitrary shift along the time axis. For $m = 1$, this implies that the marginal probability distribution at time t is the same as the marginal probability distribution at any other point in time; $p(x_t) = p(x_{t+k})$. Hence the marginal distribution does not depend on time, which in turn implies that the mean and variance of x_t must be constant, i.e.

$$E(x_1) = E(x_2) = \ldots = E(x_n) = E(x_t) = \mu,$$

and

$$V(x_1) = V(x_2) = \ldots = V(x_n) = V(x_t) = \sigma^2.$$

If $m = 2$, stationarity implies that all bivariate distributions $p(x_t, x_{t-k})$ do not depend on t; thus the covariances are functions only of the lag k, and not of time t, i.e. for all k,

$$\begin{aligned}\mathrm{Cov}(x_1, x_{1+k}) &= \mathrm{Cov}(x_2, x_{2+k}) = \ldots = \mathrm{Cov}(x_{n-k}, x_n)\\ &= \mathrm{Cov}(x_t, x_{t-k}).\end{aligned}$$

As we have seen, the stationarity assumption implies that the mean and the variance of the process are constant and that the autocovariances,

$$\gamma_k = \mathrm{Cov}(x_t, x_{t-k}) = E[(x_t-\mu)(x_{t-k}-\mu)]$$

and autocorrelations,

$$\rho_k = \frac{\mathrm{Cov}(x_t, x_{t-k})}{[V(x_t)\cdot V(x_{t-k})]^{\frac{1}{2}}} = \frac{\gamma_k}{\gamma_0}, \tag{5.1}$$

depend only on the lag (or time difference) k. Since these conditions apply only to the first- and second-order moments of the process, this is known as *second-order* or *weak stationarity*, sometimes also referred to as stationarity in the *wide sense*. If, in addition, joint normality is assumed, so that the distribution is entirely characterised by these first two moments, strict stationarity and weak stationarity are equivalent. The autocorrelations considered as a function of k are referred to as the *autocorrelation function* (ACF) or, sometimes, the *correlogram*. Note that since

$$\gamma_k = \mathrm{Cov}(x_t, x_{t-k}) = \mathrm{Cov}(x_{t-k}, x_t) = \mathrm{Cov}(x_t, x_{t+k}) = \gamma_{-k},$$

it follows that $\rho_k = \rho_{-k}$ and so only the positive half of the ACF is usually given. The ACF plays a major role in modelling the dependencies among observations since it characterises, together with the process mean $\mu = E(x_t)$ and variance $\gamma_0 = V(x_t)$, the stationary stochastic process describing the evolution of x_t. It therefore indicates, by measuring the extent to which one value of the process is correlated with previous values, the length and strength of the 'memory' of the process.

5.1.2 *Sample autocorrelation function* (*SACF*)

In general μ, γ_0 and the ρ_k will be unknown. With our stationarity and (implicit) ergodicity assumptions, μ and γ_0 can be estimated by the sample mean and sample variance respectively:

$$\bar{x} = n^{-1} \sum_{t=1}^{n} x_t,$$

$$s^2 = n^{-1} \sum_{t=1}^{n} (x_t - \bar{x})^2,$$

and an estimate of ρ_k is then given by the lag k *sample autocorrelation*

$$r_k = \frac{\sum_{t=k+1}^{n} (x_t - \bar{x})(x_{t-k} - \bar{x})}{\sum_{t=1}^{n} (x_t - \bar{x})^2}, \quad k = 1, 2, \ldots. \tag{5.2}$$

For uncorrelated observations ($\rho_k = 0$, for all $k \neq 0$), the variance of r_k is approximately given by n^{-1} (see, for example, Box and Jenkins (1976, chapter 2)). If, as well, n is large, r_k will be approximately normally distributed, so that an absolute value of r_k in excess of $2n^{-\frac{1}{2}}$ may be regarded as significantly different from zero. More generally, if $\rho_k = 0$ for $k > q$, the variance of r_k, for $k > q$, is given by (Bartlett (1946))

$$V(r_k) = n^{-1}(1 + 2\rho_1^2 + \ldots + 2\rho_q^2). \tag{5.3}$$

(a)

k	1	2	3	4	5	6	7	8
r_k	0.86	0.72	0.64	0.56	0.45	0.33	0.19	0.10
s.e.(r_k)	0.16	0.25	0.30	0.33	0.36	0.37	0.38	0.38

k	9	10	11	12	13	14	15	16
r_k	−0.01	−0.09	−0.13	−0.22	−0.32	−0.37	−0.39	−0.40
s.e.(r_k)	0.38	0.38	0.38	0.38	0.39	0.39	0.40	0.41

k	17	18	19	20	21	22	23	24
r_k	−0.38	−0.34	−0.31	−0.30	−0.28	−0.22	−0.18	−0.16
s.e.(r_k)	0.42	0.43	0.44	0.44	0.45	0.45	0.46	0.46

Here s.e.$(r_K) = [V(r_k)]^{\frac{1}{2}}$

(b)

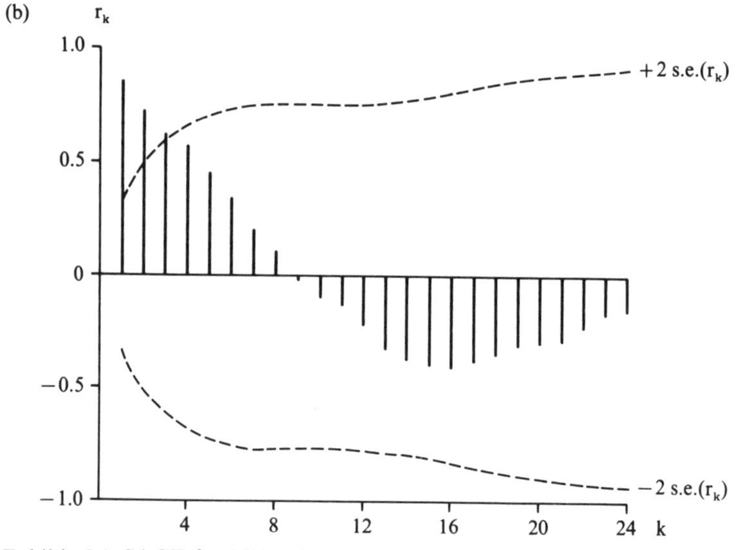

Exhibit 5.1 SACF for UK private sector housing starts series

Thus, by successively increasing the value of q and replacing the ρ_j's by their sample estimates, the variances of $r_1, r_2, \ldots, r_k$ can then be estimated as $n^{-1}, n^{-1}(1+2r_1^2), \ldots, n^{-1}(1+2r_1^2+\ldots+2r_{k-1}^2)$ and, of course, these will be larger than those calculated using the simple formula n^{-1}.

Example 5.1: SACF of annual housing starts

Consider again the annual UK private sector housing starts series shown in Exhibit 2.2. On the assumption that the stochastic process generating the series is both stationary and ergodic, the following sample moments can be calculated:

$$\bar{x} = 140; \quad s^2 = \hat{\gamma}_0 = 3872; \quad n = 39.$$

The first 24 values of the SACF, calculated using equation (5.2), are listed as part (a) of Exhibit 5.1, along with accompanying standard errors obtained from equation (5.3). The SACF is often easier to interpret when displayed graphically. This is shown as part (b) of Exhibit 5.1, where two standard error bounds are also plotted. From this Exhibit a number of points are of interest. The two standard error bounds computed using equation (5.3) are much larger than the 'naive' $2n^{-\frac{1}{2}}$ bounds, but nevertheless, r_1, r_2 and r_3 all exceed these bounds. There is a distinct 'cyclical' pattern to the r_k, although this would not have been apparent if the SACF had been truncated at values less than 15, say. In Chapter 2, we described this series as having no general trend but fluctuating widely around a mean level that itself changes quite often. We now see that although the series fluctuates widely, there is a high degree of autocorrelation between successive observations. This dependency is of obvious importance for any successful modelling of the series and it is to the development of formal models for time series that we now turn.

5.2 Stochastic difference equations

A fundamental theorem in time series analysis, known as *Wold's decomposition* (Wold, 1938), states that every weakly stationary, purely nondeterministic, stochastic process $(x_t - \mu)$ can be written as a *linear* combination (or linear filter) of a sequence of uncorrelated random variables. By purely nondeterministic we mean that any linearly deterministic components have been subtracted from x_t. Such a component is one that can be perfectly predicted from past values of itself and examples commonly found are a (constant) mean, as is implied by writing the process as $(x_t - \mu)$, periodic sequences, and polynomial or exponential sequences in t. A formal discussion of this theorem, well beyond the scope of this book, may be found in, for example, Nerlove et al. (1979, chapter 2), but Wold's decomposition underlies all the theoretical models of time series that are subsequently introduced.

This *linear filter* representation is given by

$$\begin{aligned} x_t - \mu &= a_t + \psi_1 a_{t-1} + \psi_2 a_{t-2} + \dots \\ &= \sum_{j=0}^{\infty} \psi_j a_{t-j}, \quad \psi_0 = 1. \end{aligned} \tag{5.4}$$

The $\{a_t: t = 0, \pm 1, \pm 2, \ldots\}$ are a sequence of uncorrelated random variables from a fixed distribution (often assumed to be normal) with

$$E(a_t) = 0, V(a_t) = E(a_t^2) = \sigma^2,$$

and

$$\mathrm{Cov}(a_t, a_{t-k}) = E(a_t a_{t-k}) = 0, \text{ for all } k \neq 0.$$

We will usually refer to such a sequence as a *white-noise process*, a term borrowed from the engineering literature (see Granger and Newbold (1977, page 51)). The coefficients ψ_j in the linear filter are known as ψ *(psi)-weights*: their number can be either finite or infinite.

We can easily show that the model (5.4) will lead to autocorrelation in x_t. From this equation it follows that

$$E(x_t) = \mu$$

$$\begin{aligned}
\gamma_0 &= V(x_t) = E(x_t - \mu)^2 \\
&= E(a_t + \psi_1 a_{t-1} + \psi_2 a_{t-2} + \ldots)^2 \\
&= E(a_t^2) + \psi_1^2 E(a_{t-1}^2) + \psi_2^2 E(a_{t-2}^2) + \ldots \\
&= \sigma^2 + \psi_1^2 \sigma^2 + \psi_2^2 \sigma^2 + \ldots \\
&= \sigma^2 \sum_{j=0}^{\infty} \psi_j^2,
\end{aligned}$$

by using the result that $E(a_{t-i} a_{t-j}) = 0$ for $i \neq j$. Now

$$\begin{aligned}
\gamma_k &= E(x_t - \mu)(x_{t-k} - \mu) \\
&= E(a_t + \psi_1 a_{t-1} + \ldots + \psi_k a_{t-k} + \psi_{k+1} a_{t-k-1} + \ldots) \\
&\quad \times (a_{t-k} + \psi_1 a_{t-k-1} + \ldots) \\
&= \sigma^2 (1 \cdot \psi_k + \psi_1 \psi_{k+1} + \psi_2 \psi_{k+2} + \ldots) \\
&= \sigma^2 \sum_{j=0}^{\infty} \psi_j \psi_{j+k},
\end{aligned}$$

and this implies

$$\rho_k = \frac{\sum_{j=0}^{\infty} \psi_j \psi_{j+k}}{\sum_{j=0}^{\infty} \psi_j^2}.$$

If the number of ψ-weights in (5.4) is infinite, we have to assume that the weights converge absolutely ($\sum |\psi_j| < \infty$). This condition can be shown to be equivalent to assuming that x_t is stationary, and guarantees that all moments exist and are independent of time, in particular that the variance of x_t, γ_0, is finite.

Although equation (5.4) appears complicated, many realistic models result from particular choices of these ψ-weights. Taking $\mu = 0$ without loss of generality and choosing $\psi_1 = -\theta$ and $\psi_j = 0, j \geqslant 2$, the *first-order moving average* model is obtained:

$$x_t = a_t - \theta a_{t-1}.$$

If $\psi_j = \phi^j$ is chosen, then

$$\begin{aligned} x_t &= a_t + \phi a_{t-1} + \phi^2 a_{t-2} + \dots \\ &= a_t + \phi(a_{t-1} + \phi a_{t-2} + \dots) \\ &= \phi x_{t-1} + a_t. \end{aligned}$$

This is the *first-order autoregressive* model, and both these and more general models are considered in detail in the following sections. In this and subsequent developments, we will make the further assumption that the $\{a_t\}$ are a sequence of *independent*, rather than just uncorrelated, random variables drawn from a fixed distribution. If that distribution is normal, then the two assumptions are identical, but the implications of making the assumption of independence, under fairly general conditions, are discussed in Chapter 16, where such an assumption is also relaxed, leading to classes of *nonlinear* time series models. We will also continue to assume, for ease of exposition, that the mean of x_t, μ, is zero.

5.3 Autoregressive processes

5.3.1 *First-order autoregressive process* [$AR(1)$]

As we have just seen, this model can be written as

$$x_t - \phi x_{t-1} = a_t. \tag{5.5}$$

The *backshift* (or *lag*) *operator* B is now introduced for notational convenience. This shifts time one step back, so that

$$Bx_t \equiv x_{t-1},$$

and, in general,

$$B^m x_t \equiv x_{t-m},$$

noting that $B^m \mu \equiv \mu$. The lag operator allows (possibly infinite) distributed lags to be written in a very concise way. For example, by using this notation, the AR(1) model can be written as

$$(1-\phi B)\,x_t = a_t, \tag{5.6}$$

so that

$$\begin{aligned} x_t &= (1-\phi B)^{-1} a_t \\ &= (1+\phi B+\phi^2 B^2+\ldots)\,a_t \\ &= a_t+\phi a_{t-1}+\phi^2 a_{t-2}+\ldots. \end{aligned}$$

This linear filter representation, with $\psi_j = \phi^j$, will converge as long as $|\phi| < 1$, which is therefore the stationarity condition. The ACF of the AR(1) process can be derived in the following way. Multiplying both sides of (5.5) by $x_{t-k}, k \geqslant 0$, and taking expectations, yields

$$E(x_t x_{t-k}) - \phi E(x_{t-1} x_{t-k}) = E(a_t x_{t-k})$$

or

$$\gamma_k - \phi\gamma_{k-1} = E(a_t x_{t-k}).$$

Now, for $k = 0$,

$$\begin{aligned} E(a_t x_t) &= E[a_t(a_t+\phi a_{t-1}+\phi^2 a_{t-2}+\ldots)] \\ &= E(a_t^2)+\phi E(a_t a_{t-1})+\phi^2 E(a_t a_{t-2})+\ldots \\ &= E(a_t^2) = \sigma^2. \end{aligned}$$

For $k > 0$,

$$\begin{aligned} E(a_t x_{t-k}) &= E[a_t(a_{t-k}+\phi a_{t-k-1}+\phi^2 a_{t-k-2}+\ldots)] \\ &= E(a_t a_{t-k})+\phi E(a_t a_{t-k-1})+\phi^2 E(a_t a_{t-k-2})+\ldots \\ &= 0. \end{aligned}$$

Hence

$$\gamma_0 - \phi\gamma_{-1} = \sigma^2 = \gamma_0 - \phi\gamma_1 \tag{5.7}$$

$$\gamma_k - \phi\gamma_{k-1} = 0, \quad k = 1, 2, \ldots. \tag{5.8}$$

Substituting $\gamma_1 = \phi\gamma_0$ into (5.7) yields

$$\gamma_0 = \frac{\sigma^2}{1-\phi^2},$$

and dividing (5.8) by γ_0 yields the difference equation

$$\frac{\gamma_k}{\gamma_0} = \rho_k = \phi\rho_{k-1},$$

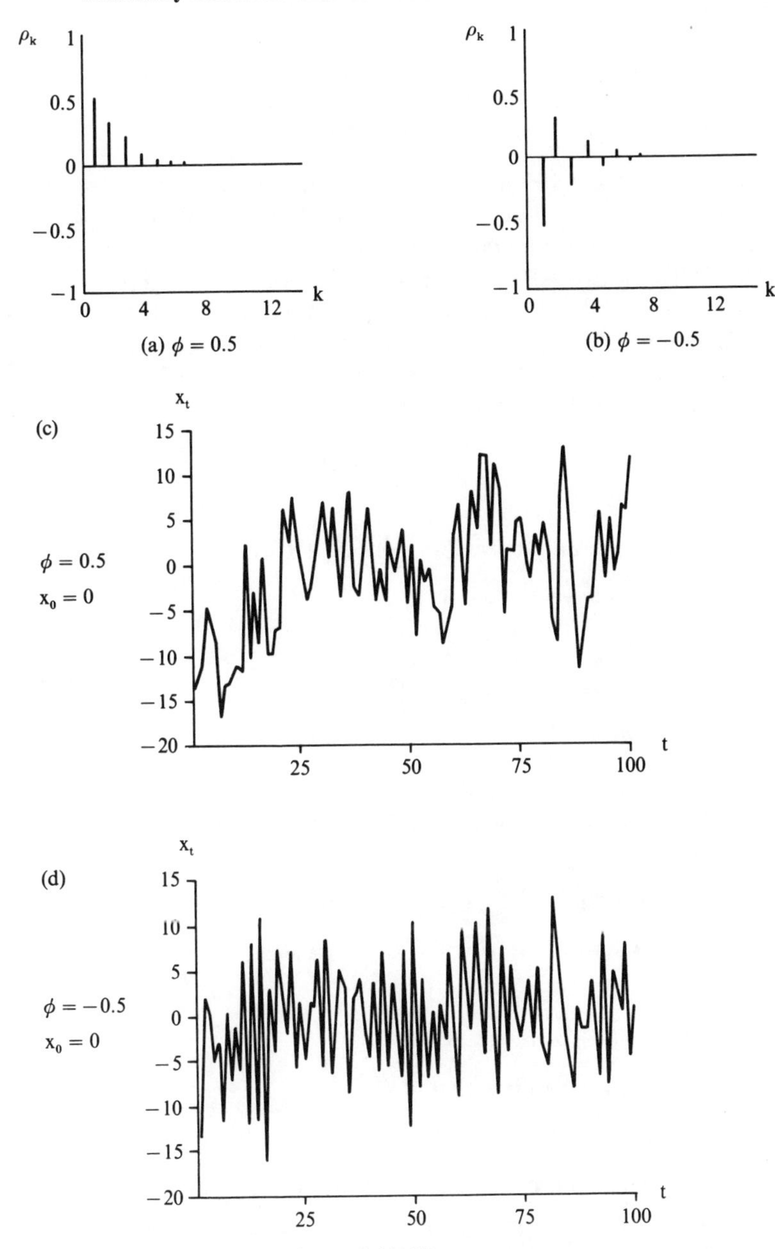

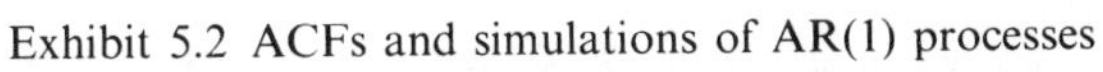
Exhibit 5.2 ACFs and simulations of AR(1) processes

which, on solving, gives

$$\rho_k = \phi\rho_{k-1} = \phi^2\rho_{k-2} = \dots$$
$$= \phi^{k-1}\rho_1 = \phi^k\rho_0 = \phi^k, \quad k = 0, 1, 2, \dots.$$

Thus, if $\phi > 0$, the ACF decays exponentially to zero, while if $\phi < 0$, the ACF decays in an oscillatory pattern, both decays being slow if ϕ is close to the nonstationary boundaries of $+1$ and -1.

The ACFs for two AR(1) processes with (a) $\phi = 0.5$, and (b) $\phi = -0.5$, are shown in Exhibit 5.2, along with generated data from the processes with a_t assumed to be $N(0, 25)$. With positive ϕ the adjacent values are positively correlated and the generated series has a tendency to exhibit 'low-frequency' trends. With negative ϕ, however, adjacent values have a negative correlation and the generated series displays violent, rapid oscillations.

5.3.2 *Second-order autoregressive process* [$AR(2)$]

This model can be written as

$$x_t = \phi_1 x_{t-1} + \phi_2 x_{t-2} + a_t$$

or, using the lag operator B, as

$$(1 - \phi_1 B - \phi_2 B^2) x_t = a_t. \tag{5.9}$$

This process can also be given a linear filter (or moving average) representation by noting that, in lag operator notation, equation (5.4) can be written

$$x_t = a_t + \psi_1 a_{t-1} + \psi_2 a_{t-2} + \dots$$
$$= (1 + \psi_1 B + \psi_2 B^2 + \dots) a_t = \psi(B) a_t,$$

where

$$\psi(B) = (1 + \psi_1 B + \psi_2 B^2 + \dots).$$

Equation (5.9) then requires that

$$(1 - \phi_1 B - \phi_2 B^2)^{-1} = \psi(B),$$

and the ψ-weights can be calculated by equating coefficients in

$$(1 - \phi_1 B - \phi_2 B^2)(1 + \psi_1 B + \psi_2 B^2 + \dots) = 1.$$

For this equality to hold, the coefficients of $B^j, j \geqslant 0$, on each side of the equation have to be the same, i.e.

$$B^1 : \psi_1 - \phi_1 = 0 \qquad \therefore \psi_1 = \phi_1$$

$$B^2 : \psi_2 - \phi_1 \psi_1 - \phi_2 = 0 \qquad \therefore \psi_2 = \phi_1^2 + \phi_2$$

$$B^3 : \psi_3 - \phi_1 \psi_2 - \phi_2 \psi_1 = 0 \quad \therefore \psi_3 = \phi_1^3 + 2\phi_1 \phi_2.$$

Noting that $\psi_3 = \phi_1 \psi_2 + \phi_2 \psi_1$, the ψ-weights can then be derived recursively for $j \geqslant 2$ from

$$\psi_j = \phi_1 \psi_{j-1} + \phi_2 \psi_{j-2}.$$

For stationarity, it is required that these ψ_j weights converge, which implies that conditions on ϕ_1 and ϕ_2 have to be imposed. Now, for the AR(1) process, stationarity required that $|\phi| < 1$, or, equivalently, that the solution to

$$1 - \phi B = 0,$$

which is ϕ^{-1}, had to be bigger than 1 in absolute value. For the AR(2) process, solutions have to be obtained to the quadratic equation (known as the *characteristic equation*)

$$(1 - \phi_1 B - \phi_2 B^2) = (1 - g_1 B)(1 - g_2 B) = 0.$$

These solutions are given by

$$g_1, g_2 = (\phi_1 \pm (\phi_1^2 + 4\phi_2)^{\frac{1}{2}})/2$$

and can both be real, or they can be a pair of complex numbers. For stationarity, it is required that the roots be such that $|g_1| < 1$ and $|g_2| < 1$, and it can be shown that these conditions imply the following set of restrictions on ϕ_1 and ϕ_2:

$$\phi_1 + \phi_2 < 1$$

$$-\phi_1 + \phi_2 < 1$$

$$-1 < \phi_2 < 1.$$

The roots will be complex if $\phi_1^2 + 4\phi_2 < 0$, although a necessary condition for complex roots is simply that $\phi_2 < 0$.

The ACF of an AR(2) process can be derived in a similar fashion to that of an AR(1). Multiplying (5.9) by x_{t-k} and taking expectations yields

$$E(x_t x_{t-k}) - \phi_1 E(x_{t-1} x_{t-k}) - \phi_2 E(x_{t-2} x_{t-k}) = E(a_t x_{t-k}),$$

or

$$\gamma_k - \phi_1 \gamma_{k-1} - \phi_2 \gamma_{k-2} = E(a_t x_{t-k}).$$

As we have already shown,

$$E(a_t x_{t-k}) = \begin{cases} \sigma^2, & \text{for} \quad k = 0, \\ 0, & \text{for} \quad k = 1, 2, \ldots \end{cases}$$

so that

$$\gamma_0 - \phi_1 \gamma_{-1} - \phi_2 \gamma_{-2} = \sigma^2 = \gamma_0 - \phi_1 \gamma_1 - \phi_2 \gamma_2 \tag{5.10}$$

$$\gamma_k - \phi_1 \gamma_{k-1} - \phi_2 \gamma_{k-2} = 0, \quad k = 1, 2, \ldots. \tag{5.11}$$

Thus the autocorrelations

$$\rho_k - \phi_1 \rho_{k-1} - \phi_2 \rho_{k-2} = 0, \quad k = 1, 2, \ldots \tag{5.12}$$

or

$$(1 - \phi_1 B - \phi_2 B^2) \rho_k = 0$$

follow a second-order difference equation. For $k = 1$ we have

$$\rho_1 = \phi_1 \rho_0 + \phi_2 \rho_{-1} = \phi_1 \rho_0 + \phi_2 \rho_1 = \frac{\phi_1}{1 - \phi_2},$$

for $k = 2$,

$$\rho_2 = \phi_1 \rho_1 + \phi_2 \rho_0 = \frac{\phi_1^2}{1 - \phi_2} + \phi_2,$$

and, for all other lags, (5.12) can be used recursively. Furthermore, equation (5.10) can be written as

$$\gamma_0 (1 - \phi_1 \rho_1 - \phi_2 \rho_2) = \sigma^2,$$

and substituting for ρ_1 and ρ_2 yields the variance of x_t as

$$\gamma_0 = \frac{1 - \phi_2}{1 + \phi_2} \cdot \frac{\sigma^2}{(1 - \phi_2)^2 - \phi_1^2} = \frac{1 - \phi_2}{1 + \phi_2} \cdot \frac{\sigma^2}{(\phi_1 + \phi_2 - 1)(\phi_2 - \phi_1 - 1)}. \tag{5.13}$$

If x_t is stationary and non-deterministic, then $0 < \gamma_0 < \infty$, and equation (5.13) shows that for this inequality to hold the stationarity conditions given earlier must be satisfied.

The behaviour of the ACF for alternative values of ϕ_1 and ϕ_2 can be studied by considering the solution of the difference equation (5.11), which is

$$\rho_k = A_1 g_1^k + A_2 g_2^k, \quad k = 0, 1, 2, \ldots$$

where A_1 and A_2 are constants determined by the initial conditions $\rho_0 = 1$ and $\rho_{-1} = \rho_1$. If the roots are equal ($g_1 = g_2 = g$), the solution is given by

$$\rho_k = (A_1 + A_2 k) g^k, \quad k = 0, 1, 2, \ldots.$$

The behaviour of the ACF for four combinations of (ϕ_1, ϕ_2) is shown in

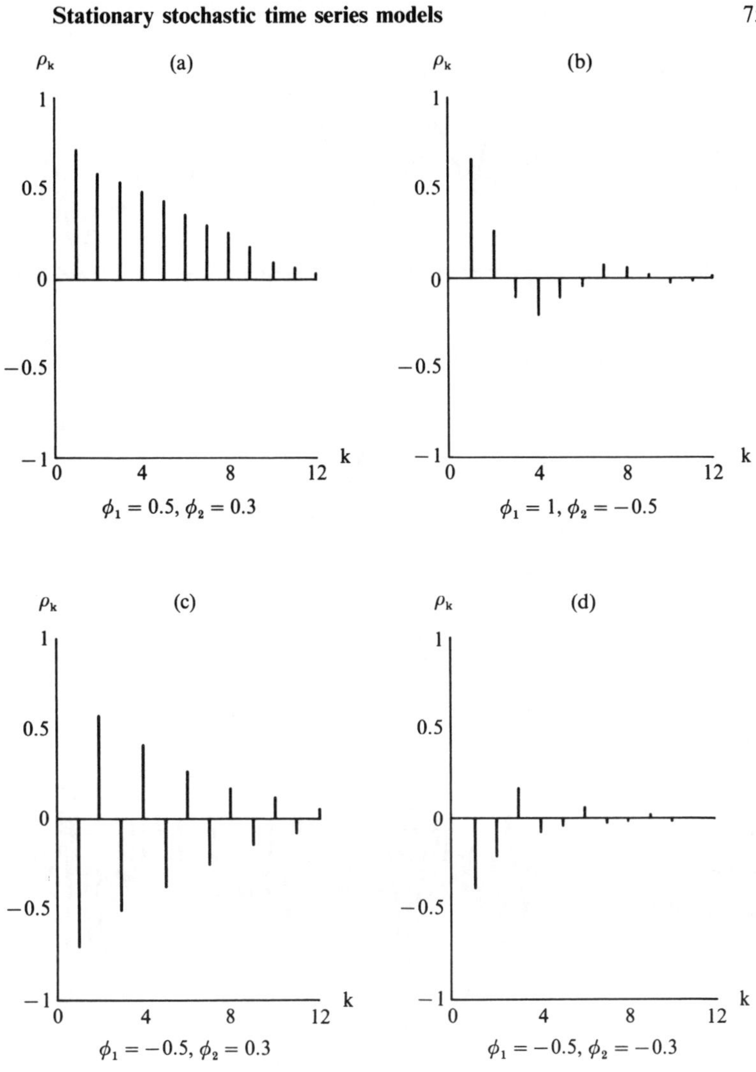

Exhibit 5.3 ACFs of various AR(2) processes

Exhibit 5.3. If g_1 and g_2 are real (cases (a) and (c)), the ACF is a mixture of two damped exponentials. Depending on their sign, the autocorrelations can also damp out in an oscillatory manner. If the roots are complex (cases (b) and (d)), the ACF follows a damped sine wave. Exhibit 5.4 shows plots of generated time series from these four AR (2) processes, in each case with $a_t \sim N(0, 25)$. Depending on the signs of the real roots, the series may be smooth or jagged, while complex roots tend to induce 'pseudo-periodic' behaviour.

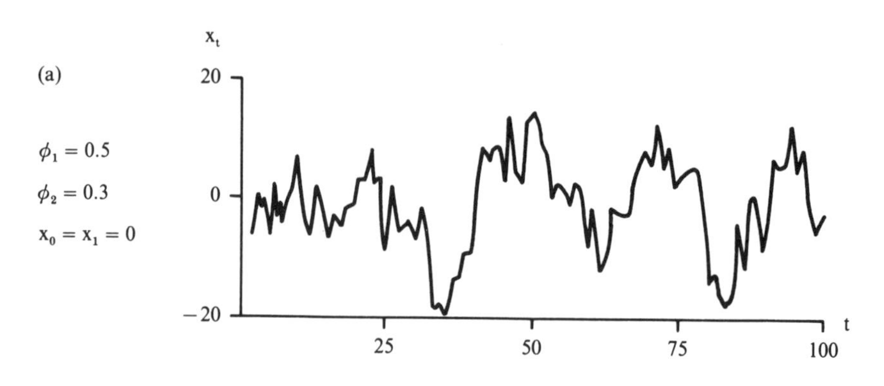

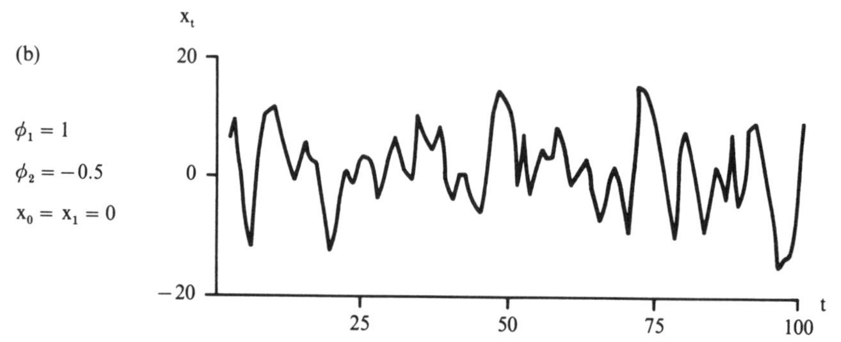

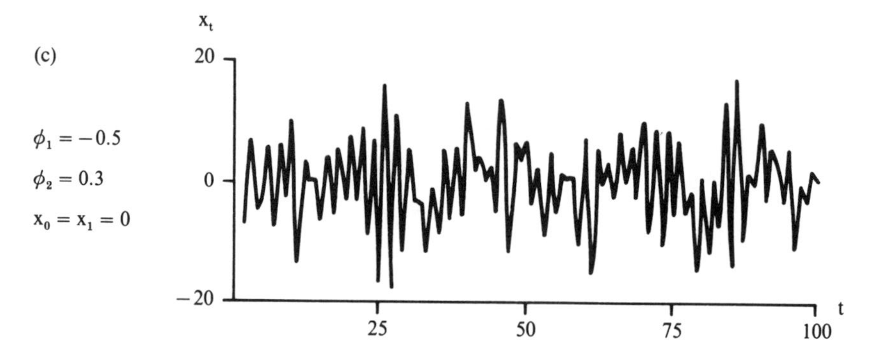

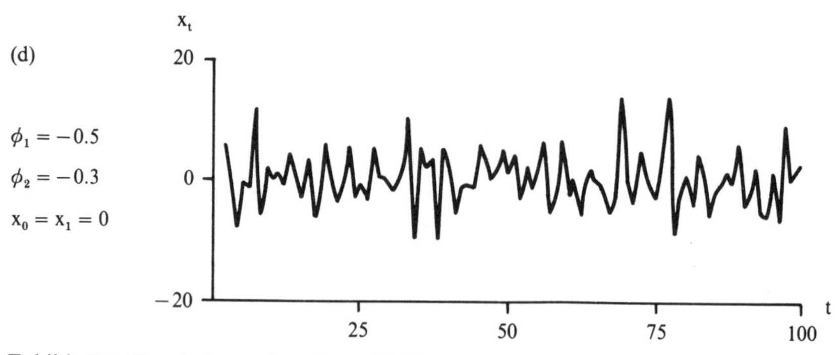

Exhibit 5.4 Simulations of various AR(2) processes

5.3.3 *Autoregressive process of order p* $[AR(p)]$

The general autoregressive model of order p can be written as

$$x_t - \phi_1 x_{t-1} - \ldots - \phi_p x_{t-p} = a_t$$

or

$$(1 - \phi_1 B - \ldots - \phi_p B^p) x_t = a_t, \tag{5.14}$$

i.e.

$$\phi(B) x_t = a_t.$$

Again, the linear filter representation $x_t = \psi(B) a_t$ can be obtained by equating coefficients in

$$\phi(B) \psi(B) = 1.$$

The stationarity conditions required for the convergence of the ψ-weights are that the roots of the characteristic equation

$$\phi(B) = (1 - g_1 B)(1 - g_2 B) \ldots (1 - g_p B) = 0$$

are such that $|g_i| < 1$ for $i = 1, 2, \ldots, p$, an alternative phrase being that the roots g_i^{-1} all lie outside the unit circle. It is straightforward to show that

$$\gamma_0 - \phi_1 \gamma_1 - \ldots - \phi_p \gamma_p = \sigma^2$$

so that

$$\gamma_0 = \frac{\sigma^2}{1 - \phi_1 \rho_1 - \ldots - \phi_p \rho_p}.$$

Furthermore, it can also be shown that the autocovariances and autocorrelations follow pth order difference equations:

$$\gamma_k = \phi_1 \gamma_{k-1} + \ldots + \phi_p \gamma_{k-p}, \quad k > 0 \tag{5.15}$$

$$\rho_k = \phi_1 \rho_{k-1} + \ldots + \phi_p \rho_{k-p}, \quad k > 0. \tag{5.16}$$

The first p equations in (5.16) are called the *Yule–Walker* equations:

$$\begin{aligned}
\rho_1 &= \phi_1 + \rho_1 \phi_2 + \ldots + \rho_{p-1} \phi_p \\
\rho_2 &= \rho_1 \phi_1 + \phi_2 + \ldots + \rho_{p-2} \phi_p \\
&\vdots \\
\rho_p &= \rho_{p-1} \phi_1 + \rho_{p-2} \phi_2 + \ldots + \phi_p
\end{aligned}$$

and these can be written in matrix form as

$$\boldsymbol{\rho} = \boldsymbol{P} \boldsymbol{\phi} \tag{5.17}$$

where $\boldsymbol{\rho} = (\rho_1, \rho_2, \ldots, \rho_p)'$, $\boldsymbol{\phi} = (\phi_1, \phi_2, \ldots, \phi_p)'$, and

$$P = \begin{bmatrix} 1 & \rho_1 & \rho_2 & \cdots & \rho_{p-1} \\ \rho_1 & 1 & \rho_1 & \cdots & \rho_{p-2} \\ \vdots & \vdots & \vdots & & \vdots \\ \rho_{p-1} & \rho_{p-2} & \rho_{p-3} & \cdots & 1 \end{bmatrix}.$$

Thus

$$\boldsymbol{\phi} = P^{-1}\boldsymbol{\rho},$$

so that the autoregressive parameters can be expressed as a function of the first p autocorrelations.

The difference equation in (5.16),

$$\phi(B)\,\rho_k = 0,$$

determines the behaviour of the ACF. Its solution is

$$\rho_k = A_1 g_1^k + \ldots + A_p g_p^k, \quad k = 0, 1, 2, \ldots$$

and since $|g_i| < 1$, the ACF is described by a mixture of damped exponentials (for real roots) and damped sine waves (for complex roots).

5.3.4 *Partial autocorrelations*

Since all autoregressive processes imply ACFs that damp out, it is sometimes difficult to distinguish between models of different orders. To help with this problem of discrimination, a new concept is introduced, the *partial autocorrelation function* (PACF). In general, the correlation between two random variables is often due to both variables being correlated with a third variable. In the context of time series, a large portion of the correlation between x_t and x_{t-k} can be due to the correlation these variables have with the intervening lags $x_{t-1}, x_{t-2}, \ldots, x_{t-k+1}$. To adjust for this correlation, the *partial autocorrelations* may be calculated.

The lag k partial autocorrelation is the partial regression coefficient ϕ_{kk} in the kth order autoregression

$$x_t = \phi_{k1} x_{t-1} + \phi_{k2} x_{t-2} + \ldots + \phi_{kk} x_{t-k} + a_t, \tag{5.18}$$

and it measures the additional correlation between x_t and x_{t-k} after adjustments have been made for the intermediate variables $x_{t-1}, \ldots, x_{t-k+1}$.

In general, the ϕ_{kk} can be obtained from the Yule–Walker equations that correspond to (5.18). These are given by (5.16) with $p = k$ and $\phi_i = \phi_{ii}$, and solving for the last coefficient ϕ_{kk} using Cramer's Rule leads to

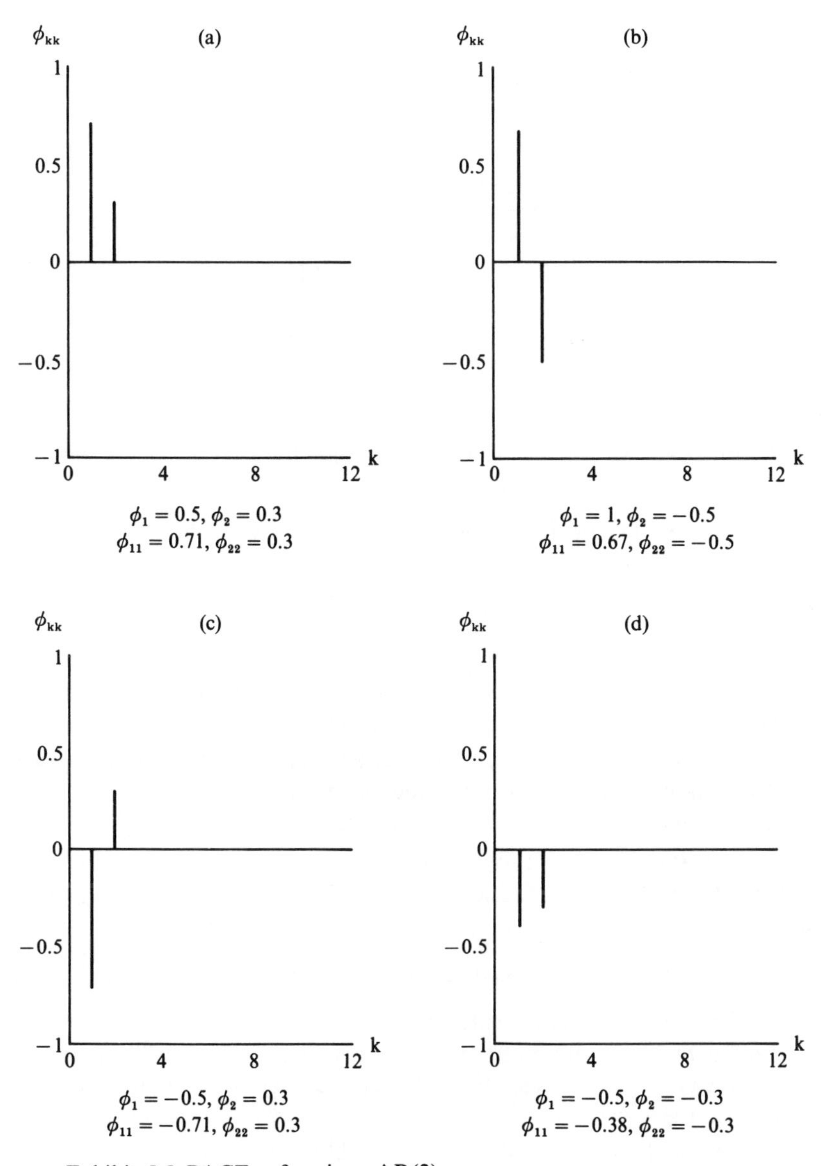

Exhibit 5.5 PACFs of various AR(2) processes

$$\phi_{kk} = \frac{\begin{vmatrix} 1 & \rho_1 & \cdots & \rho_{k-2} & \rho_1 \\ \rho_1 & 1 & \cdots & \rho_{k-3} & \rho_2 \\ \cdot & \cdot & \cdots & \cdot & \cdot \\ \cdot & \cdot & \cdots & \cdot & \cdot \\ \cdot & \cdot & \cdots & \cdot & \cdot \\ \rho_{k-1} & \rho_{k-2} & \cdots & \rho_1 & \rho_k \end{vmatrix}}{\begin{vmatrix} 1 & \rho_1 & \cdots & \rho_{k-2} & \rho_{k-1} \\ \rho_1 & 1 & \cdots & \rho_{k-3} & \rho_{k-2} \\ \cdot & \cdot & \cdots & \cdot & \cdot \\ \cdot & \cdot & \cdots & \cdot & \cdot \\ \cdot & \cdot & \cdots & \cdot & \cdot \\ \rho_{k-1} & \rho_{k-2} & \cdots & \rho_1 & 1 \end{vmatrix}}.$$

It follows from the definition of ϕ_{kk} that the PACFs of autoregressive processes are of a particular form:

$$\text{AR}(1): \phi_{11} = \rho_1 = \phi, \qquad \phi_{kk} = 0 \quad \text{for} \quad k > 1.$$

$$\text{AR}(2): \phi_{11} = \rho_1, \phi_{22} = \frac{\rho_2 - \rho_1^2}{1 - \rho_1^2}, \qquad \phi_{kk} = 0 \quad \text{for} \quad k > 2.$$

$$AR(p): \phi_{11} = 0, \phi_{22} \neq 0, \ldots, \phi_{pp} \neq 0, \quad \phi_{kk} = 0 \quad \text{for} \quad k > p.$$

Thus the partial autocorrelations for lags larger than the order of the process are zero. Hence, an AR (p) process is described by:

(i) an ACF that is infinite in extent and is a combination of damped exponentials and damped sine waves, and

(ii) a PACF that is zero for lags larger than p.

The PACFs for the four AR(2) processes shown in Exhibit 5.3 are depicted in Exhibit 5.5, and each show the distinctive cut-off at lag 2.

5.3.5 *Sample partial autocorrelation function* (*SPACF*)

The SPACF is usually calculated by fitting autoregressive models of increasing order: the estimate of the last coefficient in each model is the sample partial autocorrelation. In fact, it can be shown that the estimates, $\hat{\phi}_{kj}$, of the coefficients in (5.18) can be obtained recursively using the following updating equations (Durbin (1960))

$$\hat{\phi}_{kk} = \frac{r_k - \sum_{j=1}^{k-1} \hat{\phi}_{k-1,j} r_{k-j}}{1 - \sum_{j=1}^{k-1} \hat{\phi}_{k-1,j} r_j}$$

$$\hat{\phi}_{kj} = \hat{\phi}_{k-1,j} - \hat{\phi}_{kk} \hat{\phi}_{k-1,k-j}, \quad j = 1, 2, \ldots, k-1.$$

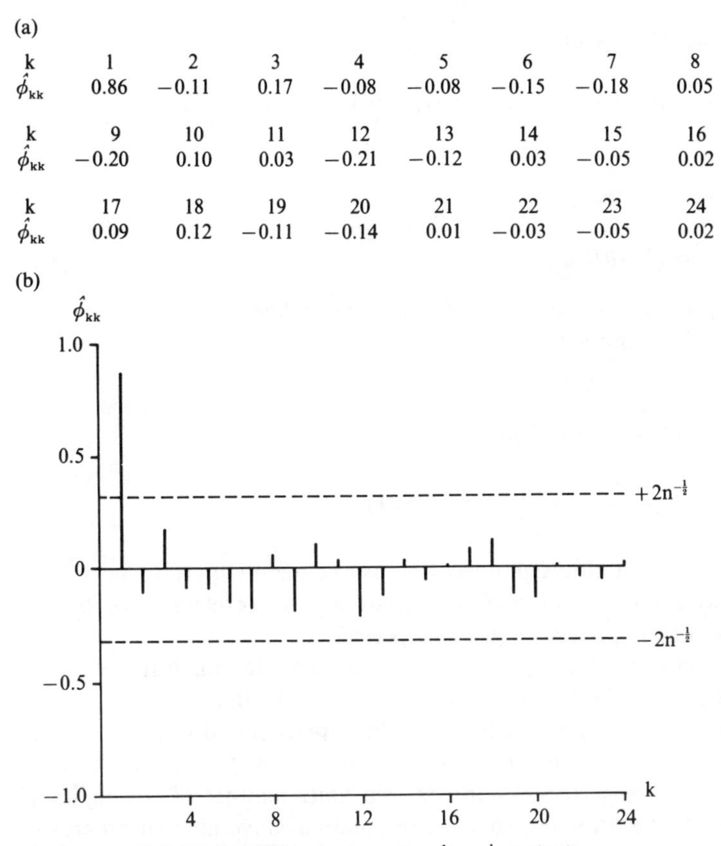

(a)

k	1	2	3	4	5	6	7	8
$\hat{\phi}_{kk}$	0.86	−0.11	0.17	−0.08	−0.08	−0.15	−0.18	0.05
k	9	10	11	12	13	14	15	16
$\hat{\phi}_{kk}$	−0.20	0.10	0.03	−0.21	−0.12	0.03	−0.05	0.02
k	17	18	19	20	21	22	23	24
$\hat{\phi}_{kk}$	0.09	0.12	−0.11	−0.14	0.01	−0.03	−0.05	0.02

Exhibit 5.6 SPACF for UK private sector housing starts

If the data follow an AR(p) process, then for lags greater than p the variance of $\hat{\phi}_{kk}$ can be approximated by

$$V(\hat{\phi}_{kk}) = n^{-1}, \quad \text{for} \quad k > p.$$

Example 5.2: SPACF of annual housing starts

The first 24 values of the SPACF of the UK private housing starts series are listed in part (a) of Exhibit 5.6 and displayed graphically in part (b). Since the $2n^{-\frac{1}{2}}$ bounds are ± 0.32, only $\hat{\phi}_{11}$ is significantly different from zero and hence the evolution of this series is consistent with it being generated by an AR(1) process. We note, however, that the SPACF of the series, shown in Exhibit 5.1, displays a 'cyclical' pattern, which would suggest that it was generated by a process having complex roots, requiring an autoregressive process of at least order 2. The resolution of such an apparent contradiction will be discussed in subsequent chapters.

5.4 Moving average processes

5.4.1 *First order moving average process* [$MA(1)$]

The MA (1) model is given by

$$x_t = a_t - \theta a_{t-1}$$

or

$$x_t = (1-\theta B)\, a_t. \tag{5.19}$$

By setting $\psi_1 = -\theta$ and $\psi_j = 0$ for $j > 1$ in the linear filter representation (5.4), it follows immediately that

$$\gamma_0 = \sigma^2(1+\theta^2), \gamma_1 = -\sigma^2\theta, \gamma_k = 0 \quad \text{for} \quad k > 1.$$

Hence the ACF is described by

$$\rho_1 = \frac{-\theta}{1+\theta^2}, \rho_k = 0 \quad \text{for} \quad k > 1,$$

and it implies that although observations one period apart are correlated, observations more than one step apart are uncorrelated, so that the 'memory' of the process is just one period.

The expression for ρ_1 can be written as the quadratic equation $\theta^2\rho_1 + \theta + \rho_1 = 0$. Since θ must be real, it follows that $-0.5 < \rho_1 < 0.5$. However, both θ and $1/\theta$ will satisfy this equation and thus two MA (1) processes can always be found that correspond to the same ACF. Since any moving average model consists of a finite number of ψ-weights, all MA models are stationary. In order to obtain a converging autoregressive representation, however, the restriction $|\theta| < 1$ must be imposed. This restriction is known as the *invertibility* condition and it implies that the process can be written in terms of an AR (∞) representation

$$x_t = \pi_1 x_{t-1} + \pi_2 x_{t-2} + \ldots + a_t,$$

where the π (*pi*)-*weights* converge, i.e. $\sum |\pi_j| < \infty$. In fact, the MA (1) model can be written as

$$(1-\theta B)^{-1} x_t = a_t,$$

and expanding $(1-\theta B)^{-1}$ yields

$$(1+\theta B + \theta^2 B^2 + \ldots)\, x_t = a_t.$$

The weights $\pi_j = -\theta^j$ will converge if $|\theta| < 1$, i.e. if the model is invertible. This implies the reasonable assumption that the effect of past observations decreases with age.

5.4.2 *Second-order moving average process* [*MA* (2)]

The MA (2) model can be written as

$$x_t = a_t - \theta_1 a_{t-1} - \theta_2 a_{t-2}$$

or

$$x_t = (1 - \theta_1 B - \theta_2 B^2) a_t. \tag{5.20}$$

By substituting $\psi_1 = -\theta_1$, $\psi_2 = -\theta_2$ and $\psi_j = 0$ for $j > 2$ into the linear filter representation (5.4), it follows that

$$\gamma_0 = \sigma^2(1 + \theta_1^2 + \theta_2^2), \gamma_1 = \sigma^2(-\theta_1 + \theta_1 \theta_2),$$

$$\gamma_2 = -\sigma^2\theta_2, \gamma_k = 0 \quad \text{for} \quad k > 2,$$

thus leading to the ACF

$$\rho_1 = \frac{-\theta_1 + \theta_1 \theta_2}{1 + \theta_1^2 + \theta_2^2}, \rho_2 = \frac{-\theta_2}{1 + \theta_1^2 + \theta_2^2}, \rho_k = 0 \quad \text{for} \quad k > 2.$$

The model therefore implies that observations more than two periods apart are uncorrelated. The AR (∞) representation's π weights are obtained by equating coefficients of B^j in

$$(1 - \pi_1 B - \pi_2 B^2 - \ldots)(1 - \theta_1 B - \theta_2 B^2) = 1$$

and are given by

$$B^1: -\pi_1 - \theta_1 = 0 \qquad \therefore \pi_1 = -\theta_1$$

$$B^2: -\pi_2 + \theta_1 \pi_1 - \theta_2 = 0 \qquad \therefore \pi_2 = \theta_1 \pi_1 - \theta_2 = -\theta_1^2 - \theta_2$$

$$\cdot \qquad\qquad \cdot$$

$$B^j: -\pi_j + \theta_1 \pi_{j-1} + \theta_2 \pi_{j-2} = 0 \quad \therefore \pi_j = \theta_1 \pi_{j-1} + \theta_2 \pi_{j-2}, j > 2.$$

For invertibility, the roots of

$$(1 - \theta_1 B - \theta_2 B^2) = (1 - h_1 B)(1 - h_2 B) = 0$$

are required to obey the restrictions $|h_1| < 1$, $|h_2| < 1$, and these are assured if, analogous to the stationarity conditions for the AR (2) process,

$$\theta_1 + \theta_2 < 1, \theta_2 - \theta_1 < 1, -1 < \theta_2 < 1.$$

5.4.3 *Moving average process of order q* [*MA* (*q*)]

This model can be written as

$$x_t = a_t - \theta_1 a_{t-1} - \ldots - \theta_q a_{t-q}$$

or

$$x_t = (1 - \theta_1 B - \ldots - \theta_q B^q) a_t = \theta(B) a_t. \tag{5.21}$$

By setting $\psi_1 = -\theta_1, \ldots, \psi_q = -\theta_q, \psi_j = 0$ for $j > q$, the following autocorrelations are obtained:

$$\gamma_0 = \sigma^2(1+\theta_1^2+\ldots+\theta_q^2),$$

$$\gamma_k = \sigma^2(-\theta_k+\theta_1\theta_{k+1}+\ldots+\theta_{q-k}\theta_q), \quad k = 1,2,\ldots,q,$$

$$\gamma_k = 0, k > q,$$

leading to the ACF

$$\rho_k = \frac{-\theta_k+\theta_1\theta_{k+1}+\ldots+\theta_{q-k}\theta_q}{1+\theta_1^2+\ldots+\theta_q^2}, \quad k = 1,2,\ldots,q,$$

$$\rho_k = 0, k > q.$$

The ACF of the MA(q) process therefore cuts off after lag q: the memory of the process extends q periods, observations more than q periods apart being uncorrelated.

The weights in the AR(∞) representation $\pi(B)x_t = a_t$ are given by $\pi(B) = \theta^{-1}(B)$ and can be obtained by equating coefficients of B^j in $\pi(B)\theta(B) = 1$. For invertibility, the roots of

$$(1-\theta_1 B-\ldots-\theta_q B^q) = (1-h_1 B)\ldots(1-h_q B) = 0$$

must satisfy $|h_1| < 1, \ldots, |h_q| < 1$.

5.4.4 *Partial autocorrelation function for MA processes*

The ACF of an MA(q) process cuts off after lag q $(\rho_k = 0; k > q)$. However, it can be shown that its PACF is infinite in extent (it tails off). Expressions for the PACF of moving average processes are complicated but, in general, they are dominated by combinations of exponential decays (for the real roots in $\theta(B)$) and/or damped sine waves (for the complex roots). Their patterns are thus very similar to the ACFs of AR processes. Indeed, an important duality between AR and MA processes should by now have been noted: while the ACF of an AR(p) process is infinite in extent, the PACF cuts off after lag p. The ACF of an MA(q) process, on the other hand, cuts off after lag q, while the PACF is infinite in extent.

The ACFs and PACFs for various MA(1) and MA(2) processes are shown in Exhibit 5.7 and generated time series for each of them, again using $a_t \sim N(0, 25)$, are plotted in Exhibit 5.8. On comparison of these plots with those for the autoregressive processes shown in Exhibits 5.2 and 5.4, it is seen that realisations from the two types of processes are often quite similar, suggesting that it may, on occasions, be difficult to distinguish between the two forms of models.

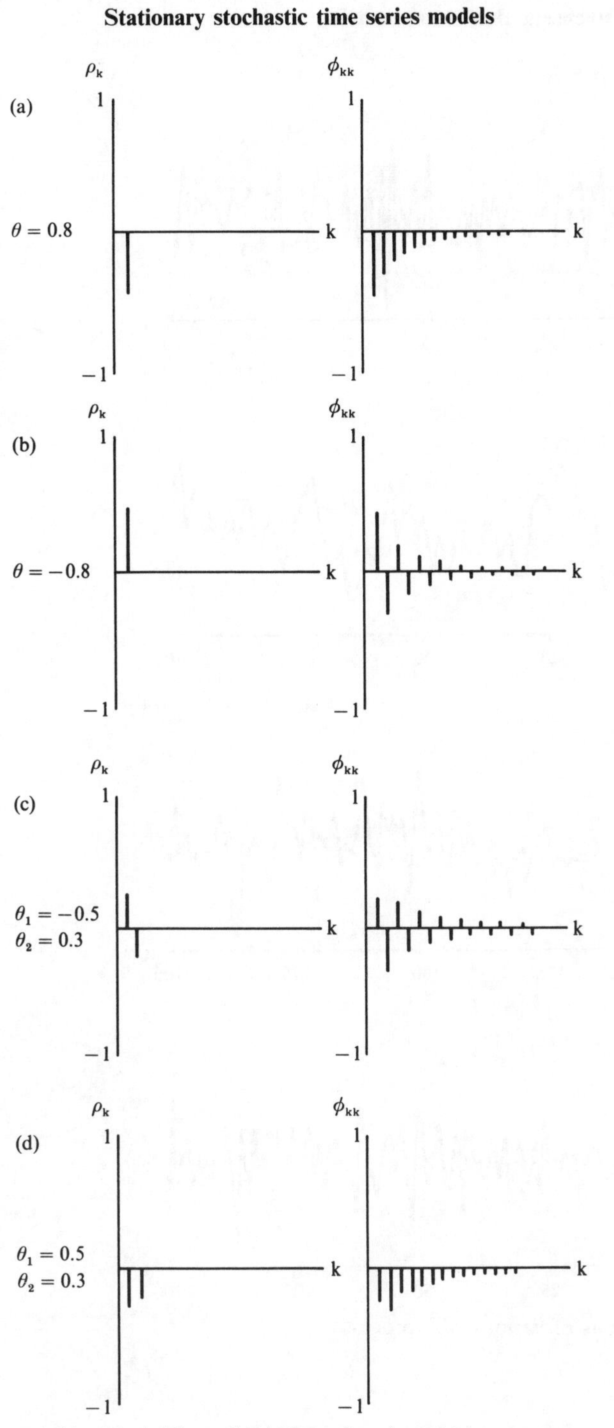

Exhibit 5.7 ACFs and PACFs of various MA processes

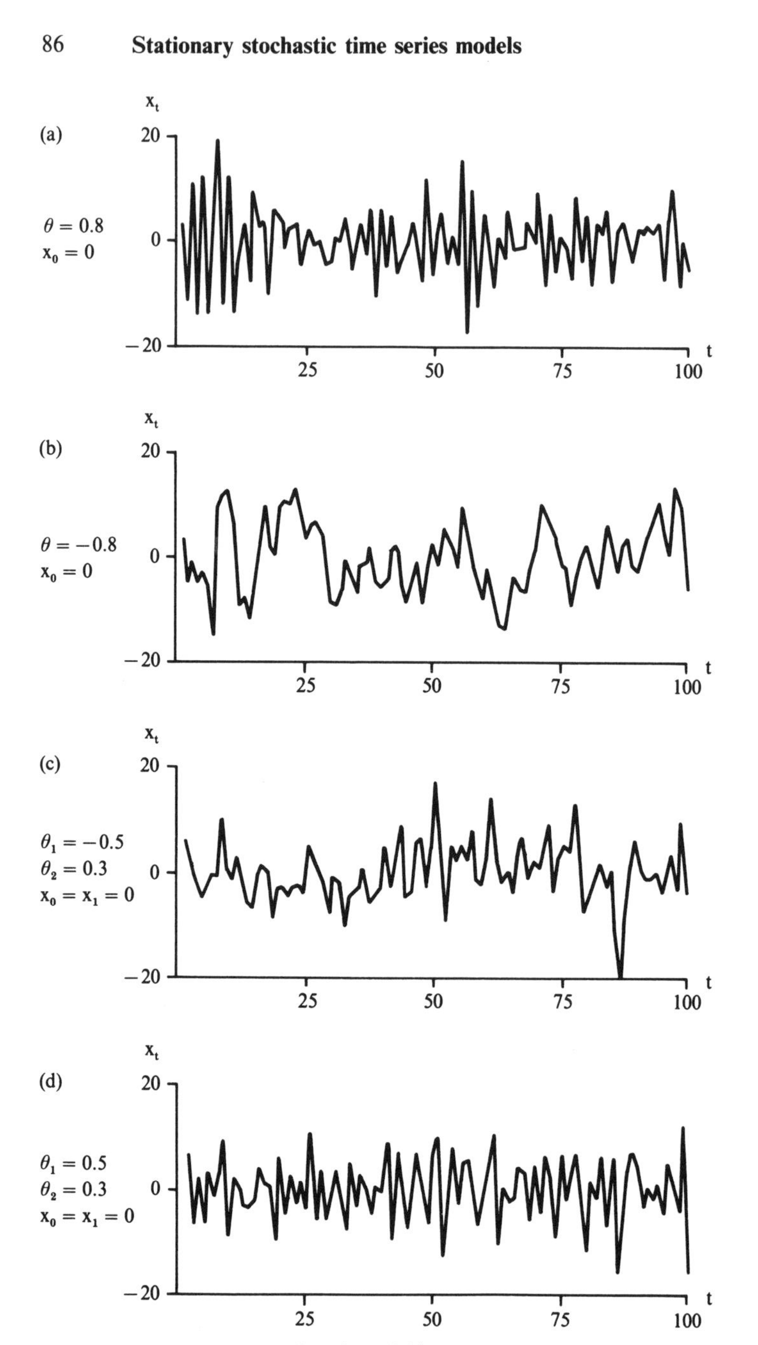

Exhibit 5.8 Simulations of various MA processes

5.5 Autoregressive moving average (ARMA) processes

A more general approach to representing the ψ-weights in the linear filter representation (5.4) is to consider $\psi(B)$ as a ratio of two finite polynomials in B:

$$\psi(B) = \frac{\theta(B)}{\phi(B)} = \frac{(1-\theta_1 B - \ldots - \theta_q B^q)}{(1-\phi_1 B - \ldots - \phi_p B^p)}.$$

This leads to the ARMA (p, q) model:

$$\phi(B)\, x_t = \theta(B)\, a_t \tag{5.22}$$

or

$$(1-\phi_1 B - \ldots - \phi_p B^p)\, x_t = (1-\theta_1 B - \ldots - \theta_q B^q)\, a_t$$

or

$$x_t - \phi_1 x_{t-1} - \ldots - \phi_p x_{t-p} = a_t - \theta_1 a_{t-1} - \ldots - \theta_q a_{t-q}.$$

In cases where a large number of parameters are needed in either the 'pure MA' or 'pure AR' models, a ratio of low-order MA and AR operators frequently leads to parsimonious representations of the ψ and π-weights.

5.5.1 *The ARMA* (1, 1) *process*

The simplest example of this class of 'mixed' models is the ARMA (1, 1):

$$(1-\phi B)\, x_t = (1-\theta B)\, a_t. \tag{5.23}$$

The ψ-weights in the MA (∞) representation

$$x_t = \psi(B)\, a_t$$

are given by $\psi(B) = (1-\theta B)/(1-\phi B)$. Equating coefficients of B^j in

$$(1-\phi B)(1+\psi_1 B + \psi_2 B^2 + \ldots) = 1 - \theta B$$

leads to

$$B^1: \psi_1 - \phi = -\theta \qquad \therefore\ \psi_1 = \phi - \theta$$

$$B^2: \psi_2 - \phi\psi_1 = 0 \qquad \therefore\ \psi_2 = \phi\psi_1 = (\phi-\theta)\,\phi$$

$$\cdot \qquad\qquad \cdot$$

$$B^j: \psi_j - \phi\psi_{j-1} = 0 \qquad \therefore\ \psi_j = \phi\psi_{j-1} = (\phi-\theta)\,\phi^{j-1}.$$

Equivalently, the π-weights in the AR (∞) representation

$$\pi(B)\, x_t = a_t$$

are given by $\pi(B) = (1-\phi B)/(1-\theta B)$. Equating coefficients of B^j in

$$(1-\theta B)(1-\pi_1 B - \pi_2 B^2 - \ldots) = 1 - \phi B$$

leads to

$$B^1: -\pi_1 - \theta = -\phi \qquad \therefore \pi_1 = \phi - \theta$$

$$B^2: -\pi_2 + \theta\pi_1 = 0 \qquad \therefore \pi_2 = \theta\pi_1 = (\phi - \theta)\,\theta$$

$$\cdot \qquad \qquad \cdot$$

$$B^j: -\pi_j + \theta\pi_{j-1} = 0 \quad \therefore \pi_j = \theta\pi_{j-1} = (\phi - \theta)\,\theta^{j-1}.$$

The ARMA (1, 1) model thus leads to both moving average and autoregressive representations having an infinite number of weights. The ψ-weights converge for $|\phi| < 1$ (the stationarity condition) and the π-weights converge for $|\theta| < 1$ (the invertibility condition). The stationarity condition for the ARMA (1, 1) model is thus the same as that of an AR (1) model: the invertibility condition is the same as that of an MA (1) model.

The model can be written as the difference equation

$$x_t - \phi x_{t-1} = a_t - \theta a_{t-1}. \tag{5.24}$$

Multiplying (5.24) by x_{t-k} and taking expectations obtains

$$\gamma_k - \phi\gamma_{k-1} = E(a_t\, x_{t-k}) - \theta E(a_{t-1}\, x_{t-k}).$$

If $k > 1$, $E(a_t\, x_{t-k}) = E(a_{t-1}\, x_{t-k}) = 0$, since a_t and a_{t-1} are uncorrelated with x_{t-k}. Therefore

$$\gamma_k - \phi\gamma_{k-1} = 0, \quad \text{for} \quad k > 1.$$

Furthermore, since

$$E(a_t\, x_t) = E(a_{t-1}\, x_{t-1}) = \sigma^2$$

and

$$E(a_{t-1}\, x_t) = E[a_{t-1}(a_t + \psi_1\, a_{t-1} + \psi_2\, a_{t-2} + \ldots)]$$

$$= \psi_1\, \sigma^2 = (\phi - \theta)\, \sigma^2,$$

it follows that, for $k = 0$,

$$\gamma_0 - \phi\gamma_1 = \sigma^2 - \theta(\phi - \theta)\, \sigma^2$$

and, for $k = 1$,

$$\gamma_1 - \phi\gamma_0 = -\theta\sigma^2.$$

Solving these two equations leads to

$$\gamma_0 = \frac{1 + \theta^2 - 2\phi\theta}{1 - \phi^2}\sigma^2,$$

$$\gamma_1 = \frac{(1 - \phi\theta)(\phi - \theta)}{1 - \phi^2}\sigma^2.$$

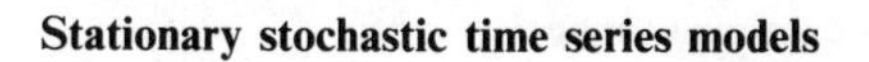

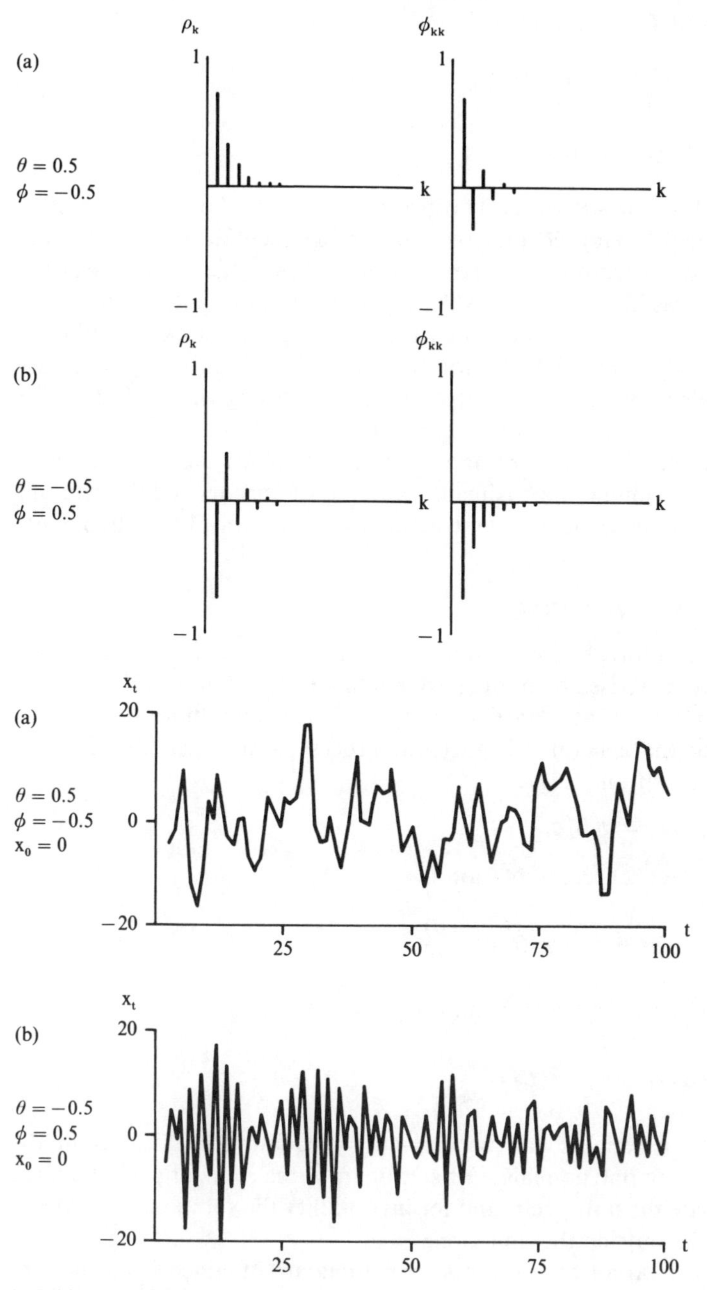

Exhibit 5.9 ARMA(1, 1) processes

Hence the ACF is given by

$$\rho_1 = \frac{(1-\phi\theta)(\phi-\theta)}{1+\theta^2-2\phi\theta}, \tag{5.25}$$

$$\rho_k = \phi\rho_{k-1}, \quad \text{for} \quad k > 1.$$

The ACF of an ARMA (1, 1) process is therefore similar to that of an AR (1) process, being characterised by an exponential decay. However, this decay starts from ρ_1, and not from $\rho_0 = 1$ as in the AR (1) case. The PACF behaves like that of an MA (1) process after an initial value $\phi_{11} = \rho_1$ and will therefore be dominated by an exponential decay. Exhibit 5.9 shows the ACFs and PACFs for two ARMA (1, 1) processes and also provides plots of series generated from them using $a_t \sim N(0, 25)$. The series have distinct patterns, a consequence of the autoregressive and moving average operators being very different. When the operators have roots that are similar, approximate 'cancellation' occurs and the resulting series will have little discernible structure (see Anderson (1976, chapter 6)).

5.5.2 *ARMA* (p, q) *process*

Unlike pure autoregressive or moving average models, the mixed ARMA model is characterised by both an ACF and a PACF that tail off to infinity rather than cut off at a particular lag. This is shown more formally by considering the general ARMA (p, q) process, which has already been defined in equation (5.22):

$$\phi(B)\, x_t = \theta(B)\, a_t.$$

The pure MA (∞) representation is

$$x_t = \psi(B)\, a_t, \quad \psi(B) = \frac{\theta(B)}{\phi(B)},$$

and the pure AR (∞) representation is

$$\pi(B)\, x_t = a_t, \quad \pi(B) = \frac{\phi(B)}{\theta(B)} = \psi^{-1}(B).$$

The ψ and π weights can be derived by equating coefficients in the appropriate lag polynomials. For stationarity, the roots of $\phi(B) = 0$ must all lie outside the unit circle, and for invertibility the solutions of $\theta(B) = 0$ must also lie outside the unit circle.

The autocovariances and autocorrelations are obtained from the pth-order difference equations

$$\gamma_k - \phi_1 \gamma_{k-1} - \ldots - \phi_p \gamma_{k-p} = 0, \quad k > q$$

and

$$\rho_k - \phi_1 \rho_{k-1} - \ldots - \phi_p \rho_{k-p} = 0, \quad k > q, \tag{5.26}$$

thus implying that eventually the ACF of an ARMA(p,q) process will follow the same pattern as that of an AR(p) process, being described by combinations of damped exponentials and/or damped sine waves. Since the p starting values $\rho_q, \rho_{q-1}, \ldots, \rho_{q-p+1}$ have to lie on the solution of equation (5.26), the solution itself holds for $k > q-p$. The initial $q-p+1$ starting values $\rho_0, \rho_1, \ldots, \rho_{q-p}$, however, will not follow this general pattern.

A number of special cases are of interest, and these may be summarised as follows:

ARMA(1, 1): $q-p+1 = 1$; one initial value ρ_0;
exponential decay from $k = 1$.
ARMA(2, 2): $q-p+1 = 1$; one initial value ρ_0;
exponential decay or damped sine wave from $k = 1$.
ARMA(1, 2): $q-p+1 = 2$; two initial values ρ_0, ρ_1;
exponential decay from $k = 2$.
ARMA(2, 1): $q-p+1 = 0$; no initial values:
all autocorrelations (for $k \geqslant 0$) are described by either a sum of two damped exponentials or a damped sine wave.

Eventually (for $k > p-q$), the PACF of an ARMA(p,q) process behaves like that of an MA(q) process. For $k \leqslant p-q$, however, the PACF does not follow this general pattern.

In summary, both the ACF and PACF of a mixed ARMA process will be infinite in extent and tail off as k increases. Eventually (for $k > q-p$), the ACF is determined from the AR part of the model, while for $k > p-q$ the PACF is determined from the MA part of the model.

Although more complicated than the pure autoregressive and moving average models, there are a number of reasons why mixed ARMA processes should occur frequently when modelling economic time series. These and further interpretations of the class of ARMA processes developed here are discussed in future chapters.

6 Modelling nonstationary processes

6.1 Introduction

The class of ARMA models developed in the previous chapter relies on the assumption that the underlying process is stationary, thus implying that the mean, variance and autocovariances of the process are invariant under time translations. As we have seen, this restricts the mean and variance to be constant and requires the autocovariances to depend only on the time lag.

Many economic time series, however, are certainly not stationary and, in particular, have a tendency to exhibit time-changing levels and/or variances. These features are clearly seen in a number of the series introduced in Chapter 2; for example, the nominal and real TFE series plotted in Exhibit 2.1 show very pronounced trends in their mean levels, private sector housing starts in Exhibit 2.2 has a changing, but not trending, mean level, and the exchange rates and interest rates shown in Exhibits 2.8 and 2.9 display marked changes in both level and variance.

6.2 Nonstationarity in variance

We begin by assuming that a time series can be decomposed into a non-stochastic mean level and a random error component:

$$x_t = \mu_t + \varepsilon_t, \tag{6.1}$$

and we suppose that the variance of the errors, ε_t, is functionally related to the mean level μ_t by

$$V(x_t) = V(\varepsilon_t) = h^2(\mu_t)\,\sigma^2,$$

where h is some known function. Our objective is to find a transformation of the data, $g(x_t)$, that will stabilise the variance, i.e. the variance of the transformed variable $g(x_t)$ should be constant. Expanding $g(x_t)$ in a first-order Taylor series around μ_t yields

$$g(x_t) \simeq g(\mu_t) + (x_t - \mu_t)\, g'(\mu_t),$$

where $g'(\mu_t)$ is the first derivative of $g(x_t)$ evaluated at μ_t. The variance of $g(x_t)$ can then be approximated as

$$\begin{aligned} V[g(x_t)] &\simeq V[g(\mu_t) + (x_t - \mu_t)\, g'(\mu_t)] \\ &= [g'(\mu_t)]^2\, V(x_t) \\ &= [g'(\mu_t)]^2\, h^2(\mu_t)\, \sigma^2. \end{aligned}$$

Thus, in order to stabilise the variance, we have to choose the transformation $g(\cdot)$ such that

$$g'(\mu_t) = \frac{1}{h(\mu_t)}.$$

For example, if the standard deviation of x_t is proportional to its level, then $h(\mu_t) = \mu_t$ and the variance-stabilising transformation $g(\mu_t)$ has to satisfy $g'(\mu_t) = \mu_t^{-1}$. This implies that $g(\mu_t) = \ln \mu_t$, and thus (natural) logarithms of x_t should be used to stabilise the variance. If the variance of x_t is proportional to its level, $h(\mu_t) = \mu_t^{\frac{1}{2}}$, so that $g'(\mu_t) = \mu_t^{-\frac{1}{2}}$. Thus since $g(\mu_t) = 2\mu_t^{\frac{1}{2}}$, the square root transformation $x_t^{\frac{1}{2}}$ will stabilise the variance. These two examples are special cases of the class of power transformations introduced in Chapter 4, here modified to (see Box and Cox (1964))

$$g(x_t) = \frac{x_t^{\lambda} - 1}{\lambda},$$

the modification being employed since $\lim_{\lambda \to 0}[(x_t^{\lambda} - 1)/\lambda] = \ln x_t$.

6.3 Nonstationarity in mean

A nonconstant mean level in a time series can occur in a variety of ways. One possibility is that the mean can evolve as a polynomial of order d in time. This will arise if x_t can be decomposed into a trend component, given by the polynomial, and a stochastic, stationary zero mean error component. This is always possible given Cramer's (1961) extension of Wold's decomposition theorem to nonstationary processes. Thus we may have

$$\begin{aligned} x_t &= \mu_t + \varepsilon_t \\ &= \sum_{j=0}^{d} \beta_j t^j + \psi(B)\, a_t. \end{aligned} \tag{6.2}$$

Since

$$E(\varepsilon_t) = \psi(B)\,E(a_t) = 0, \quad E(x_t) = E(\mu_t) = \sum_{j=0}^{d} \beta_j t^j,$$

and as the β_j coefficients remain constant through time, such a trend in the mean is said to be *deterministic*. Deterministic trends of this type can be removed by a simple transformation. Consider the linear trend model obtained by setting $d = 1$, where, for simplicity, the error component is assumed to be a white-noise sequence:

$$x_t = \beta_0 + \beta_1 t + a_t. \tag{6.3}$$

Lagging (6.3) one period and subtracting this from (6.3) yields

$$x_t - x_{t-1} = \beta_1 + a_t - a_{t-1}. \tag{6.4}$$

The result is a difference equation following an ARMA(1,1) process where, since $\phi = \theta = 1$, both autoregressive and moving average roots are unity and the model is neither stationary nor invertible. If we consider the *first-differences* of x_t, w_t say, then

$$w_t = x_t - x_{t-1} = (1-B)\,x_t = \nabla x_t,$$

where $\nabla = 1 - B$ is known as the *first difference operator*, first introduced in Chapter 4. Equation (6.4) can then be written as

$$w_t = \nabla x_t = \beta_1 + \nabla a_t$$

and w_t is then generated by a stationary, since $E(w_t) = \beta_1$ is a constant, but not invertible, MA(1) process.

In general, if the trend polynomial is of order d, and ε_t is characterised by the ARMA process $\phi(B)\,\varepsilon_t = \theta(B)\,a_t$, then the *dth-difference* of x_t,

$$\nabla^d x_t = (1-B)^d\,x_t,$$

which we note is *not* the same as $\nabla_d\,x_t = x_t - x_{t-d}$, will follow the process

$$\nabla^d x_t = \theta_0 + \frac{\nabla^d \theta(B)}{\phi(B)}\,a_t,$$

where $\theta_0 = d!\,\beta_d$. Thus the MA part of the process generating $\nabla^d x_t$ will have d unit roots. Note also that the variance of x_t will be the same as the variance of ε_t, which will be constant for all t. Exhibit 6.1 shows plots of generated data for both linear and quadratic trend models. Because the variance of the error component, here assumed to be white noise and distributed as $N(0,9)$, is constant and independent of the level, the variability of the series is bounded about its expected value, and the trend components are clearly observed in the plots.

An alternative way of generating a nonstationary mean level is to

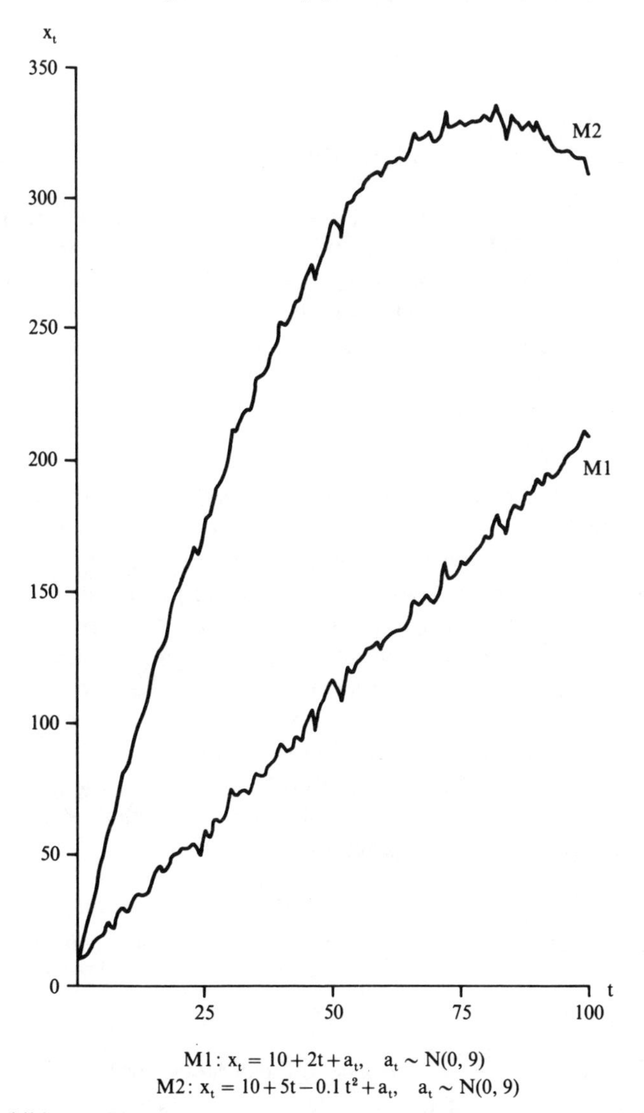

Exhibit 6.1 Linear and quadratic trends

consider ARMA models whose autoregressive parameters do not satisfy stationarity conditions. For example, consider the AR(1) model

$$x_t = \phi x_{t-1} + a_t, \tag{6.5}$$

where $\phi > 1$. If the process is assumed to have started at time $t = -N$, the difference equation (6.5) has the solution

$$x_t = x_{-N}\,\phi^{t+N} + \sum_{i=0}^{t+N} \phi^i a_{t-i}. \tag{6.6}$$

The 'complementary function' $x_{-N}\,\phi^{t+N}$ can be regarded as the *conditional expectation of x_t at time $-N$* (Box and Jenkins (1976, chapter 4)), and is an increasing function of t. The conditional expectation of x_t at times $-N+1, -N+2, \ldots, t-2, t-1$ depend on the random shocks $a_{-N}, a_{-N+1}, \ldots, a_{t-3}, a_{t-2}$, and hence, since this conditional expectation may be regarded as the trend of x_t, the trend changes stochastically.

It is possible for $x_{-N} = 0$, in which case the conditional expectation will be zero for all t. However, even in this unlikely case, the variance of x_t is still given by

$$V(x_t) = \sigma^2 \frac{\phi^{2(t+N+1)} - 1}{\phi^2 - 1},$$

which is an increasing function of time and becomes infinite as $N \rightarrow \infty$. In general, then, x_t will have a trend in both mean and variance, and such processes are said to be *explosive*. A plot of generated data from the process (6.5) with $\phi = 1.05$ and $a_t \sim N(0, 9)$, with starting value $x_0 = 10$, is shown in Exhibit 6.2. We see that, after a short 'induction period', the series follows essentially an exponential curve with the generating a_t's playing almost no further part. The same behaviour would be observed if further autoregressive and moving average terms were added to the model, as long as the stationarity condition is violated.

As we can see from (6.6), the solution of (6.5) is explosive if $\phi > 1$ but stationary if $\phi < 1$. The case $\phi = 1$ provides a process that is neatly balanced between the two. If x_t is generated by the model

$$x_t = x_{t-1} + a_t, \tag{6.7}$$

then x_t is said to follow a *random walk*. If we allow a constant, θ_0, to be included,

$$x_t = x_{t-1} + \theta_0 + a_t, \tag{6.8}$$

then x_t will follow a *random walk with drift*. If the process starts at $t = -N$, then

$$x_t = x_{-N} + (t+N)\,\theta_0 + \sum_{i=0}^{t+N} a_{t-i},$$

so that

$$\mu_t = E(x_t) = x_{-N} + (t+N)\,\theta_0,$$

$$\gamma_{0,t} = V(x_t) = (t+N)\,\sigma^2,$$

$$\gamma_{k,t} = \operatorname{Cov}(x_t, x_{t-k}) = (t+N-k)\,\sigma^2, \quad k \geqslant 0.$$

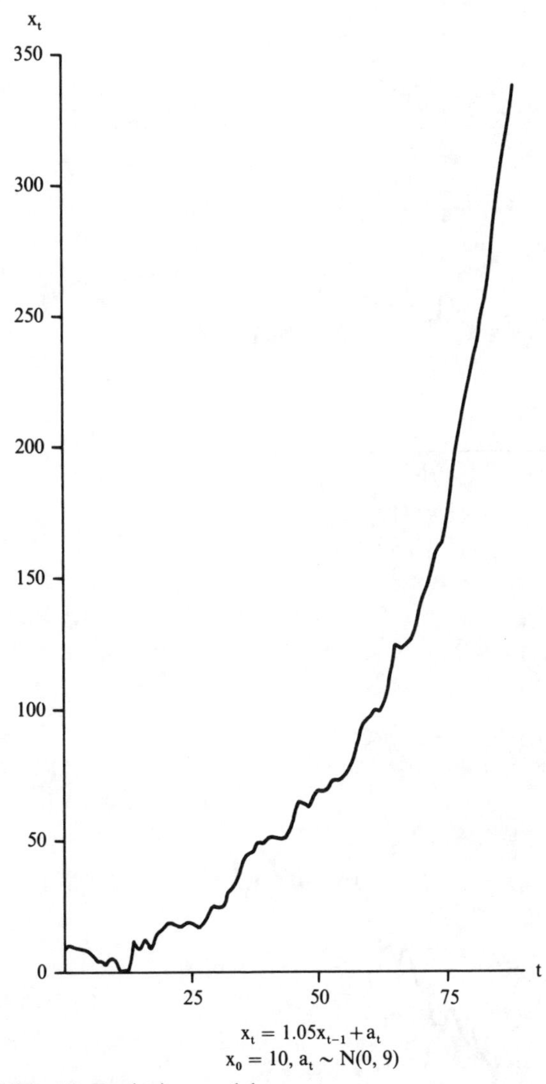

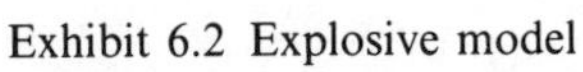
Exhibit 6.2 Explosive model

Thus the correlation between x_t and x_{t-k} is given by

$$\rho_{k,t} = \frac{t+N-k}{\sqrt{(t+N)(t+N-k)}} = \sqrt{\frac{t+N-k}{t+N}}.$$

If $t+N$ is large compared to k, all $\rho_{k,t}$ will approximate unity. The

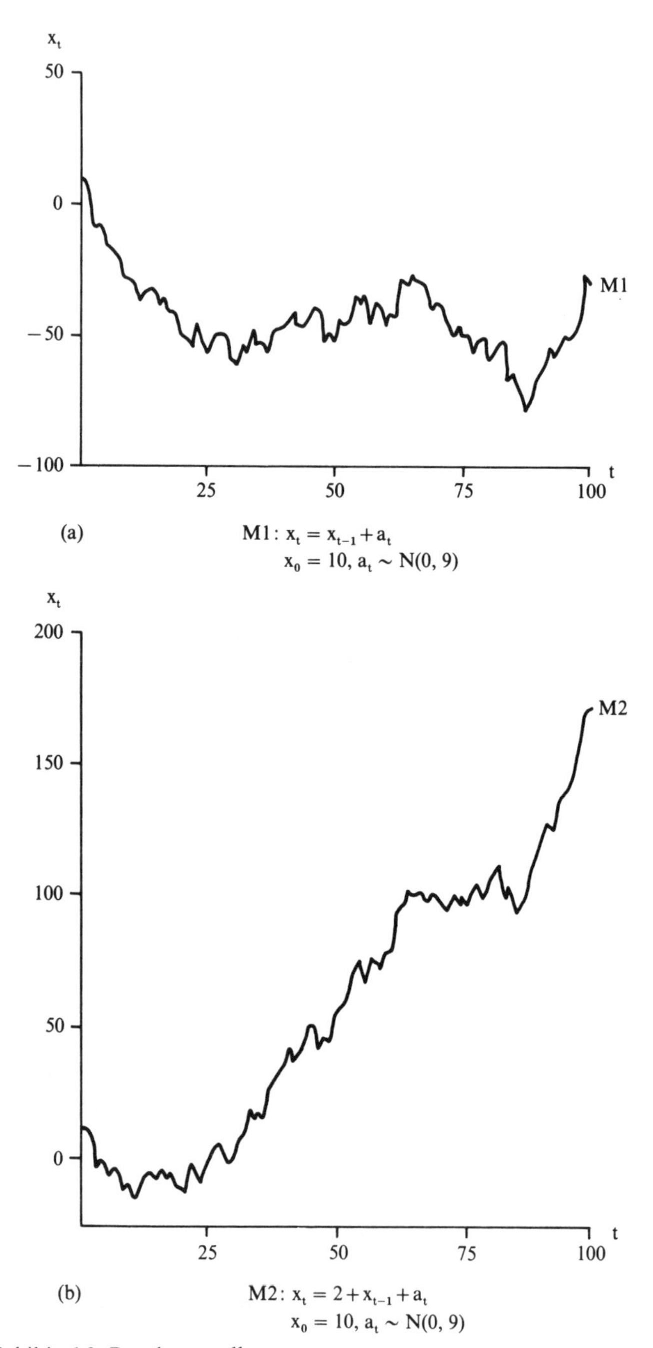

Exhibit 6.3 Random walks

sequence of x_t values will therefore be very smooth, but will also be nonstationary since its variance increases with t. Exhibit 6.3 shows generated plots of the random walks (6.7) and (6.8) with $x_0 = 10$ and $a_t \sim N(0, 9)$. In part (a) of the Exhibit the drift parameter, θ_0, is set to zero while in part (b) we have set $\theta_0 = 2$. The two plots differ considerably, with the drift model showing no affinity whatsoever with the initial value x_0, while the model with $\theta_0 = 0$ only shows some slight affinity. In fact, the expected length of time for a random walk without drift to pass again through an arbitrary value is infinite.

The random walk is an example of a class of nonstationary processes known as *integrated processes*. Equation (6.8) can be written as

$$\nabla x_t = \theta_0 + a_t,$$

and so first-differencing the series leads to a stationary model, in this case the white noise process a_t. Generally, a series may need dth-differencing to attain stationarity, and the series so obtained may itself be autocorrelated. If this autocorrelation is modelled by an ARMA (p, q) process, then the model for the original series x_t is of the form

$$\phi(B)\, \nabla^d x_t = \theta_0 + \theta(B)\, a_t, \tag{6.9}$$

where $\phi(B)$ and $\theta(B)$ are polynomials in B of orders p and q, the solutions of $\phi(B) = 0$ and $\theta(B) = 0$ all being less than unity. The model (6.9) is said to be an *autoregressive-integrated-moving average* process of orders, p, d and q, or ARIMA (p, d, q), and x_t is said to be integrated of order d, $I(d)$. It will usually be the case that the degree of differencing, d, will be 0, 1 or 2. Again it will be the case that the autocorrelations of an ARIMA process will be near one for all nonlarge k. For example, consider the (stationary) ARMA (1, 1) process

$$x_t - \phi x_{t-1} = a_t - \theta a_{t-1},$$

whose ACF has been shown to be (see equation 5.25)

$$\rho_1 = \frac{(1-\phi\theta)(\phi-\theta)}{1+\theta^2-2\phi\theta}, \quad \rho_k = \phi\rho_{k-1}, \quad \text{for } k > 1.$$

As $\phi \rightarrow 1$, the ARIMA (0, 1, 1) process

$$\nabla x_t = a_t - \theta a_{t-1}$$

results, and all the ρ_k tend to unity.

A number of points concerning the ARIMA class of models are of importance. Consider again (6.9), with $\theta_0 = 0$ for simplicity:

$$\phi(B)\, \nabla^d x_t = \theta(B)\, a_t. \tag{6.10}$$

This process can equivalently be defined by the two equations

$$\phi(B)\, w_t = \theta(B)\, a_t \tag{6.11}$$

and

$$w_t = \nabla^d x_t, \tag{6.12}$$

so that, as we have noted above, the model corresponds to assuming that the dth difference of the series can be represented by a stationary, invertible ARMA process. Alternatively, for $d \geqslant 1$, (6.12) can be inverted to give

$$x_t = S^d w_t, \tag{6.13}$$

where S is the infinite summation operator defined by

$$S = (1+B+B^2+\ldots) = (1-B)^{-1} = \nabla^{-1}.$$

Equation (6.13) implies that the process (6.10) can be obtained by summing, or 'integrating', the stationary process (6.11) d times: hence the term integrated process.

Box and Jenkins (1976, chapter 4) refer to this type of nonstationary behaviour as *homogeneous nonstationarity*, and it is important to discuss why this form of nonstationarity is felt to be useful in describing the behaviour of economic time series. Consider again the first order autoregressive process (6.5). A basic characteristic of the AR(1) model is that, for both $|\phi| < 1$ and $|\phi| > 1$, the local behaviour of a series generated from the model is heavily dependent upon the level of x_t. For many (nonseasonal) economic time series, such as those shown in Exhibits 2.8 and 2.9, for example, local behaviour appears to be roughly independent of level, and this is what we mean by homogeneous nonstationarity.

If we want to use ARMA models for which the behaviour of the process is indeed independent of its level, then the autoregressive operator $\phi(B)$ must be chosen so that

$$\phi(B)\,(x_t+c) = \phi(B)\, x_t,$$

where c is any constant. Thus

$$\phi(B)\, c = 0,$$

implying that $\phi(1) = 0$, so that $\phi(B)$ must be able to be factorised as

$$\phi(B) = \phi_1(B)\,(1-B) = \phi_1(B)\,\nabla,$$

in which case the class of processes that need to be considered will be of the form

$$\phi_1(B)\, w_t = \theta(B)\, a_t,$$

where $w_t = \nabla x_t$. Since the requirement of homogeneous nonstationarity

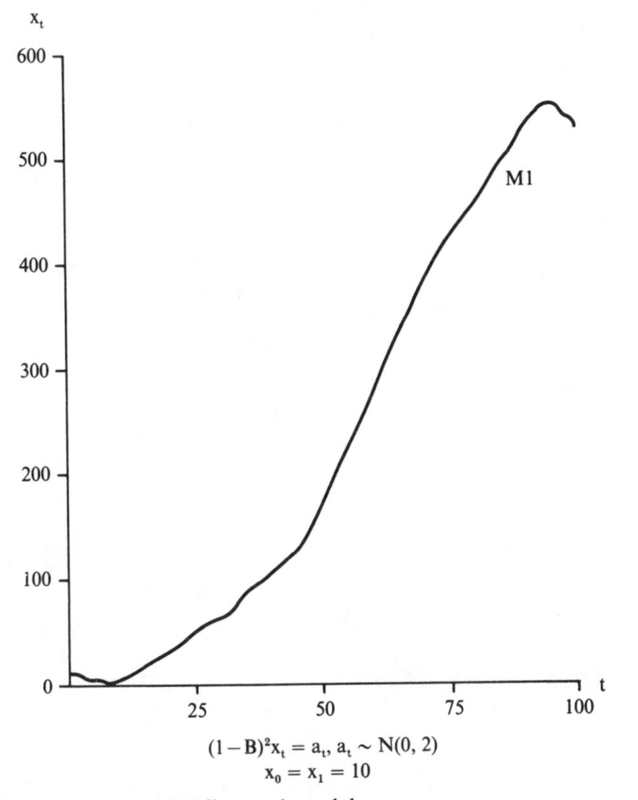

$(1-B)^2x_t = a_t,\ a_t \sim N(0, 2)$
$x_0 = x_1 = 10$

Exhibit 6.4 'Second difference' model

precludes w_t increasing explosively, either $\phi_1(B)$ is a stationary autoregressive operator, or $\phi_1(B) = \phi_2(B)(1-B)$, so that $\phi_2(B)\,w_t^* = \theta(B)\,a_t$, where $w_t^* = \nabla^2 x_t$. Since this argument can be used recursively, it follows that for time series that are homogeneously nonstationary, the autoregressive operator must be of the form $\phi(B)\,\nabla^d$, where $\phi(B)$ is a stationary autoregressive operator. Exhibit 6.4 plots generated data from the model $\nabla^2 x_t = a_t$, where $a_t \sim N(0,2)$ and $x_0 = x_1 = 10$, and such a series is seen to display random movements in both level and slope.

We see from Exhibits 6.3(a) and 6.4 that ARIMA models without the constant θ_0 in (6.9) are capable of representing series that have *stochastic* trends, which typically will consist of random changes in both the level and slope of the series. As seen from Exhibit 6.3(b) and equation (6.8), however, the inclusion of a nonzero drift parameter introduces a deterministic linear trend into the generated series, since $\mu_t = E(x_t) = \beta_0 + \theta t$ if we set $\beta_0 = x_{-N} + N\theta$. In general, if a constant is included in the

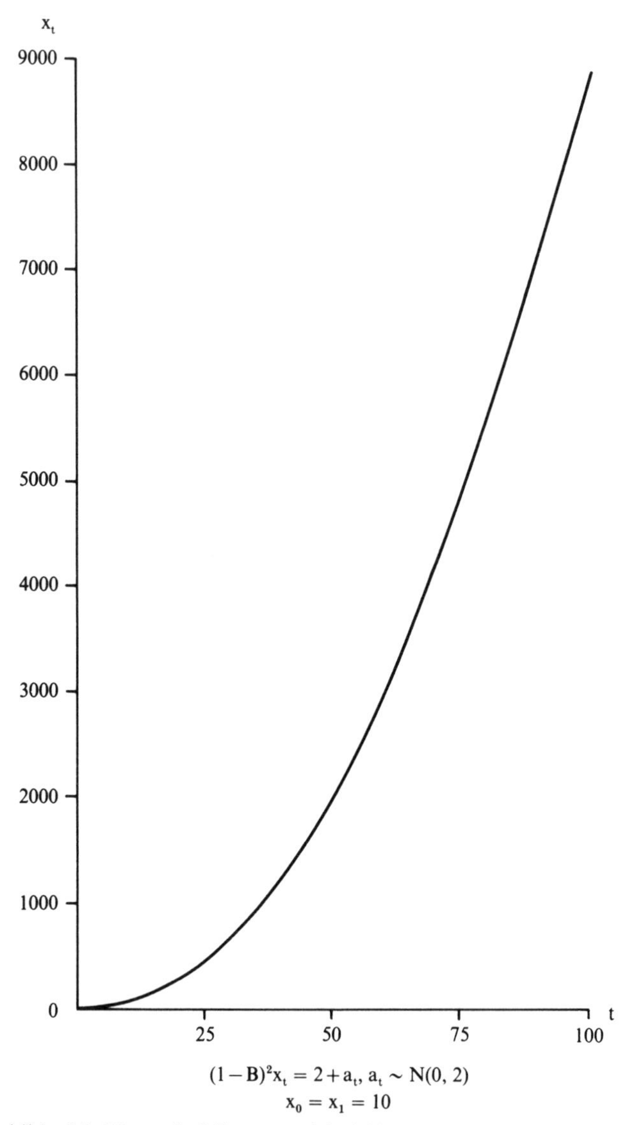

$(1-B)^2 x_t = 2 + a_t, \; a_t \sim N(0, 2)$

$x_0 = x_1 = 10$

Exhibit 6.5 'Second difference with drift' model

model for dth differences, then a deterministic polynomial trend of degree d is automatically allowed for. Equivalently, if θ_0 is allowed to be nonzero, then

$$E(w_t) = E(\nabla^d x_t) = \mu_w = \theta_0/(1-\phi_1-\phi_2-\ldots-\phi_p)$$

is nonzero, so that an alternative way of expressing (6.9) is as

$$\phi(B)\,\tilde{w}_t = \theta(B)\,a_t,$$

where $\tilde{w}_t = w_t - \mu_w$. Exhibit 6.5 plots generated data for $\nabla^2 x_t = 2 + a_t$, where again $a_t \sim N(0, 2)$ and $x_0 = x_1 = 10$. The inclusion of the deterministic quadratic trend has a dramatic effect on the evolution of the series, with the nonstationary 'noise' being completely swamped after a few periods.

Model (6.9) therefore allows both stochastic and deterministic trends to be modelled. When $\theta_0 = 0$, a stochastic trend is incorporated, while if $\theta_0 \neq 0$, the model may be interpreted as representing a deterministic trend (a polynomial in time of order d) buried in nonstationary noise, which will typically be correlated. The models presented earlier in this section could be described as deterministic trends buried in *stationary* noise, since they can be written as

$$\phi(B)\nabla^d x_t = \phi(1)\beta_d d! + \nabla^d \theta(B) a_t,$$

the stationary nature of the noise in the level of x_t being manifested in d roots of the moving average operator being unity. Further discussion of the relationships existing between stochastic and deterministic trends will be found in Chapter 11.

6.4 Nonstationarity and nonlinear models

The discussion so far has concentrated on explaining nonstationarity by linear models with fixed parameters. However, we can expand the class of ARMA processes to include both nonlinear models and time varying parameters. Such extensions are very much a current research area and we will only touch upon certain ideas later in the book in Chapters 15 and 16: nonlinear ARMA models have been studied by Granger and Andersen (1978) and extensively utilised for modelling financial time series by Taylor (1986), while models with time varying parameters have typically been analysed by application of the Kalman filter, as discussed, for example, by Harvey (1981, 1987).

7 Forecasting using ARIMA models

7.1 Introduction

Although a major use of ARIMA models is in their application to forecasting (see, for example, Box and Jenkins (1976) and Granger and Newbold (1977)), forecasting, per se, is not our primary aim in this book. Nevertheless, certain aspects of the theory of forecasting using ARIMA models are extremely important for understanding the full implications and properties of fitting such models to economic time series.

Thus, suppose our observed series $(x_1, x_2, \ldots, x_n)$ is regarded as a realisation from the general ARIMA (p, d, q) process

$$\phi(B)\nabla^d x_t = \theta(B)\,a_t, \tag{7.1}$$

and we wish to forecast a future value x_{n+h}. The linear filter representation of x_{n+h} is

$$x_{n+h} = a_{n+h} + \psi_1 a_{n+h-1} + \ldots + \psi_{h-1} a_{n+1} + \psi_h a_n + \psi_{h+1} a_{n-1} + \ldots, \tag{7.2}$$

where $\psi(B) = \phi^{-1}(B)\nabla^{-d}\theta(B)$. Our forecast of $x_{n+h}(h \geqslant 1)$ is to be made at time n, so that only the set of values $x_n, x_{n-1}, \ldots$ are known. We often refer to n as the *origin* and h as the *lead time*. The forecast of x_{n+h} will therefore be a linear combination of the past and present values of x, so that it can, in fact, be regarded as the conditional expectation of x_{n+h}, given $x_n, x_{n-1}, \ldots$. If the *h-step ahead forecast* made at origin n is denoted $f_{n,h}$, then

$$\begin{aligned} f_{n,h} &= E(x_{n+h}|x_n, x_{n-1}, \ldots) \\ &= E((a_{n+h} + \psi_1 a_{n+h-1} + \ldots + \psi_{h-1} a_{n+1} \\ &\qquad + \psi_h a_n + \psi_{h+1} a_{n-1} + \ldots)\,|x_n, x_{n-1}, \ldots). \end{aligned}$$

Now

$$E(a_{n+j}|x_n, x_{n-1}, \ldots) = \begin{cases} a_{n+j}, & j \leqslant 0 \\ 0, & j > 0, \end{cases}$$

since past values of a_{n+j} (i.e. $j \leqslant 0$) are known, and future values, although unknown, have zero expectation. Hence

$$f_{n,h} = \psi_h a_n + \psi_{h+1} a_{n-1} + \ldots$$

and this can be shown to be the *minimum mean square error* (MMSE) forecast of x_{n+h} made at origin n (Box and Jenkins (1976, page 128)). Furthermore, any linear function $\sum_h w_h f_{n,h}$ of the forecasts will be a MMSE forecast of the corresponding linear function $\sum_h w_h x_{n+h}$ of future observations. Note that here we have regarded $f_{n,h}$ as a function of h for fixed n, and when this is done, $f_{n,h}$ is called the *forecast function* for origin n.

The h-step ahead *forecast error* for origin n is

$$e_{n,h} = x_{n+h} - f_{n,h} = a_{n+h} + \psi_1 a_{n+h-1} + \ldots + \psi_{h-1} a_{n+1}.$$

It is immediately seen that, since $E(e_{n,h}|x_n, x_{n-1}, \ldots) = 0$, $f_{n,h}$ is an unbiased forecast. The variance of the forecast error is then

$$V(e_{n,h}) = \sigma^2(1 + \psi_1^2 + \psi_2^2 + \ldots + \psi_{h-1}^2).$$

The forecast error is therefore a linear combination of the unobservable future shocks entering the system after time n and, in particular, the one-step ahead forecast error ($h = 1$) is

$$e_{n,1} = x_{n+1} - f_{n,1} = a_{n+1}.$$

Thus, for a MMSE forecast, the one-step ahead forecast errors must be uncorrelated. However, it is not the case that h-step ahead forecasts made at different origins will be uncorrelated, nor will be forecasts for different lead times made at the same origin. From Box and Jenkins (1976, Appendix A5.1), the autocorrelations between h-step ahead forecasts made at origins n and $n-j$, where $j \geqslant 0$, are given by

$$\rho[e_{n,h}, e_{n-j,h}] = \frac{\sum_{i=j}^{h-1} \psi_i \psi_{i-j}}{\sum_{i=0}^{h-1} \psi_i^2}, \quad 0 \leqslant j < h$$

and 0 otherwise. The correlation coefficient between the n-origin forecast errors at lead times h and $h+j$ is

$$\rho[e_{n,h}, e_{n,h+j}] = \frac{\sum_{i=0}^{h-1} \psi_i \psi_{h+i}}{\left(\sum_{l=0}^{h-1} \psi_l^2 \sum_{m=0}^{h+j-1} \psi_m^2\right)^{\frac{1}{2}}}.$$

For example, setting $h = 2$ and $j = 1$ in the above formulae yield

$$\rho[e_{n,2}, e_{n-1,2}] = \frac{\psi_1}{1+\psi_1^2}$$

and

$$\rho[e_{n,2}, e_{n,3}] = \frac{\psi_2 + \psi_1 \psi_3}{\{(1+\psi_1^2)(1+\psi_1^2+\psi_2^2)\}^{\frac{1}{2}}}.$$

As a consequence, there will often be a tendency for the forecast function to lie either wholly above or below the values of the series when they eventually become available.

7.2 A general procedure for computing forecasts

Consider again the ARIMA (p, d, q) process (7.1), now written more generally to include a constant:

$$\phi(B)\nabla^d x_{n+h} = \theta_0 + \theta(B)\, a_{n+h}. \tag{7.3}$$

Letting

$$\alpha(B) = \phi(B)\nabla^d = (1 - \alpha_1 B - \alpha_2 B^2 - \ldots - \alpha_{p+d} B^{p+d}),$$

we have

$$x_{n+h} = \alpha_1 x_{n+h-1} + \alpha_2 x_{n+h-2} + \ldots + \alpha_{p+d} x_{n+h-p-d} + \theta_0 + a_{n+h} - \theta_1 a_{n+h-1} - \ldots - \theta_q a_{n+h-q},$$

so that

$$f_{n,h} = E\{\alpha_1 x_{n+h-1} + \alpha_2 x_{n+h-2} + \ldots + \alpha_{p+d} x_{n+h-p-d} + \theta_0 + a_{n+h} - \theta_1 a_{n+h-1} - \ldots - \theta_q a_{n+h-q} | x_n, x_{n-1}, \ldots\}.$$

Now

$$E(x_{n+j} | x_n, x_{n-1}, \ldots) = \begin{cases} x_{n+j}, & j \leqslant 0 \\ f_{n,j}, & j > 0 \end{cases}$$

and

$$E(a_{n+j} | x_n, x_{n-1}, \ldots) = \begin{cases} a_{n+j}, & j \leqslant 0 \\ 0 & j > 0, \end{cases}$$

so that, to evaluate $f_{n,h}$, all we need to do is: (i) replace past expectations ($j \leqslant 0$) by known values, x_{n+j} and a_{n+j}, and (ii) replace future expectations ($j > 0$) by forecast values, $f_{n,j}$ and 0.

Example 7.1: Computing forecasts from ARIMA models

Consider the following models (a) $(1-\phi B)x_t = \theta_0 + a_t$, (b) $\nabla x_t = (1-\theta B)a_t$, and (c) $\nabla^2 x_t = (1-\theta_1 B - \theta_2 B^2)a_t$.

(a) Here $\alpha(B) = (1-\phi B)$ and so

$$x_{n+h} = \phi x_{n+h-1} + \theta_0 + a_{n+h}.$$

Hence

$$\begin{aligned} f_{n,h} &= \phi f_{n,h-1} + \theta_0 \\ &= \phi^h x_n + \theta_0(1+\phi+\phi^2+\ldots+\phi^{h-1}), \end{aligned}$$

by repeated substitution. Thus, for stationary processes ($|\phi| < 1$), as $h \rightarrow \infty$,

$$f_{n,h} = \frac{\theta_0}{1-\phi} = E(x_t) = \mu,$$

so that for large lead times the best forecast of a future observation is eventually the mean of the process. Since $\psi_j = \phi^j$, the variance of the h-step ahead forecast is given by

$$\begin{aligned} V(e_{n,h}) &= \sigma^2(1+\phi^2+\phi^4+\ldots+\phi^{2(h-1)}) \\ &= \sigma^2 \frac{(1-\phi^{2h})}{(1-\phi^2)}. \end{aligned}$$

Thus, as h tends to infinity, the variance increases to a constant value $\sigma^2/(1-\phi^2)$, this being the variation of the process about the ultimate forecast, μ.

(b) Here $\alpha(B) = (1-B)$ and so

$$x_{n+h} = x_{n+h-1} + a_{n+h} - \theta a_{n+h-1}.$$

For $h = 1$, we have

$$f_{n,1} = x_n - \theta a_n,$$

for $h = 2$,

$$f_{n,2} = f_{n,1} = x_n - \theta a_n,$$

and, in general,

$$f_{n,h} = f_{n,h-1}, h > 1.$$

Hence, for all lead times, the forecasts from origin n will follow a straight line parallel to the time axis passing through $f_{n,1}$. Note that, since

$$f_{n,h} = x_n - \theta a_n$$

and

$$a_n = (1-B)(1-\theta B)^{-1} x_n,$$

the h-step ahead forecast can be written as

$$\begin{aligned} f_{n,h} &= (1-\theta)(1-\theta B)^{-1} x_n \\ &= (1-\theta)[x_n + \theta x_{n-1} + \theta^2 x_{n-2} + \ldots], \end{aligned}$$

i.e. the forecast for all future values of x is an exponentially weighted moving average of current and past values. Since, for this model, $\psi_j = 1-\theta, (j = 1, 2, \ldots)$, the expression for the variance of the h-step ahead forecast error is

$$V(e_{n,h}) = \sigma^2(1+(h-1)(1-\theta)^2),$$

which increases with h.

(c) Here $\alpha(B) = (1-B)^2 = (1-2B+B^2)$ and so

$$x_{n+h} = 2x_{n+h-1} - x_{n+h-2} + a_{n+h} - \theta_1 a_{n+h-1} - \theta_2 a_{n+h-2}.$$

For $h = 1$, we have

$$f_{n,1} = 2x_n - x_{n-1} - \theta_1 a_n - \theta_2 a_{n-1},$$

for $h = 2$,

$$f_{n,2} = 2f_{n,1} - x_n - \theta_2 a_n,$$

for $h = 3$,

$$f_{n,3} = 2f_{n,2} - f_{n,1},$$

and thus, for $h \geqslant 3$,

$$f_{n,h} = 2f_{n,h-1} - f_{n,h-2}.$$

Hence, for all lead times, the forecasts from origin n will follow a straight line passing through the forecasts $f_{n,1}$ and $f_{n,2}$. For this model, $\psi_j = 1+\theta_2+j(1-\theta_1-\theta_2), (j = 1, 2, \ldots)$ and the variance of the h-step ahead forecast error is (Box and Jenkins (1976, page 149))

$$\begin{aligned} V(e_{n,h}) = \sigma^2(1 &+ (h-1)(1+\theta_2)^2 + \tfrac{1}{6}h(h-1)(2h-1)(1-\theta_1-\theta_2)^2 \\ &+ h(h-1)(1+\theta_2)(1-\theta_1-\theta_2)), \end{aligned}$$

which again increases with h.

These examples show that the degree of differencing, or 'order of integration', determines not only how successive forecasts are related to each other, but also the behaviour of the associated forecast error variances. These ideas are formalised in the next section.

7.3 Eventual forecast functions

The procedure just described is the most convenient for actually calculating forecasts. However, from the point of view of studying the nature of these forecasts, it is useful to consider an alternative form of the forecast function. At time $n+h$ the model (7.3) can be written

$$x_{n+h}-\alpha_1 x_{n+h-1}-\ldots-\alpha_{p+d} x_{n+h-p-d} = \theta_0+a_{n+h}-\theta_1 a_{n+h-1}-\ldots-\theta_q a_{n+h-q}. \quad (7.4)$$

The difference equation (7.4) has, for origin n, the solution,

$$x_{n+h} = \sum_{i=1}^{p+d} b_i^{(n)} f_i(h)+\xi \sum_{j=n+1}^{n+h} \psi_{n+h-j}+\sum_{j=n+1}^{n+h} \psi_{n+h-j}\, a_j, \quad (7.5)$$

where $\xi = \theta_0/(1-\theta_1-\ldots-\theta_q)$, the ψ's are as in (7.2), and $f_1(h), f_2(h), \ldots, f_{p+d}(h)$ are functions of the lead time h. By taking conditional expectations at time n of (7.5), we have, for $h > q-p-d$,

$$f_{n,h} = \sum_{i=1}^{p+d} b_i^{(n)} f_i(h)+\xi \sum_{j=n+1}^{n+h} \psi_{n+h-j} \quad (7.6)$$

and, as Box and Jenkins (1976, chapter 4) show, this conditional expectation is also the *complementary function* of the difference equation (7.4), the *particular integral* being the term

$$\sum_{j=n+1}^{n+h} \psi_{n+h-j}\, a_j,$$

which shows how the expectation (or forecast) is modified by *subsequent* events represented by the shocks $a_{n+1}, a_{n+2}, \ldots, a_{n+h}$.

The functions $f_1(h), f_2(h), \ldots, f_{p+d}(h)$ may be polynomials, exponentials, sines and cosines, and also products of these functions. For example, if $d = 0$ and $\alpha(B)$ factorises as

$$\alpha(B) = (1-g_1 B)(1-g_2 B)\ldots(1-g_p B),$$

and if all $g_i, (i = 1, 2, \ldots, p)$ are distinct, then if g_1, say, is real, $f_1(h) = g_1^h$. If g_2 and g_3, say, are a pair of complex roots $(\alpha \pm \beta i)^h$, then they will contribute a damped sine wave function of the form $\delta^h \sin(2\pi f_0(n+h)+F)$, where δ is the damping factor, f_0 the frequency, and F the phase, all being functions of the process parameters (Box and Jenkins (1976, pages 58–63)).

If $\alpha(B)$ has d equal roots g_0^{-1}, so that it factorises as

$$\alpha(B) = (1-g_1 B)(1-g_2 B)\ldots(1-g_p B)(1-g_0 B)^d$$

then $f_{p+1}(h) = g_0, f_{p+2}(h) = hg_0, \ldots, f_{p+d}(h) = h^{d-1}g_0$. If these roots are equal to unity, so that $\alpha(B) = \phi(B)(1-B)^d$, then equation (7.6) becomes

$$f_{n,h} = b_0 + \sum_{j=1}^{p} b_j^{(n)} f_j(h) + \sum_{j=p+1}^{p+d} b_j^{(n)} h^{j-p-1} \tag{7.7}$$

where

$$b_0 = \xi \sum_{j=n+1}^{n+h} \psi_{n+h-j},$$

thus introducing a polynomial in h of order $d-1$ into the solution. For a given origin n, the coefficients $b_j^{(n)}$ are constants applying for all lead times h, but they change from one origin to the next, adapting themselves to the particular observations of the series being considered. We note that equation (7.7) only provides forecasts for lead times $h > q-p-d$. Because it is sometimes the case that $q > p+d$, this equation is known as the *eventual forecast function*, and further discussion of its usefulness in forecasting is to be found in McKenzie (1988).

Box and Jenkins (1976, Appendix A4.1) show that the expression

$$\sum_{j=n+1}^{n+h} \psi_{n+h-j} a_j$$

in equation (7.5) can be regarded as the 'particular integral' and we can therefore write the general solution (7.5) as

$$x_{n+h} = b_0 + \boldsymbol{f}'(h)\boldsymbol{\beta}^{(n)} + \sum_{j=n+1}^{n+h} \psi_{n+h-j} a_j \tag{7.8}$$

where

$$\boldsymbol{f}'(h) = (f_1(h), \ldots, f_{p+d}(h))$$

and

$$\boldsymbol{\beta}^{(n)} = (b_1^{(n)}, \ldots, b_{p+d}^{(n)})'.$$

We can express the solution (7.8) with respect to origin $n-1$ as well as origin n:

$$\begin{aligned} x_{n+h} &= b_0 + \boldsymbol{f}'(h)\boldsymbol{\beta}^{(n)} + a_{n+h} + \psi_1 a_{n+h-1} + \ldots + \psi_{h-1} a_{n+1} \\ &= b_0 + \boldsymbol{f}'(h+1)\boldsymbol{\beta}^{(n-1)} + a_{n+h} + \psi_1 a_{n+h-1} + \ldots + \psi_{h-1} a_{n+1} + \psi_h a_n, \end{aligned}$$

so that

$$\boldsymbol{f}'(h)\boldsymbol{\beta}^{(n)} = \boldsymbol{f}'(h+1)\boldsymbol{\beta}^{(n-1)} + \psi_h a_n. \tag{7.9}$$

Since the solution holds for $h > q-p-d = v$, the updating equation (7.9) can be obtained for $h = v+1, v+2, \ldots, v+p+d$, and solved for $\boldsymbol{\beta}^{(n)}$, yielding

$$\boldsymbol{\beta}^{(n)} = (\boldsymbol{F}_v^{-1}\boldsymbol{F}_{v+1})\boldsymbol{\beta}^{(n-1)} + (\boldsymbol{F}_v^{-1}\boldsymbol{\psi}_v)\,a_n \tag{7.10}$$

where

$$\boldsymbol{\psi}_v = (\psi_{v+1}, \psi_{v+2}, \ldots, \psi_{v+p+d})'$$

and

$$\boldsymbol{F}_v = (\boldsymbol{f}'(v+1), \ldots, \boldsymbol{f}'(v+p+d))'.$$

Example 7.2: Computing eventual forecast functions

Consider again the models (a) $(1-\phi B)\,x_t = \theta_0 + a_t$, (b) $\nabla x_t = (1-\theta B)\,a_t$, and (c) $\nabla^2 x_t = (1-\theta_1 B - \theta_2 B^2)\,a_t$.

(a) The solution of the difference equation

$$x_{n+h} - \phi x_{n+h-1} = \theta_0 + a_t$$

is, from (7.5),

$$x_{n+h} = \theta_0(1+\phi+\phi^2+\ldots+\phi^{h-1}) + b_1^{(n)}\phi^h,$$

so that the eventual forecast function is, for $h > 0$,

$$f_{n,h} = \theta_0(1+\phi+\phi^2+\ldots+\phi^{h-1}) + b_1^{(n)}\phi^h$$

$$= \frac{\theta_0(1-\phi^h)}{1-\phi} + b_1^{(n)}\phi^h.$$

Note that $f_{n,1} = \theta_0 + b_1^{(n)}\phi$, but from Example 7.2(a), $f_{n,1} = \theta_0 + \phi x_n$, so that $b_1^{(n)} = x_n$. Alternatively, we have

$$\boldsymbol{F}_v = \phi, \boldsymbol{F}_{v+1} = \phi^2, \boldsymbol{\psi}_v = \phi,$$

so that substitution into the updating equations (7.10) yields

$$b_1^{(n)} = \phi b_1^{(n-1)} + a_n.$$

(b) The solution of the difference equation $(1-B)f_{n,h} = 0$ for $h > 1$ is given by

$$f_{n,h} = b_1^{(n)}, h > 0$$

and substituting $F_v = 1$ and $\psi_v = 1-\theta$ into the updating equation (7.10) yields

$$b_1^{(n)} = b_1^{(n-1)} + (1-\theta)\,a_n.$$

Hence, for *any fixed origin*, $b_1^{(n)}$ is a constant and the forecasts for all lead times will follow a straight line parallel to the time axis (see Example 7.1). However, the coefficient $b_1^{(n)}$ will be updated as a new observation becomes available and the origin advances, the updating being a fraction $(1-\theta)$ of the latest shock a_n. The forecast function can be thought of as a polynomial of degree zero in the lead time h with a coefficient that is adaptive with respect to the origin n.

(c) The solution of the difference equation $(1-B)^2 f_{n,h} = 0$ for $h > 2$ is given by

$$f_{n,h} = b_1^{(n)} + b_2^{(n)} h, h > 0.$$

Here

$$\boldsymbol{F}_v = \begin{bmatrix} 1 & 1 \\ 1 & 2 \end{bmatrix}, \boldsymbol{F}_{v+1} = \begin{bmatrix} 1 & 1 \\ 1 & 2 \end{bmatrix}, \boldsymbol{\psi}_v = \begin{bmatrix} 2-\theta_1 \\ 3-2\theta_1-\theta_2 \end{bmatrix}$$

and substituting these into the updating equation (7.10) yields

$$b_1^{(n)} = b_1^{(n-1)} + b_2^{(n-1)} + (1-\theta_2)\, a_n$$

$$b_2^{(n)} = \qquad b_2^{(n-1)} + (1-\theta_1-\theta_2)\, a_n.$$

The presence of the slope term $b_2^{(n-1)}$ in the updating formula for $b_1^{(n)}$ allows the location parameter b_1 to be adjusted to a value appropriate to the new origin. With this adjustment, both coefficients are updated by (different) fractions of the latest shock a_n. Here the forecast function can be thought of as a polynomial of degree one in h, with both location and slope parameters that are adaptive with respect to the origin n.

Example 7.3: The ARIMA (0, 1, 1) process with drift

A model that is often found to be useful is the ARIMA (0, 1, 1) process 'with deterministic drift'

$$\nabla x_t = \theta_0 + (1-\theta_1 B)\, a_t,$$

which has, from (7.7), the complementary function

$$f_{n,h} = b_0 + b_1^{(n)}, h > 0.$$

Now, $b_0 = \xi((h-1)(1-\theta_1)+1)$, since $\psi_0 = 1$ and $\psi_j = 1-\theta_1, j \geqslant 1$, and with $\xi = \theta_0/(1-\theta_1)$, we have

$$f_{n,h} = (h-1)\,\theta_0 + \frac{\theta_0}{1-\theta_1} + b_1^{(n)}, h > 0,$$

where again

$$b_1^{(n)} = b_1^{(n-1)} + (1-\theta_1)\, a_n.$$

Thus x_{n+h} contains a deterministic slope or drift due to the term $(h-1)\,\theta_0$. An important special case of this model, introduced in Chapter 6, is the random walk with drift, obtained when $\theta_1 = 0$. In this case the eventual forecast function becomes

$$f_{n,h} = h\theta_0 + b_1^{(n)}$$

with

$$b_1^{(n)} = b_1^{(n-1)} + a_n$$

i.e.

$$f_{n,h} = h\theta_0 + x_n, h > 0.$$

These examples lead us to the following summarisations. For the ARIMA$(0, d, q)$ process, the eventual forecast function satisfies $(1-B)^d f_{n,h} = 0$, and has for its solution a polynomial in h of degree $d-1$:

$$f_{n,h} = b_0 + b_1^{(n)} + b_2^{(n)} h + \ldots + b_d^{(n)} h^{d-1},$$

which provides forecasts $f_{n,h}$ for $h > q-d$. The coefficients $b_1^{(n)}, \ldots, b_d^{(n)}$ are progressively updated as the origin advances. The forecast for origin n makes $q-d$ initial 'jumps', which depend upon $a_n, a_{n-1}, \ldots, a_{n-q+1}$, before following this polynomial, whose position is uniquely determined by the 'pivotal' values $f_{n,q}, f_{n,q-1}, \ldots, f_{n,q-d+1}$, where $f_{n,j} = x_{n-j}$ for $j \leqslant 0$.

For the ARIMA$(p, d, 0)$ process, the eventual forecast function satisfies $\phi(B)(1-B)^d f_{n,h} = 0$ and has for its solution equation (7.7):

$$f_{n,h} = b_0 + \sum_{j=1}^{p} b_j^{(n)} f_j(h) + \sum_{j=p+1}^{p+d} b_j^{(n)} h^{j-p-1}.$$

This solution provides forecasts $f_{n,h}$ for all $h > 0$ and passes through the last $p+d$ available values of the series; $x_n, x_{n-1}, \ldots, x_{n-p-d+1}$, these being the pivotal values.

For mixed ARIMA(p, d, q) processes, equation (7.7) holds for $h > q-p-d$ if $q > p+d$, and for $h > 0$ if $q < p+d$. In both cases the forecast function is uniquely determined by the pivotal values $f_{n,q}, f_{n,q-1}, \ldots, f_{n,q-p-d+1}$.

Example 7.4: The eventual forecast function of a mixed ARIMA process

Consider the nonstationary ARIMA$(1, 1, 1)$ process

$$(1-\phi B)\nabla x_t = (1-\theta B) a_t.$$

Forecasts are readily obtained from

$$f_{n,1} = (1+\phi) x_n - \phi x_{n-1} - \theta a_n$$

$$f_{n,h} = (1+\phi) f_{n,h-1} - \phi f_{n,h-2}, h > 1.$$

Since $q < p+d$, the eventual forecast function for all $h > 0$ is the solution of $(1-\phi B)(1-B) f_{n,h} = 0$, which is

$$f_{n,h} = b_1^{(n)} + b_2^{(n)} \phi^h.$$

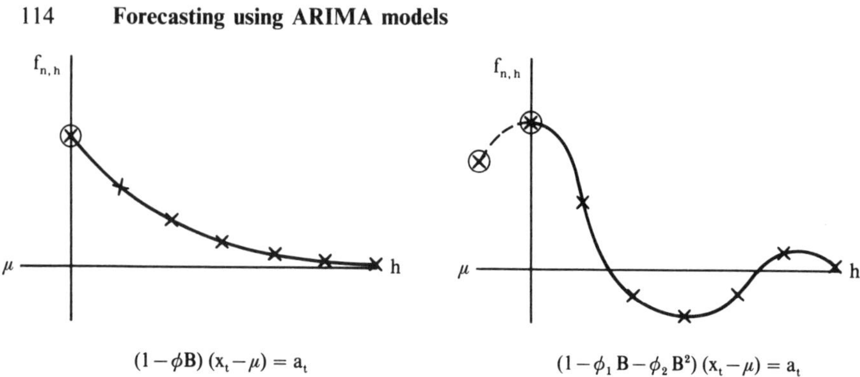

$(1-\phi B)(x_t-\mu) = a_t$ $\qquad$ $(1-\phi_1 B-\phi_2 B^2)(x_t-\mu) = a_t$

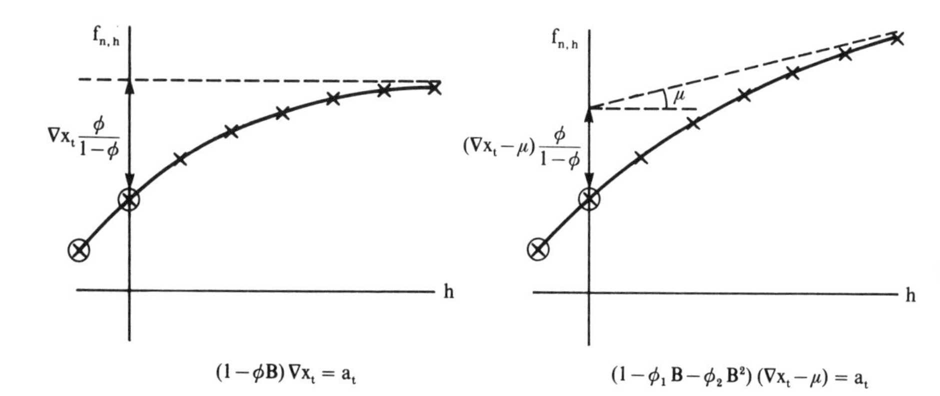

$(1-\phi B)\nabla x_t = a_t$ $\qquad$ $(1-\phi_1 B-\phi_2 B^2)(\nabla x_t-\mu) = a_t$

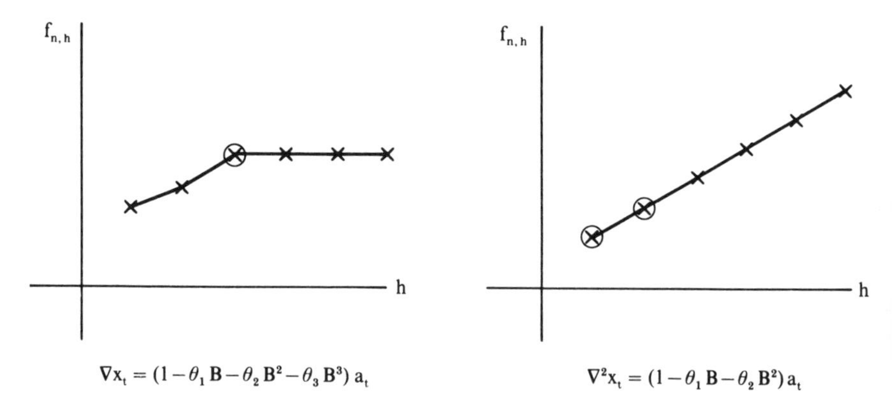

$\nabla x_t = (1-\theta_1 B-\theta_2 B^2-\theta_3 B^3)\,a_t$ $\qquad$ $\nabla^2 x_t = (1-\theta_1 B-\theta_2 B^2)\,a_t$

$\otimes$: denotes a 'pivotal value'.

Exhibit 7.1 Examples of eventual forecast functions

Here

$$\boldsymbol{F}_v = \begin{bmatrix} 1 & \phi \\ 1 & \phi^2 \end{bmatrix}, \quad \boldsymbol{F}_{v+1} = \begin{bmatrix} 1 & \phi^2 \\ 1 & \phi^3 \end{bmatrix}, \quad \boldsymbol{\psi}_v = \begin{bmatrix} \dfrac{1-\theta}{1-\phi}+\dfrac{\theta-\phi}{1-\phi}\cdot\phi \\ \dfrac{1-\theta}{1-\phi}+\dfrac{\theta-\phi}{1-\phi}\cdot\phi^2 \end{bmatrix},$$

and after substitution into the updating equations (7.10) we obtain

$$b_1^{(n)} = b_1^{(n-1)} + \frac{(1-\theta)}{(1-\phi)} a_n,$$

$$b_2^{(n)} = \phi b_2^{(n-1)} + \frac{(\theta-\phi)}{(1-\phi)} a_n.$$

When $\phi = 0$, the eventual forecast function and updating equations reduce to those given in Example 7.2(b) for the model $\nabla x_t = (1-\theta B)\, a_t$. Substituting for $f_{n,1}$ and $f_{n,2}$ in terms of $b_1^{(n)}$ and $b_2^{(n)}$, we obtain explicitly

$$b_1^{(n)} = x_n + \frac{\phi}{1-\phi}(x_n - x_{n-1}) - \frac{\theta}{1-\phi} a_n,$$

$$b_2^{(n)} = \frac{\theta a_n - \phi(x_n - x_{n-1})}{1-\phi},$$

so that

$$f_{n,h} = x_n + \phi\frac{(1-\phi^h)}{1-\phi}(x_n - x_{n-1}) - \theta\frac{(1-\phi^h)}{1-\phi} a_n.$$

As $h \to \infty$, we see that $f_{n,h} \to b_1^{(n)}$.

The forecast functions for various commonly encountered ARIMA models are depicted graphically in Exhibit 7.1, where the dependence of the initial forecasts on the pivotal values and the eventual behaviour of the forecast function are clearly seen.

8 ARIMA model building

8.1 Introduction

In the previous three chapters of the book we have developed the class of ARIMA models for analysing economic time series. We now discuss how such models can be fitted to observed series. In Box and Jenkins (1976), this process is known as the iterative strategy of 'identification–estimation–diagnostic checking' and is based on a sequence of procedures, many somewhat informal, which are designed to arrive, eventually, at the most appropriate specification from the general class of ARIMA (p, d, q) models. It is often argued that this strategy requires a good deal of experience and expertise and much recent research has been undertaken on the design of more formal testing procedures and the use of selection criteria, which enable a more standard set of 'rules' to be followed in model specification.

Since the 'Box–Jenkins' methodology is well documented in many text-books on time series, we focus greater attention on these newer approaches, believing them to be extremely useful aids in model building. Nevertheless, we do not wish our discussion to be seen as a downgrading of the traditional approach, for, as with any model building exercise, all useful methods should be examined in arriving at an appropriate specification.

8.2 Estimation of ARIMA models

The class of ARIMA models that has been developed in previous chapters may be written generally as

$$\phi(B)\,\nabla^d x_t = \theta_0 + \theta(B)\,a_t, \tag{8.1}$$

where x_t can be some transformation of the actually observed series, z_t say, so that $x_t = g(z_t)$. If this transformation is known, and the degree of differencing, d, and the orders of the polynomials $\phi(B)$ and $\theta(B)$, denoted p and q respectively, are also specified, then the estimation problem is that of estimating the parameters $\boldsymbol{\phi} = (\phi_1, \ldots, \phi_p)'$, $\boldsymbol{\theta} = (\theta_0, \theta_1, \ldots, \theta_q)'$, and σ^2, the variance of the white noise error process a_t. Let $\boldsymbol{x} = (x_1, x_2, \ldots, x_n)'$ be the vector of original (possibly transformed) observations, and let $\boldsymbol{w} = (w_1, w_2, \ldots, w_T)$ be the vector of $T = n-d$ stationary differences, $w_t = \nabla^d x_t$. The ARIMA (p, d, q) model (8.1) can then be written as

$$a_t = \theta_0 + \theta_1 a_{t-1} + \ldots + \theta_q a_{t-q} + w_t - \phi_1 w_{t-1} - \ldots - \phi_p w_{t-p}. \tag{8.2}$$

On the assumption that the a_t s are independent and come from a normal distribution with mean zero and variance σ^2, the joint probability density function of $\boldsymbol{a} = (a_1, a_2, \ldots, a_T)'$ is given by

$$p(\boldsymbol{a} \mid \boldsymbol{\phi}, \boldsymbol{\theta}, \sigma^2) = (2\pi\sigma^2)^{-T/2} \exp\left[-\frac{1}{2\sigma^2} \sum_{t=1}^{T} a_t^2\right].$$

The joint probability density function of $\boldsymbol{w}$ (or, equivalently, the likelihood function of the parameters $(\boldsymbol{\phi}, \boldsymbol{\theta}, \sigma^2)$) can then be written in the form (Newbold (1974))

$$L(\boldsymbol{\phi}, \boldsymbol{\theta}, \sigma^2 \mid \boldsymbol{w}) = g_1(\boldsymbol{\phi}, \boldsymbol{\theta}, \sigma^2) \exp\left[-\frac{1}{2\sigma^2} S(\boldsymbol{\phi}, \boldsymbol{\theta})\right], \tag{8.3}$$

where g_1 is a function of the parameters $(\boldsymbol{\phi}, \boldsymbol{\theta}, \sigma^2)$ and

$$S(\boldsymbol{\phi}, \boldsymbol{\theta}) = \sum_{t=1-p-q}^{T} E(u_t \mid \omega)^2.$$

Here $\omega = (\boldsymbol{w}, \boldsymbol{\phi}, \boldsymbol{\theta}, \sigma^2)$ and hence $E(u_t \mid \omega)$ is the conditional expectation of u_t given $\boldsymbol{w}$, $\boldsymbol{\phi}$, $\boldsymbol{\theta}$ and σ^2, where

$$u_t = \begin{cases} a_t & t = 1, 2, \ldots, T \\ g_2(\boldsymbol{a}^*, \boldsymbol{w}^*) & t \leqslant 0, \end{cases}$$

g_2 being a linear function of the initial unobservable values $\boldsymbol{a}^* = (a_{1-q}, \ldots, a_{-1}, a_0)'$ and $\boldsymbol{w}^* = (w_{1-p}, \ldots, w_{-1}, w_0)'$ needed for the evaluation of $\boldsymbol{a}$ in equation (8.2). The functions g_1 and g_2 depend on the particular ARIMA model, and the exact likelihood functions for a variety of cases may be found, for example, in Abraham and Ledolter (1983, chapter 5).

Maximum likelihood (ML) estimates of the parameters $(\boldsymbol{\phi}, \boldsymbol{\theta}, \sigma^2)$ can be obtained by maximising the likelihood function (8.3). In general, closed-form solutions cannot be found, but various algorithms are available to compute ML estimates, or close approximations to them, numerically: see, for example, Ansley (1979), Ljung and Box (1979). ML estimation can be difficult and is often expensive in computing time. This is because the presence of the function $g_1(\boldsymbol{\phi}, \boldsymbol{\theta}, \sigma^2)$ in the likelihood function can make maximisation difficult, and the conditional expectations $E(u_t \mid \omega)$, $(t = 1-p-q, \ldots, -1, 0)$, are required, which in turn need the calculation of $E(a_t \mid \omega)$, $(t = 1-q, \ldots, -1, 0)$ and $E(w_t \mid \omega)$, $(t = 1-p, \ldots, -1, 0)$.

In view of these difficulties, a number of approximations to ML have been proposed. A standard approach is to ignore the function $g_1(\boldsymbol{\phi}, \boldsymbol{\theta}, \sigma^2)$ and maximise $\exp[-(1/2\sigma^2) S(\boldsymbol{\phi}, \boldsymbol{\theta})]$ or, equivalently, to minimise $S(\boldsymbol{\phi}, \boldsymbol{\theta})$. This, of course, leads to *least squares* (LS) estimates and such an approximation will be satisfactory unless the parameters are close to the invertibility boundaries for, except in this case, the likelihood function is dominated by the exponential part of (8.3) and the removal of g_1 has only a negligible effect.

Nevertheless, even with LS estimation, the conditional expectations $E(u_t \mid \omega)$ must still be evaluated, and the way in which this is done leads to two forms of LS estimates. In Chapter 7 we found that the MMSE forecast of x_{n+h} made at time n is given by the conditional expectation $E(x_{n+h} \mid x_n, x_{n-1}, \ldots)$. This result may be used to compute the unknown quantities $E(a_t \mid \omega)$, $(t = 1-q, \ldots, -1, 0)$, and $E(w_t \mid \omega)$, $(t = 1-p, \ldots, -1, 0)$, needed for the evaluation of $E(u_t \mid \omega)$. The MMSE forecasts of a_t and w_t, $(t \leqslant 0)$, can be found by considering the 'backward' model

$$e_t = \theta_0 + \theta_1 e_{t+1} + \ldots + \theta_q e_{t+q} + w_t - \phi_1 w_{t+1} - \ldots - \phi_p w_{t+p}. \qquad (8.4)$$

Since this model has the same ACF as that of equation (8.2), and e_t is also a white noise sequence with variance σ^2, this method of evaluating $E(a_t \mid \omega)$ and $E(w_t \mid \omega)$, $(t \leqslant 0)$, is called *backforecasting* (or simply *backcasting*). Having obtained these backcasts for a given set of starting $(\boldsymbol{\phi}, \boldsymbol{\theta})$ values, a nonlinear least squares procedure can then be used to find the set of parameter values $(\hat{\boldsymbol{\phi}}, \hat{\boldsymbol{\theta}}, \hat{\sigma}^2)$ that minimises $S(\boldsymbol{\phi}, \boldsymbol{\theta})$; see, for example, Abraham and Ledolter (1983, chapter 5) for details. These values are usually called *unconditional least squares* (ULS), or *exact least squares*, estimates.

Computationally simpler estimates can be obtained by minimising the *conditional sum of squares*

$$S_c(\boldsymbol{\phi}, \boldsymbol{\theta}) = \sum_{t=p+1}^{T} a_t^2,$$

where the starting values $a_p, a_{p-1}, \ldots, a_{p+1-q}$ are set equal to their expected value of zero. The resulting estimates are known as *conditional least squares* (CLS) estimates.

The ML estimates will be consistent, asymptotically efficient and normally distributed and the inverse of the information matrix

$$I(\tilde{\boldsymbol{v}}) = -E\left[\frac{\partial^2 \ln L}{\partial \boldsymbol{v}\, \partial \boldsymbol{v}'}\bigg|_{\boldsymbol{v}=\tilde{\boldsymbol{v}}}\right]$$

can be used as an estimator of the asymptotic covariance matrix of the ML estimator $\tilde{\boldsymbol{v}} = (\tilde{\boldsymbol{\phi}}, \tilde{\boldsymbol{\theta}})$. For ULS and CLS estimates, the covariance matrix obtained at the final iteration of the nonlinear least squares procedure may be used, and this will be asymptotically equivalent to $I^{-1}(\tilde{\boldsymbol{v}})$.

Thus there are various, asymptotically equivalent, estimation procedures available and a choice has to be made as to which one should be used. This will usually depend upon the available software, but if different estimation algorithms are available, the choice should be based on the small-sample properties of the estimators. Evidence on such properties is somewhat limited, but the results of Ansley and Newbold (1980), Davidson (1981) and Osborn (1982) suggest that ML is usually preferable in small samples and particularly so when the parameter values approach the invertible boundaries. CLS is comparable to ML when values are away from the invertibility boundaries and ULS does not appear to be recommended by any authors. In terms of estimating the error variance, it seems to be the case that CLS has a tendency to overestimate σ^2, while use of ULS leads to an understatement of σ^2.

8.3 Choosing a transformation

The previous section assumed that the observed series x_t was itself some (instantaneous) transformation of the series z_t, so that $x_t = g(z_t)$. As we have discussed earlier, such a transformation is often employed to stabilise the variance of the series and, more generally, to induce normality. The family of transformations typically used for this purpose is the Box–Cox power family

$$g(z_t) = z_t^{(\lambda)} = (z_t^{\lambda} - 1)/\lambda,$$

with $z_t^{(0)} = \ln z_t$, as introduced in Chapter 6. The value of the index λ can either be chosen beforehand or estimated along with the other parameters $(\boldsymbol{\phi}, \boldsymbol{\theta}, \sigma^2)$; see Nelson and Granger (1979) for details of joint estimation procedures. The former approach is the most usual and the exploratory methods of choosing a power transformation, most notably range-median plots, outlined in Chapter 4 may be employed to determine a value of λ, usually restricting attention to a limited range of transformations such as those indexed by 0, $\frac{1}{2}$ and 1. The examples given in Chapter 4 provide illustrations of how these techniques operate in practice.

We are not necessarily restricted to power transformations for stabilising the variance. Granger and Hughes (1971), for example, found that the transformation

$$g(z_t) = \frac{z_t}{\frac{1}{2m+1}\sum_{j=-m}^{m} z_{t-j}}$$

was useful when both the mean and variance trend together.

Although the logarithmic transformation ($\lambda = 0$) is, in practice, the most widely used, since it often succeeds in stabilising the variance of the observed series and also reflects the fact that many economic variables are, by definition, positive valued, it may not be the most appropriate when variables are bounded both above and below. For a variable known to lie between 0 and 1, Wallis (1987) suggests using the logistic transformation

$$g(z_t) = \ln\left[\frac{z_t}{1-z_t}\right],$$

and shows that this has a number of advantages over the logarithmic. In particular, an attractive feature of the logistic transformation is its symmetry. An example of a series bounded as $0 \leqslant z_t \leqslant 1$ is the unemployment rate, and hence $1-z_t$ is its complement, the employment rate. Since $\ln\{z_t/(1-z_t)\} = -\ln\{(1-z_t)/(1-(1-z_t))\}$, the ACFs for the logistic transformations of z_t and $1-z_t$ are identical and it is of no consequence whether one chooses to work with the unemployment or employment rate, which is not true for the logarithmic transformation.

8.4 Determining the degree of differencing

Having decided upon the appropriate transformation of the data, the next stage is to determine the degree of differencing, d. If the series is nonstationary then d will be positive, and this will usually be apparent from a plot of the series. Determining the actual value of d, however, is not so easy from just a visual inspection, and examination of the SACFs for various differences will also be required.

To see why this is so, recall from Chapter 5 that a stationary AR(p) process requires that all the roots g_i in

$$\phi(B) = (1-g_1 B)(1-g_2 B)\ldots(1-g_p B)$$

are such that $|g_i| < 1$. Now suppose that one of them, say g_1, approaches

1, i.e. $g_1 = 1-\delta$, where δ is a small positive number. The autocorrelations

$$\rho_k = A_1 g_1^k + A_2 g_2^k + \ldots + A_p g_p^k \simeq A_1 g_1^k$$

will then be dominated by $A_1 g_1^k$, since all other terms will go to zero more rapidly. Furthermore, as g_1 is close to 1, the exponential decay $A_1 g_1^k$ will be slow and almost linear, since

$$A_1 g_1^k = A_1(1-\delta)^k = A_1(1-\delta k + \delta^2 k^2 - \ldots) \simeq A_1(1-\delta k).$$

Hence, failure of the SACF to die down quickly is therefore an indication of nonstationarity, its behaviour tending to be a slow, linear decline. As a further illustration, consider the ARMA (1, 1) model

$$(1-\phi B)\, x_t = (1-\theta B)\, a_t.$$

The ACF of this process has been shown to be (equation (5.25))

$$\rho_1 = \frac{(1-\phi B)(\phi-\theta)}{1+\theta^2-2\phi\theta},$$

$$\rho_k = \phi^{k-1}\rho_1, \quad k > 1.$$

As $\phi \to 1, \rho_k \to 1$ for any finite $k > 0$ and again the autocorrelations will decay very slowly and approximately linearly.

Thus a slow and almost linear decay in the SACF may be taken as an indication of nonstationarity and hence of the need for differencing. It should also be pointed out that the slow decay in the SACF can start at values of r_1 considerably smaller than 1, sometimes 0.5 or even less. If the original series x_t is found to be nonstationary, the first difference ∇x_t is then analysed. If ∇x_t is still nonstationary, the second difference $\nabla^2 x_t = (x_t - x_{t-1})^2$ is analysed: the procedure being repeated until a stationary difference is found, although it is seldom the case in practice that d exceeds 2.

Sole reliance on the SACF can sometimes lead to problems of *overdifferencing*. Although further differences of a stationary series will still be stationary, overdifferencing can lead to serious difficulties. Consider the stationary MA (1) process $x_t = (1-\theta B)\, a_t$. The first difference of this process is

$$\begin{aligned}(1-B)\, x_t &= (1-B)(1-\theta B)\, a_t \\ &= (1-(1+\theta)\, B + \theta B^2)\, a_t \\ &= (1-\theta_1 B - \theta_2 B^2)\, a_t.\end{aligned}$$

We now have a more complicated model containing two parameters rather than one and, moreover, one of the roots of the $\theta(B)$ polynomial is unity since $\theta_1+\theta_2 = 1$. The model is therefore not invertible, so that the AR(∞) representation does not exist and, as noted in section 8.2, attempts to estimate this model will almost surely run into difficulties.

Note also that the variance of x_t is given by

$$\gamma_0(x) = (1+\theta^2)\,\sigma^2,$$

whereas the variance of $w_t = \nabla x_t$ is given by

$$\gamma_0(w) = (1+(1+\theta)^2+\theta^2)\,\sigma^2$$
$$= 2(1+\theta+\theta^2)\,\sigma^2.$$

Hence

$$\gamma_0(w)-\gamma_0(x) = (1+\theta)^2\sigma^2 > 0,$$

thus showing that the variance of the overdifferenced process will always be larger than that of the original process.

As discussed by Anderson (1976), the behaviour of the sample variances associated with different values of d can provide a useful means of deciding the appropriate level of differencing: the sample variances will decrease until a stationary sequence has been found, but will tend to increase on overdifferencing. However, this is not always the case, and a comparison of sample variances for successive differences of a series is best employed as a useful auxiliary method of determining the appropriate value of d. In any case, as we shall shortly find, it may be difficult to decide sensibly between two alternative values of d. This will necessarily lead to alternative models, discrimination between which must be made on other criteria.

Example 8.1: Determining the degree of differencing for housing starts

Consider again the UK housing starts series originally analysed in Examples 5.1 and 5.2. The SACF of the original series x_t (the appropriateness of working with untransformed data being confirmed by range-median plots) was shown in Exhibit 5.1. Although it starts at the large value of $r_1 = 0.86$, it quickly damps in a nonlinear fashion, the cyclical pattern suggesting complex roots associated with an ARIMA (2, 0, 0) model (AR (2) in our simpler terminology), although examination of the SPACF suggested the simpler ARIMA (1, 0, 0) model. Note that $V(x_t) = 3872$. Calculation of the first differences $\nabla x_t = x_t - x_{t-1}$ led to the SACF shown in Exhibit 8.1. None of the r_k are now significant, thus suggesting that ∇x_t is white noise, i.e.

$$\nabla x_t = a_t,$$

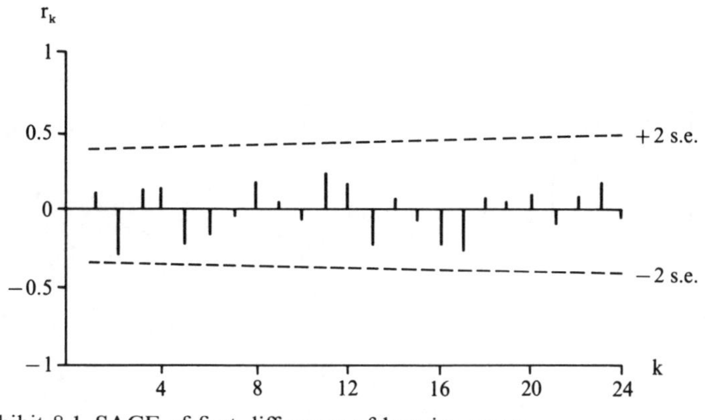

Exhibit 8.1 SACF of first difference of housing starts

the random walk model introduced in Chapter 6. For this first differenced series,

$$V(\nabla x_t) = 923 < V(x_t),$$

and hence yet a third model, the random walk (or ARIMA (0, 1, 0)), is suggested by the 'min $V(\nabla^d x_t)$' criterion, already showing that it is often difficult to isolate a single model for fitting to a series that has a relatively small number of observations.

Example 8.2: Determining the degree of differencing for nominal TFE

Consider now the logarithms of the nominal TFE series originally analysed in Chapter 4. Exhibit 8.2 displays the SACFs for x_t, ∇x_t and $\nabla^2 x_t$, along with the associated sample variances. The SACF for the undifferenced series x_t clearly reveals nonstationary behaviour, but that for ∇x_t is rather more difficult to interpret. Although the decline is fairly slow, linearity is not particularly apparent, although we must always bear in mind that the rather short series available ($T = 38$) may make precise discrimination difficult. $\nabla^2 x_t$ is certainly stationary, and $d = 2$ is also suggested by the 'min $V(\nabla^d x_t)$' criterion.

Example 8.3: Determining the degree of differencing for real TFE

Exhibit 8.3 shows various SACFs for the logarithms of the real TFE series, again originally analysed in Chapter 4. Here, both examination of the SACFs and use of the 'min $V(\nabla^d x_t)$' criterion point to ∇x_t being the appropriately differenced series to analyse.

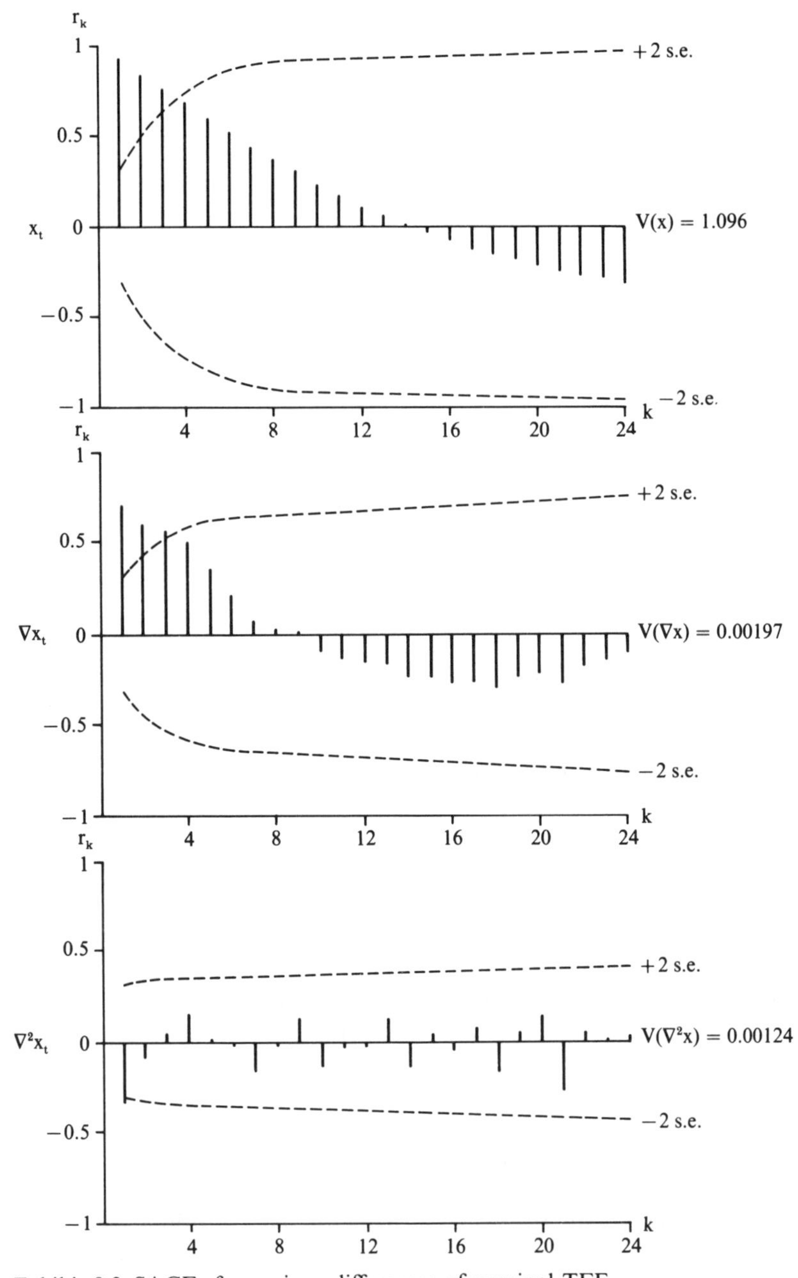

Exhibit 8.2 SACFs for various differences of nominal TFE

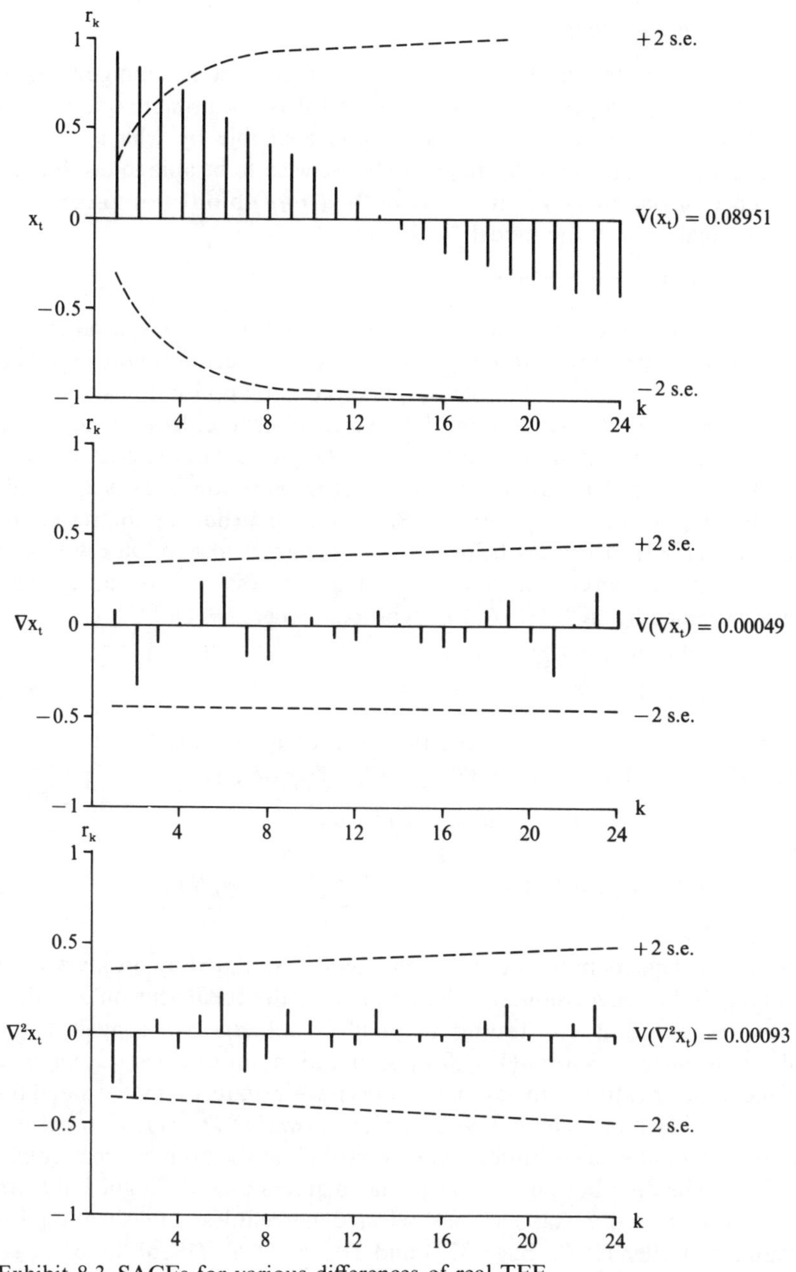

Exhibit 8.3 SACFs for various differences of real TFE

8.4.1 *Testing for unit roots*

We note that in two of the above examples, there is some ambiguity as to the appropriate degree of differencing, and it is a legitimate question to ask whether there are more formal tests available by which we may determine the value of d. More precisely, we wish to be able to test for the presence of one or more unit roots in the $(p+d)$th order autoregressive polynomial $\alpha(B)$ in the model

$$\alpha(B)\,x_t = \theta_0 + \theta(B)\,a_t. \tag{8.5}$$

The theory and practice of testing for unit roots has produced a voluminous literature in recent years, and this has been formally reviewed in Fuller (1985) and conveniently surveyed for applied researchers by Dickey et al. (1986). To develop this testing procedure, let us concentrate on autoregressive models, which may not be particularly restrictive since (8.5) will always have an autoregressive representation if, as is assumed, the moving average polynomial $\theta(B)$ is invertible (but see the results of Schwert (1987), discussed below). In particular, Said and Dickey (1984) argue that an unknown ARIMA$(p,1,q)$ process can be adequately approximated by an ARIMA$(l,0,0)$ process, where $l = O(T^{1/3})$. Consider, for example, the process

$$\phi(B)(1-\alpha B)\,x_t = \theta_0 + a_t \tag{8.6}$$

where $\theta_0 = \phi(1)(1-\alpha)\mu$, μ being the mean of x_t. Noting that $(1-\alpha B) = [(1-B)+(1-\alpha)B]$, equation (8.6) can be rewritten as

$$\phi(B)\,\nabla x_t + \phi(B)(1-\alpha)\,x_{t-1} = \theta_0 + a_t$$

or

$$\nabla x_t = \theta_0 + \phi(1)(\alpha-1)\,x_{t-1} + \phi_1^*\,\nabla x_{t-1} + \ldots + \phi_p^*\,\nabla x_{t-p} + a_t, \tag{8.7}$$

where $\phi_i^* = \alpha\phi_i$, $i = 1,\ldots,p$.

The coefficients in (8.7) can be consistently estimated by ordinary least squares (OLS) regression and the estimate of the coefficient on x_{t-1} thus provides a means for testing the null hypothesis $\alpha = 1$ against the alternative $\alpha < 1$. Since $\phi(1) \neq 0$ by assumption, an obvious test statistic is the usual 't-ratio' of the estimate of $\phi(1)(\alpha-1)$ to its estimated standard error. Dickey and Fuller (1979), however, show that this statistic does not have a Student's t distribution, even in the limit as the sample size becomes infinite. The distribution of this statistic, denoted τ_μ to distinguish it from the conventional t statistic, has selected percentiles published in, for example, Fuller (1976, page 373) and Dickey et al. (1986, table 1). To appreciate the differences in the two distributions, these percentiles show that for large sample sizes, using a 0.05 significance level would require a critical τ_μ value of -2.86, rather than -1.96 for the normal approximation to Student's t.

If the τ_μ statistic calculated from (8.7) does not allow the rejection of the null hypothesis $\alpha = 1$, i.e. one unit root, the presence of a second unit root may be tested by estimating the regression of $\nabla^2 x_t$ on 1, $\nabla x_{t-1}, \nabla^2 x_{t-1}, \ldots, \nabla^2 x_{t-p}$ and comparing the 't-ratio' of the coefficient of ∇x_{t-1} to the τ_μ distribution. Alternatively, the presence of two unit roots may be tested jointly by estimating the regression of $\nabla^2 x_t$ on $x_{t-1}, \nabla x_{t-1}, \nabla^2 x_{t-1}, \ldots, \nabla^2 x_{t-p}$ and computing the usual F statistic for testing the joint significance of x_{t-1} and ∇x_{t-1}. Again, though, this test statistic has a distribution under the null hypothesis of a double unit root that is not Snedecor's F, but is one labelled $\Phi_1(2)$ by Hasza and Fuller (1979), who provide various percentiles of this distribution as part of their table 4.1.

Example 8.4: Examples of testing for unit roots

This testing approach was used on the three series analysed in Examples 8.1 to 8.3. For housing starts, there was a debate over whether the series should be differenced or not. The appropriate regression is

$$\nabla x_t = \underset{(12.50)}{31.03} - \underset{(0.080)}{0.188} x_{t-1},$$

where numbers in parentheses below the coefficients are standard errors. No lags of ∇x_t are required and the τ_μ statistic is -2.36. The approximate critical values of this statistic are -2.95 for a 0.05 significance level and -2.60 for a 0.10 significance level, and hence we cannot reject the hypothesis of a unit root. Note, of course, that if we had incorrectly compared our τ_μ value with the t distribution we would have rejected the null hypothesis at a significance level of less than 0.025. We may go further and test whether a second unit root is required. From preliminary analysis, one lagged value of $\nabla^2 x_t$ is required, and the appropriate regression is

$$\nabla^2 x_t = \underset{(5.06)}{4.76} - \underset{(0.236)}{1.397} \nabla x_{t-1} + \underset{(0.163)}{0.313} \nabla^2 x_{t-1}$$

and now $\tau_\mu = -5.92$, clearly rejecting the hypothesis of a second unit root.

Similar conclusions are drawn from the regressions with real TFE, these being

$$\nabla x_t = \underset{(0.174)}{0.236} - \underset{(0.014)}{0.017} x_{t-1}$$

and

$$\nabla^2 x_t = \underset{(0.007)}{0.032} - \underset{(0.229)}{1.232} \nabla x_{t-1} + \underset{(0.169)}{0.342} \nabla^2 x_{t-1},$$

respectively, with τ_μ statistics of -1.20 and -5.28. For nominal TFE, examination of the SACF suggested first differencing while the 'min $V(\nabla^d x_t)$' criterion pointed towards second differencing. Here the appropriate regressions are

$$\underset{}{\nabla x_t} = \underset{(0.070)}{-0.054} + \underset{(0.007)}{0.008 x_{t-1}} + \underset{(0.151)}{0.587 \nabla x_{t-1}}$$

and

$$\nabla^2 x_t = \underset{(0.013)}{0.029} - \underset{(0.124)}{0.307 \nabla x_{t-1}}.$$

In this second regression, the τ_μ statistic is -2.48 and thus we cannot reject the hypothesis that second differencing of the nominal TFE series is appropriate. This is also confirmed by the joint test calculated from the regression

$$\nabla^2 x_t = \underset{(0.135)}{-0.361 \nabla x_{t-1}} + \underset{(0.0013)}{0.0031 x_{t-1}},$$

which is $F = 3.58$. From Hasza and Fuller (1979, page 1116), the 95th percentile of the appropriate $\Phi_1(2)$ distribution is approximately 3.7 and hence we cannot reject the null hypothesis of a double unit root.

We are thus led to the conclusion that both the housing starts and real TFE series require first differencing to induce stationarity, while the nominal TFE series requires second differencing.

8.4.2 *Extensions of unit root tests*

Some further extensions to these unit root tests have recently been developed. Said and Dickey (1985) show that the τ_μ statistic from an ARIMA $(p, 0, q)$ process calculated from the regression

$$\nabla x_t = \theta_0 + (\alpha - 1) x_{t-1} + \sum_{i=1}^{p} \phi_i \nabla x_{t-i} + a_t - \sum_{j=1}^{q} \theta_j a_{t-j}$$

has the same asymptotic distribution as the τ_μ statistic calculated from equation (8.7). The problem here is that p and q are assumed known, and this is unlikely to be the case in practice. Alternatively, Phillips (1987a, b) has shown that, irrespective of the orders p and q, a test for a unit root can be carried out by estimating (8.7) with $p = 0$, i.e.

$$\nabla x_t = \theta_0 + (\alpha - 1) x_{t-1} + a_t,$$

where a_t may now be both serially correlated and heteroskedastic, and adjusting the 't-ratio' τ_μ statistic to

$$Z_{\tau\mu} = \tau_\mu(\hat{\sigma}_a/\hat{\sigma}_{\tau l}) - \tfrac{1}{2}(\hat{\sigma}^2_{\tau l} - \hat{\sigma}^2_a)\{\hat{\sigma}^2_{\tau l}\, T^{-\frac{1}{2}} \sum_{t=2}^{T} (x_{t-1} - \bar{x}_{-1})^2\}^{-\frac{1}{2}},$$

where $\hat{\sigma}^2_a$ is the sample variance of a_t,

$$\bar{x}_{-1} = \sum_{t=1}^{T-1} x_t$$

and

$$\hat{\sigma}^2_{\tau l} = T^{-1} \sum_{t=1}^{T} u_t^2 + 2T^{-1} \sum_{j=1}^{l} \omega_{jl} \sum_{t=j+1}^{T} u_t u_{t-j},$$

the weights $\omega_{jl} = \{1 - j/(l+1)\}$ ensuring that the variance estimate $\hat{\sigma}^2_{\tau l}$ is positive. Schwert (1987) suggests setting l equal to the integer parts of either $\{4(T/100)^{0.25}\}$ or $\{12(T/100)^{0.25}\}$, and the $Z_{\tau l}$ statistic can again be compared to the τ_μ tables.

Schwert also presents simulation and empirical evidence suggesting that all of these test statistics can be affected by the process actually generating the data and, in particular, it is important to consider whether the underlying process contains a moving average component, since the test statistics can then be very different to the τ_μ distributions reported by Fuller (1976). Schwert concludes that unit root tests may depend critically on the assumption that the underlying process is a pure autoregression, and one should therefore consider the correct specification of the ARIMA process before testing for the presence of unit roots in the autoregressive polynomial, a problem that we now turn to.

8.5 Determining the orders of $\phi(B)$ and $\theta(B)$

Having determined the order of differencing, the orders of the autoregressive ($\phi(B)$) and moving average ($\theta(B)$) polynomials must be specified. In the traditional 'Box–Jenkins' approach this is done by matching the patterns in the sample autocorrelations and partial autocorrelations with the theoretical patterns of known models. The orders p and q will usually be small and in Exhibit 8.4 we summarise the theoretical properties of several commonly used models.

We shall not go into great detail in discussing this approach since, along with the original work of Box and Jenkins (1976), there are a number of books devoted to it; see, for example, Anderson (1976), Nelson (1973), Pankratz (1983), Vandaele (1983), McLeod (1983), and O'Donovan (1983). We shall simply concentrate on providing examples illustrating the

Exhibit 8.4 *Properties of the ACF and the PACF for various ARIMA models*

Model	ACF	PACF
$(1, d, 0)$	Exponential or oscillatory decay.	$\phi_{kk} = 0$ for $k > 1$.
$(2, d, 0)$	Exponential or sine wave decay.	$\phi_{kk} = 0$ for $k > 2$.
$(p, d, 0)$	Exponential and/or sine wave decay.	$\phi_{kk} = 0$ for $k > p$.
$(0, d, 1)$	$\rho_k = 0$ for $k > 1$.	Dominated by damped exponential.
$(0, d, 2)$	$\rho_k = 0$ for $k > 2$.	Dominated by damped exponential or sine wave.
$(0, d, q)$	$\rho_k = 0$ for $k > q$.	Dominated by linear combination of damped exponentials and/or sine waves.
$(1, d, 1)$	Tails off. Exponential decay from lag 1.	Tails off. Dominated by exponential decay from lag 1.
(p, d, q)	Tails off after $q-p$ lags. Exponential and/or sine wave decay after $q-p$ lags.	Tails off after $p-q$ lags. Dominated by damped exponentials and/or sine waves after $p-q$ lags.

approach and pointing towards the various difficulties that may be encountered with it.

Example 8.5: Examples of order identification using the traditional approach

We begin by considering the three series previously analysed in this chapter. For the housing starts series, first differencing is required and the associated SACF is shown in Exhibit 8.1. This shows no significant r_k's and, indeed, the sample PACF, not shown here, is of a similar pattern. However, r_2 is marginally significant, with an estimate of -0.30 and a

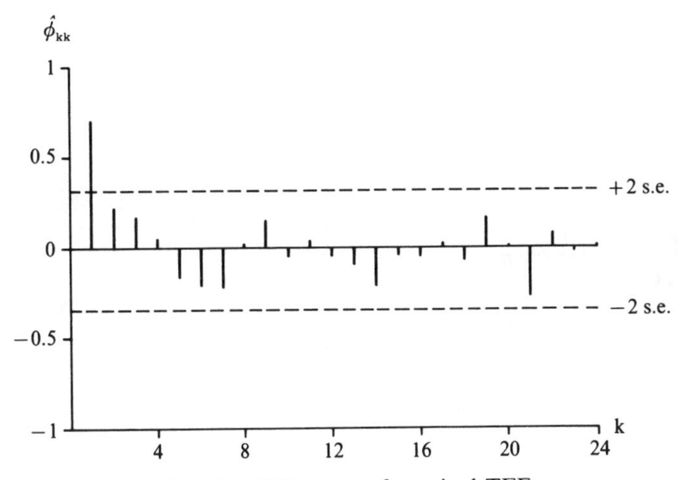

Exhibit 8.5 PACF for first differences of nominal TFE

standard error of 0.16, and hence, rather than a random walk, the following (restricted) ARIMA (0, 1, 2) model might be 'tentatively entertained', in the phraseology of Box and Jenkins,

$$\nabla x_t = \theta_0 + (1 - \theta_2 B^2) a_t. \tag{8.8}$$

First differencing was also required for real TFE. From the SACF shown in Exhibit 8.3, we see that again the only significant r_k is at lag 2, with the PACF also showing a similar pattern. Thus we identify the restricted ARIMA (0, 1, 2) model given by equation (8.8) for this series as well.

For nominal TFE, second differencing is needed, and the associated SACF shown in Exhibit 8.2 shows a single significant autocorrelation at r_1. Exhibit 8.5 displays the corresponding PACF, which shows a tendency to 'tail off', although with no particularly striking pattern, thus suggesting that the process is a moving average. Hence we identify, on grounds of parsimony, an ARIMA (0, 2, 1) model.

Example 8.6: Identification of income and consumption series

Two further annual macroeconomic series introduced in Chapter 3 are real personal disposable income and real consumption, their growth rates being shown as Exhibit 3.1. Initial analysis found that logarithmic transformations were satisfactory for both series, and that both needed first differencing to induce stationarity, so that growth rates are indeed the appropriate transformation to make. The SACFs and PACFs for the first differenced series are shown in Exhibit 8.6. Although few, if any, of the

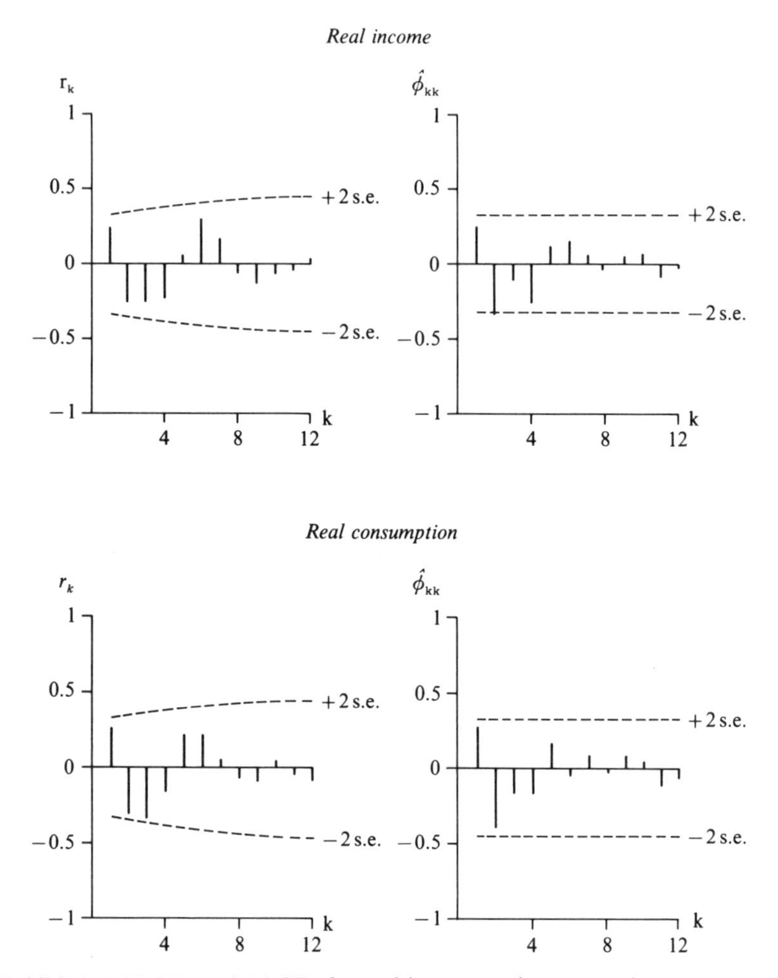

Exhibit 8.6 SACFs and PACFs for real income and consumption

autocorrelations are individually significant, both SACFs show something of a sine wave pattern, perhaps suggestive of an autoregressive process. Their PACFs, on the other hand, have large ϕ_{22}s, with little else at higher lags, thus pointing towards ARIMA(2, 1, 0) models for both series, although these identifications should be regarded as extremely tentative and, as we shall see later, various other models may also be entertained.

The series analysed in the above examples have all been annual and of around 40 observations in length. It is sometimes suggested that series of such limited extent are unsuitable for ARIMA modelling, but economists often have to work with such data and it is therefore important to provide

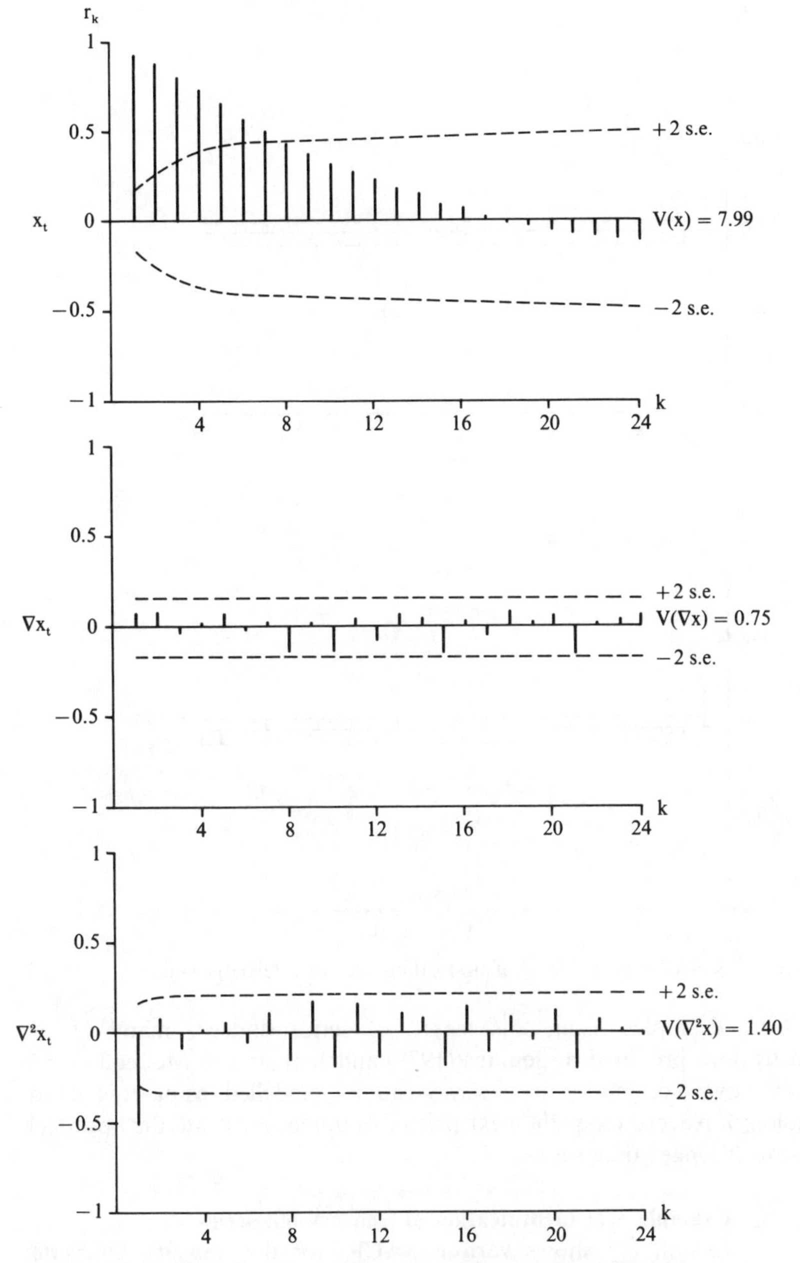

Exhibit 8.7 SACFs of various differences of the treasury bill yield

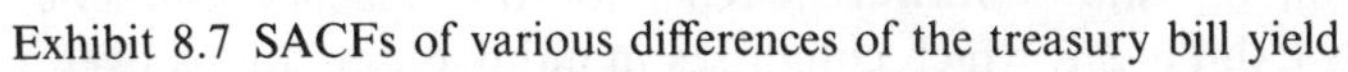

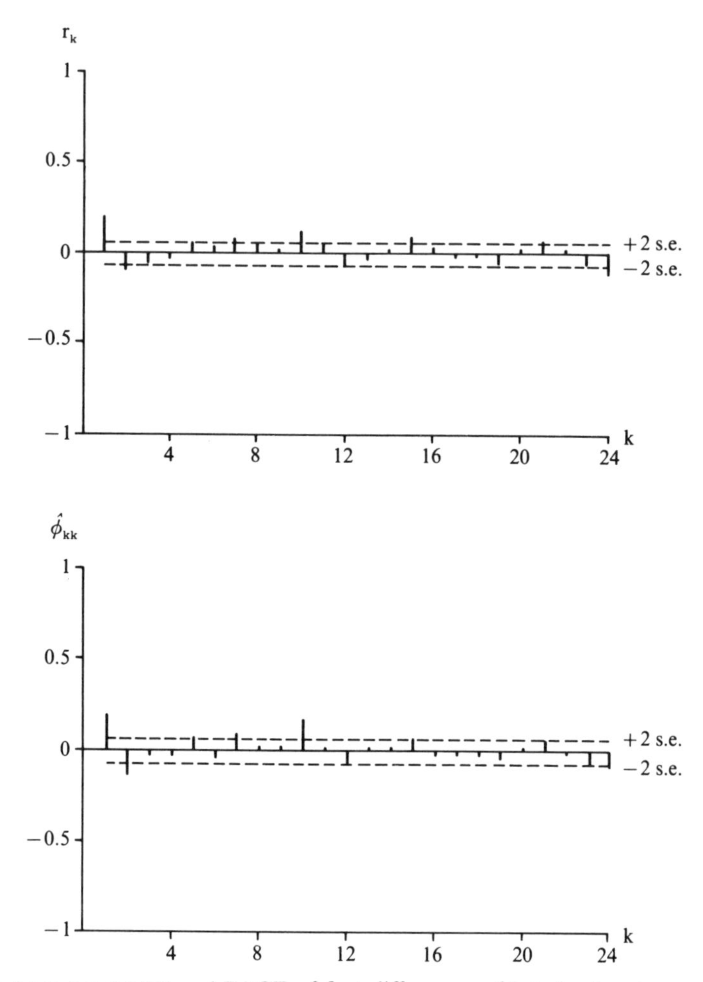

Exhibit 8.8 SACF and PACF of first differences of interbank rate

realistic examples using this length of series (indeed, many of the illustrations provided by Jenkins (1979) and Jenkins and McLeod (1983) show how successfully such series can be modelled using this methodology). Nevertheless, the next pair of examples illustrate the approach on much longer time series.

Example 8.7: Identification of treasury bill series

Exhibit 8.7 shows various SACFs for the treasury bill series introduced in Exhibit 2.7, which comprises monthly observations over a period of 13 years. It is clear from all criteria that first differencing is appropriate and, furthermore, that since the SACF for ∇x_t is completely flat, a random walk, ARIMA (0, 1, 0), model is identified. Note also that for $\nabla^2 x_t$, the only significant autocorrelation is at lag 1, suggesting the possibility of an ARIMA (0, 2, 1) model. However, this autocorrelation is estimated to be -0.50 and from Chapter 5.3, this corresponds to θ in the model

$$\nabla^2 x_t = (1-\theta B)\, a_t \tag{8.9}$$

taking on the value 1, thus confirming that the series has indeed been overdifferenced.

Example 8.8: Identification of 3-month interbank rate series

This example continues the theme of modelling interest rates and examines the daily three-month interbank rate shown in Exhibit 2.9, and analysed in an exploratory fashion in Chapter 3, where its first differences were shown as Exhibit 3.8. Standard analysis confirms that first differencing is indeed appropriate and the SACF and PACF of ∇x_t are shown as Exhibit 8.8. Only the first two auto and partial correlations are significant (apart from one at lag 10, the modelling of which is discussed in Chapter 15) and this points to three possible (low order) ARIMA models: (2, 1, 0), (0, 1, 2) and (1, 1, 1).

8.5.1 *Examples of model estimation*

Having identified a model for a particular time series or, as we have often seen, a number of models, the next stage is to estimate the model parameters using the estimation methods outlined in Section 8.2. The following set of examples thus discuss the estimation of the models identified in Examples 8.5 to 8.8.

Example 8.9: Estimation of housing starts model

The ARIMA (0, 1, 2) model shown as equation (8.8) was identified for the housing starts series. Using CLS estimation the following results were obtained:

$$\nabla x_t = \underset{(3.610)}{2.488} + (1 - \underset{(0.164)}{0.268} B^2)\, a_t, \hat{\sigma} = 29.88.$$

ML and ULS estimation produced almost identical results, the estimates of θ_2 being 0.267 and 0.277 respectively. The estimate of θ_0 is clearly insignificant, thus implying that there is no deterministic linear trend, which confirms the visual inspection of the series provided by Exhibit 2.2. Re-estimation without θ_0 obtained, using CLS for example,

$$\nabla x_t = (1 - \underset{(0.162)}{0.256} B^2)\, a_t, \hat{\sigma} = 29.66.$$

The estimate of θ_2 is rather imprecisely estimated, having a 't-ratio' of only 1.58 and, indeed, the random walk model obtained by setting this parameter to zero yields an estimate of the residual standard error of $\hat{\sigma} =$ 30.46, only marginally greater than the above model.

The same model was also identified for the real TFE series. Here CLS estimation obtained

$$\nabla x_t = \underset{(0.003)}{0.027} + (1 - \underset{(0.165)}{0.301} B^2)\, a_t, \hat{\sigma} = 0.0217.$$

The constant θ_0 is here significant, implying an annual trend growth rate of 2.7%, and θ_2 is fairly precisely estimated, with ML and ULS estimates of this parameter being 0.293 and 0.306 respectively.

Example 8.10: Estimation of nominal TFE model

Nominal TFE was identified as an ARIMA(0, 2, 1) process, estimation of which by CLS produced

$$\nabla^2 x_t = (1 - \underset{(0.150)}{0.437} B)\, a_t, \hat{\sigma} = 0.0327.$$

A constant was found to be insignificant, which should be the case for, from Chapter 7, this would imply that the series contains a deterministic quadratic trend, which is unlikely on visual inspection of the data.

Example 8.11: Estimation of income and consumption models

Real income and consumption were both tentatively identified as ARIMA(2, 1, 0) processes:

$$(1 - \phi_1 B - \phi_2 B^2) \nabla x_t = \theta_0 + a_t.$$

Exhibit 8.9 shows the estimates obtained for both series using the three alternative estimation methods. Typically, the two least squares techniques give virtually identical results, with ML producing slightly different estimates of the autoregressive parameters ϕ_1 and ϕ_2. Both series appear to be well fitted by this particular model, in the (limited) sense that the ϕ parameters are precisely estimated.

Example 8.12: Estimation of treasury bill model

In Example 8.7 the treasury bill series was identified as a random walk and on estimation no constant was found to be necessary. We noted that second differencing was inappropriate because the value of $\hat{r}_1$ suggested a parameter value on the invertibility boundary. Indeed, estimation of (8.9) by both CLS and ML produced values for $\hat{\theta}$ of unity, thus confirming the inappropriateness of second differencing.

Exhibit 8.9 *Alternative estimates for real income model*

	CLS	ULS	ML
ϕ_1	0.324	0.324	0.316
	(0.163)	(0.163)	(0.163)
ϕ_2	−0.347	−0.347	−0.329
	(0.163)	(0.163)	(0.163)
θ	0.026	0.026	0.026
σ	0.0230	0.0230	0.0230

Alternative estimates for real consumption model

	CLS	ULS	ML
ϕ_1	0.359	0.361	0.351
	(0.160)	(0.160)	(0.159)
ϕ_2	−0.407	−0.410	−0.388
	(0.162)	(0.162)	(0.161)
θ	0.022	0.023	0.023
σ	0.0180	0.0180	0.0180

Example 8.13: Estimation of alternative 3-month interbank models

Three possible models were identified for the daily three-month interbank series and estimation of these by CLS produced the following results.

$$\underset{}{(1-\underset{(0.036)}{0.221}B+\underset{(0.036)}{0.135}B^2)}\nabla x_t = a_t, \hat{\sigma} = 0.1558$$

$$\nabla x_t = (1+\underset{(0.037)}{0.221}B-\underset{(0.037)}{0.093}B^2)\,a_t, \hat{\sigma} = 0.1559$$

$$(1+\underset{(0.125)}{0.330}B)\nabla x_t = (1+\underset{(0.110)}{0.556}B)\,a_t, \hat{\sigma} = 0.1559.$$

All parameter estimates are significant and the three models fit equally well. This is not surprising, as all models are remarkably similar, for the linear filter representations of the ARIMA (2, 1, 0) and ARIMA (1, 1, 1) models are, respectively,

$$\nabla x_t = (1+0.221B-0.086B^2-\ldots)\,a_t$$

and

$$\nabla x_t = (1+0.226B-0.075B^2-\ldots)\,a_t,$$

which correspond closely to the fitted ARIMA (0, 1, 2) model.

8.6 The use of model selection criteria

The previous examples make it clear that the 'Box–Jenkins' approach to model identification requires a good deal of discretion and experience from the model builder and, while this is by no means a bad thing, a substantial amount of research has been undertaken on devising selection criteria which allow models to be chosen by sets of 'rules' and, indeed, in some automatic computer programs control of model identification is taken more or less completely out of the hands of the analyst; see, for example, Hill and Fildes (1984) and Poulos et al. (1987).

A number of model selection criteria have been proposed, most of which were originally developed for purely autoregressive schemes but have now been extended to the more general class of ARIMA models. In all cases it is assumed that the degree of differencing has been decided and that the object of the procedure is to determine the most appropriate values of p and q. Perhaps the earliest selection criterion is AIC (Akaike (1974)), which is defined as

$$\mathrm{AIC}(p,q) = \ln \hat{\sigma}^2 + 2(p+q)\,T^{-1}, \tag{8.10}$$

where $\hat{\sigma}^2$ is the estimate of the error variance σ^2 of the ARMA (p,q) model fitted to $w_t = \nabla^d x_t$, $(t = 1, 2, \ldots, T)$. A second criterion is BIC (Rissanen (1978), Schwarz (1978)):

$$\mathrm{BIC}(p,q) = \ln \hat{\sigma}^2 + (p+q)\,T^{-1} \ln T, \tag{8.11}$$

while yet a third is Φ (Hannan (1980)), defined for $c \geqslant 2$, by

$$\Phi(p,q) = \ln \hat{\sigma}^2 + (p+q)\,cT^{-1} \ln \ln T. \tag{8.12}$$

A number of other criteria have been proposed (see Shibata (1985) for a survey), but these three are the most commonly employed. All are structured in terms of the estimated variance $\hat{\sigma}^2$ (which is also, from Chapter 7, the estimated one-step ahead prediction error variance of the model), plus a penalty adjustment involving the number of estimated parameters, and it is in the extent of this penalty that the criteria differ. (Poskitt and Tremayne (1983) present a Bayesian analysis of the differences between these criteria.) The criteria are used in the following way. Upper bounds, say P and Q, are fixed for the polynomial operators $\phi(B)$ and $\theta(B)$, and with $\bar{P} = \{0, 1, \ldots, P\}$ and $\bar{Q} = \{0, 1, \ldots, Q\}$, orders p_1 and q_1 are selected such that, for example,

$$\mathrm{AIC}(p_1, q_1) = \min \mathrm{AIC}(p,q), p \in \bar{P}, q \in \bar{Q}$$

with parallel strategies obviously being employed in conjunction with BIC, Φ, or any other criterion. One possible difficulty with the application

Exhibit 8.10 *Model selection criteria for real income*

(i) *AIC*

		q			
		0	1	2	3
	0	−7.415	−7.455	−7.426	−7.371
	1	−7.390	−7.395	−7.422	−7.272
p	2	−7.433	−7.383	−7.174	−7.221
	3	−7.360	−7.296	−7.228	−7.219

(ii) *BIC*

	0	−7.415	−7.411	−7.338	−7.239
	1	−7.346	−7.251	−6.998	−7.001
p	2	−7.345	−7.251	−6.998	−7.001
	3	−7.228	−7.120	−7.008	−6.955

of this strategy is that no specific guidelines on how to determine $\bar{P}$ and $\bar{Q}$ seem to be available, although they are tacitly assumed to be sufficiently large for the range of models to contain the true model, which we may denote as having orders (p_0, q_0), and which, of course, will not necessarily be the same as (p_1, q_1), the orders chosen by the criterion under consideration.

Given these alternative criteria, are there reasons for preferring one to another? Hannan (1980) shows that, if the true orders (p_0, q_0) are contained in the set (p, q), $p \in \bar{P}$, $q \in \bar{Q}$, then for all these criteria, $p_1 \geqslant p_0$ and $q_1 \geqslant q_0$, almost surely, as $T \to \infty$. However, BIC and Φ are *strongly consistent* in that they determine the true model asymptotically, whereas for AIC, an overparameterised model will always emerge no matter how long the available realisation. Thus it would appear that either the BIC or Φ criterion should be used in preference to AIC.

Example 8.14: Use of criteria for selecting models for income and consumption

To illustrate the performance of these model selection criteria, one strongly consistent criterion, BIC, and the inconsistent criterion AIC were used to determine appropriate values of p and q for the first differences of the real income and consumption series, both of which we have already identified, albeit tentatively, as having orders (2, 0). Exhibits 8.10 and 8.11 show the AIC and BIC values obtained by setting $\bar{P} = \bar{Q} = 3$, bearing in mind the short series available, $T = 36$ in both cases. For neither series is the (2, 0) model selected. For real income, AIC selects the orders (0, 1), i.e. a MA(1) model, while BIC selects the random walk

Exhibit 8.11 *Model selection criteria for real consumption*

(i) *AIC*

		q			
		0	1	2	3
p	0	−7.848	−7.858	−7.853	−7.711
	1	−7.774	−7.833	−7.730	−7.586
	2	−7.808	−7.694	−7.582	−7.517
	3	−7.687	−7.388	−7.389	−7.372

(ii) *BIC*

		0	1	2	3
p	0	−7.848	−7.870	−7.876	−7.746
	1	−7.785	−7.856	−7.764	−7.632
	2	−7.831	−7.729	−7.628	−7.575
	3	−7.722	−7.434	−7.446	−7.442

model, (0, 0). Moving average models are again selected for real consumption, AIC once more choosing (0, 1) but BIC this time selecting (0, 2).

Example 8.15: Use of criteria for selecting models for 3-month interbank rate

As a second example, consider the daily three-month interbank rate series. Here we were unable to choose between models indexed by orders (2, 0), (0, 2) and (1, 1). Exhibit 8.12 shows the AIC and BIC values obtained by setting $\bar{P} = \bar{Q} = 6$, the series being of length $T = 740$. BIC selects the model indexed as (0, 1), close to those chosen previously, whereas AIC reaches a minimum at (5, 5), thus confirming that this criterion does tend to select an overparameterised model as T increases.

8.6.1 *Constructing a model portfolio*

From an examination of Exhibits 8.10 to 8.12, we see that there are a number of models for each series that are 'close to' the selected model in terms of their criterion value. Using the concept of 'grades of evidence' advanced by Jeffreys (1961), Poskitt and Tremayne (1987) introduce the idea of a model portfolio, with models being compared to the selected (p_1, q_1) specification by way of the 'posterior odds ratio' defined, using BIC for illustration, as

$$\mathbb{R} = \exp[-\tfrac{1}{2}T\{\mathrm{BIC}(p_1, q_1) - \mathrm{BIC}(p, q)\}]. \tag{8.13}$$

Although $\mathbb{R}$ has no physical meaning, its value may be used to 'grade the

Exhibit 8.12 *Model selection criteria for 3-month interbank rate*

(i) *AIC*

		q						
		0	1	2	3	4	5	6
	0	−3.664	−3.709	−3.712	−3.709	−3.707	−3.707	−3.703
	1	−3.698	−3.711	−3.709	−3.705	−3.706	−3.712	−3.708
	2	−3.713	−3.709	−3.705	−3.718	−3.713	−3.698	−3.704
	3	−3.709	−3.707	−3.704	−3.705	−3.715	−3.695	−3.711
p	4	−3.706	−3.706	−3.713	−3.713	−3.709	−3.702	−3.724
	5	−3.705	−3.703	−3.705	−3.706	−3.715	−3.724	−3.720
	6	−3.702	−3.699	−3.693	−3.707	−3.719	−3.714	−3.706

(ii) *BIC*

	0	−3.664	−3.703	−3.700	−3.690	−3.682	−3.676	−3.666
	1	−3.692	−3.699	−3.690	−3.680	−3.674	−3.675	−3.665
	2	−3.700	−3.690	−3.680	−3.687	−3.676	−3.654	−3.655
	3	−3.690	−3.682	−3.673	−3.668	−3.671	−3.645	−3.654
p	4	−3.681	−3.674	−3.676	−3.669	−3.659	−3.646	−3.662
	5	−3.674	−3.666	−3.662	−3.657	−3.659	−3.662	−3.652
	6	−3.664	−3.656	−3.644	−3.650	−3.657	−3.645	−3.631

decisiveness of the evidence' against a particular model. Poskitt and Tremayne suggest that a value of $\mathbb{R}$ less than 10 would not be sufficient to warrant discarding the model in favour of that chosen through the criterion minimising procedure, while any (p, q) satisfying $1 < \mathbb{R} < \sqrt{10}$ may be thought of as a close competitor to (p_1, q_1). The set of models closely competing with (p_1, q_1) would then be taken as the 'model portfolio'.

Example 8.16: Examples of constructing a model portfolio

The $\mathbb{R}$ values calculated using (8.13) for real income and consumption are shown as Exhibit 8.13. Taking $\sqrt{10}$ as an approximate upper bound leads to the model portfolio for real income containing the six specifications; (0, 0), (1, 0), (2, 0), (0, 1), (0, 2) and (1, 1), suggesting that any of these models should provide an adequate description of the series. For real consumption an identical model portfolio is obtained.

Exhibit 8.14 shows the model portfolio for the interbank rate series, along with associated $\mathbb{R}$ values. This contains four specifications, that minimising BIC plus the three which had been identified earlier as possible contenders.

Exhibit 8.13 *Grades of evidence for real income and consumption models*

(i) $\mathbb{R}$ *values for real income*

		q			
		0	1	2	3
p	0	0	1.09	4.02	24.1
	1	3.46	7.02	9.59	312
	2	3.54	19.2	1832	1739
	3	29.1	203	1530	3930

(ii) $\mathbb{R}$ *values for real consumption*

p	0	1.66	1.12	0	10.4
	1	5.10	1.44	7.44	81
	2	2.22	14.1	86	224
	3	16.0	2837	2274	2466

Exhibit 8.14 *Model portfolio for interbank rate*

Model	$\mathbb{R}$
(0, 1)	0
(2, 0)	2.62
(0, 2)	3.15
(1, 1)	4.23

8.7 Other model identification procedures

A number of other procedures for use in model identification have been proposed, but since they have not proved popular and are not in common use, we will only briefly discuss them here.

The traditional 'Box–Jenkins' identification procedure relies heavily on the SACF and PACF of the appropriately differenced series $w_t = \nabla^d x_t$. These can be supplemented by an examination of the *inverse autocorrelation function* (IACF). If w_t is generated by the ARMA(p, q) process

$$\phi(B)\, w_t = \theta(B)\, a_t, \tag{8.14}$$

then the ARMA(q, p) model

$$\theta(B)\, w_t = \phi(B)\, a_t \tag{8.15}$$

is known as the *dual* of (8.14). The ACF of this model is the IACF of the original model, and thus if the original model is a pure autoregression,

Exhibit 8.15 *Portfolio of estimated models for real consumption*

p	q	$\hat{\phi}_1$	$\hat{\phi}_2$	$\hat{\theta}_0$	$\hat{\theta}_1$	$\hat{\theta}_2$	$\hat{\sigma}$
0	0	—	—	0.022	—	—	0.0198
				(0.003)			
1	0	0.253	—	0.022	—	—	0.0194
		(0.166)		(0.004)			
2	0	0.359	−0.407	0.022	—	—	0.0180
		(0.160)	(0.162)	(0.003)			
0	1	—	—	0.021	−0.520	—	0.0186
				(0.005)	(0.152)		
0	2	—	—	0.023	−0.407	0.593	0.0176
				(0.002)	(0.148)	(0.150)	
1	1	−0.555	—	0.022	−1	—	0.0178
		(0.171)		(0.004)	(0.07)		

then the IACF is an ACF corresponding to a pure moving average. It will therefore cut off sharply at lag p, behaviour which is similar to that of the PACF.

The *sample inverse autocorrelation function* (SIACF) (see Chatfield (1979) and Bhansali (1980)) plays much the same role in model identification as does the SPACF. Additionally, if the series is non-stationary, the SIACF has the characteristics of a noninvertible moving average, i.e. it will have properties similar to that of an ARMA(0, 1) process with ρ_1 close to 0.5. Similarly, if the series has been overdifferenced, the SIACF will appear as an SACF from a nonstationary process. Abraham and Ledolter (1984) provide simulation evidence, however, to show that the SIACF is less powerful than the SPACF for appropriately identifying autoregressive processes.

Various other techniques have been proposed for order selection other than information criteria such as AIC and BIC. These include the '*R*- and *S*-array' approach of Gray et al. (1978), the 'generalised partial autocorrelations' of Woodward and Gray (1981), the 'corner' method of Beguin et al. (1980), and the 'extended sample autocorrelation function' method of Tsay and Tiao (1984), details of which can be found in the above references.

8.8 Diagnostic checking

The use of model selection criteria, when extended to incorporate a portfolio of competing specifications, has led to a number of alternative models being specified for the three series investigated in the set of

Exhibit 8.16 *Portfolio of estimated models for real income*

p	q	$\hat{\phi}_1$	$\hat{\phi}_2$	$\hat{\theta}_0$	$\hat{\theta}_1$	$\hat{\theta}_2$	$\hat{\sigma}$
0	0	—	—	0.025	—	—	0.0245
				(0.004)			
1	0	0.241	—	0.025	—	—	0.0242
		(0.167)		(0.005)			
2	0	0.324	−0.347	0.025	—	—	0.0230
		(0.163)	(0.163)	(0.004)			
0	1	—	—	0.025	−0.451	—	0.0234
				(0.006)	(0.153)		
0	1	—	—	0.025	−0.242	0.328	0.0231
				(0.004)	(0.165)	(0.169)	
1	1	−0.307	—	0.025	−0.690	—	0.0234
		(0.351)		(0.005)	(0.270)		

examples discussed above. The estimated parameters for the real income and consumption models are shown in Exhibits 8.15 and 8.16 respectively, and certain of these models are immediately seen to be over-specified; for example, the (1, 1) model for real income has an insignificant autoregressive parameter, a consequence of high correlation between the estimates of ϕ_1 and θ_1, this being a manifestation of the 'common factor' phenomenon discussed by Box and Jenkins (1976, page 195), while the corresponding model for real consumption has a moving average parameter on the invertibility boundary.

Are there any further checks on the competing models that are available to the analyst? Indeed there are, for we have yet to discuss one of the most important aspects of model building, that of diagnostic checking. Box et al. (1978) argue that the purpose of model building is to transform the, presumably autocorrelated, observed series to a structureless white noise process, i.e.

$$a_t = \frac{\phi(B)}{\theta(B)} \nabla^d x_t. \tag{8.16}$$

Hence, a check on whether a particular model is adequate or not revolves around ascertaining whether the calculated residuals,

$$\hat{a}_t = \frac{\hat{\phi}(B)}{\hat{\theta}(B)} \nabla^d x_t, \tag{8.17}$$

mimic to a reasonable degree the assumed properties of the error process a_t. This implies that (i) the mean of the residuals should be close to zero, (ii) the variance of the residuals should be approximately constant, and

(iii) the autocorrelations of the residuals should be negligible. To check whether the mean of the residuals is zero, the sample mean $\bar{\hat{a}}$ can be compared with its standard error, and a residual plot may be examined to check whether the variance is constant.

To check whether the residuals are uncorrelated, we may calculate their sample autocorrelations

$$\hat{r}_k = \frac{\sum_{t=k+1}^{T} (\hat{a}_t - \bar{\hat{a}})(\hat{a}_{t-k} - \bar{\hat{a}})}{\sum_{t=1}^{T} (\hat{a}_t - \bar{\hat{a}})^2}. \tag{8.18}$$

These can be compared with their standard errors, which are usually approximated as $T^{-\frac{1}{2}}$. However, for small k the true standard error can be much smaller, but since these true standard errors depend on the form of the fitted model, the true parameter values, and the value of k, they are difficult to calculate. Although improved approximations to the standard errors are available (see McLeod (1978)), most attention has been focused on the construction of 'portmanteau' tests. The original test of this type was developed by Box and Pierce (1970), who showed that if the stationary series $w_t = \nabla^d x_t$ was correctly generated by an ARMA(p, q) process, then the statistic

$$Q^* = T \sum_{k=1}^{m} \hat{r}_k^2 \tag{8.19}$$

would be asymptotically distributed as χ^2 with $(m-p-q)$ degrees of freedom as long as m was $0(T^{\frac{1}{2}})$. (If a constant θ_0 is included in the model then the degrees of freedom are reduced by one.)

Unfortunately, it is often the case that significant values of this statistic are rarely observed and that it is of little use in discriminating between possible models (see, for example, Prothero and Wallis (1976)). In fact, Davies et al. (1977) showed that, even for quite large samples, the true significance levels of the Q^* statistic could be much smaller than those given by asymptotic theory, so that the chance of incorrectly rejecting the null hypothesis of model adequacy is smaller than the chosen significance level. Ljung and Box (1978) argue that a better approximation could be obtained if tests were based upon the modified portmanteau statistic

$$Q = T(T+2) \sum_{k=1}^{m} (T-k)^{-1} \hat{r}_k^2. \tag{8.20}$$

Thus, if the calculated Q statistic exceeds the tabulated value $\chi^2_\alpha(m-p-q)$, then the adequacy of the fitted ARMA model would be questioned. However, the power of this modified statistic can still be quite low, even

Exhibit 8.17 *Portmanteau statistics for model portfolios*

(p, q)	Q	$df = m-p-q-1$	$P[\chi^2(df) > Q]$
(i) *Real consumption*			
(0, 0)	15.92	5	0.007
(1, 0)	11.82	4	0.019
(2, 0)	4.12	3	0.249
(0, 1)	6.94	4	0.139
(0, 2)	4.53	3	0.210
(1, 1)	7.94	3	0.047
(ii) *Real income*			
(0, 0)	14.93	5	0.011
(1. 0)	10.68	4	0.030
(2, 0)	4.28	3	0.233
(0, 1)	6.30	4	0.178
(0, 2)	6.14	3	0.105
(1, 1)	6.74	3	0.081

in the presence of severe misspecification; see Davies and Newbold (1979) and Godfrey (1979).

Example 8.17: Portmanteau tests for the consumption and income model portfolios

Exhibit 8.17 shows the Q statistics, along with associated degrees of freedom and marginal significance levels of rejecting the null hypothesis of model adequacy, for both the real income and consumption model portfolios. As $T = 36$ for both series, the number of residual autocorrelations used in the calculation of the statistics was set at $m = T^{\frac{1}{2}} = 6$, this being a commonly used rule for choosing m (see Poskitt and Tremayne (1981) for further discussion and alternatives). For real consumption the (0, 0), (1, 0) and (1, 1) specifications are clearly inadequate, leaving only the models (2, 0), (0, 1) and (0, 2) as competitors. The same three models are also left as competitors for real income if the (1, 1) model, marginally adequate on the Q test, is ruled out through having a noninvertible moving average part.

Example 8.18: Portmanteau tests for the interbank rate model portfolio

As a second example, highlighting some of the difficulties faced with using the portmanteau test, consider the Q values obtained for the four specifications contained in the model portfolio for the interbank rate, shown as Exhibit 8.18. As $T = 740$, the setting of m was taken to be 24,

Exhibit 8.18 *Portmanteau statistics for interbank model portfolio*

(p,q)	Q	$df = m-p-q-1$	$P[\chi^2(df) > Q]$
(2, 0)	51.4	22	0.0001
(0, 1)	57.2	23	0.0001
(0, 2)	50.7	22	0.0001
(1, 1)	50.8	22	0.0001

and all models appear to be inadequate since their marginal significance levels are all less than 0.0001! Unfortunately, inspection of the individual $\hat{r}_k$s for all models does not show any 'large' values, particularly at low orders of k, and so it is difficult to see how the models could be respecified (but see Example 15.1).

8.8.1 *Lagrange multiplier tests*

An explanation for this anomaly may be found by considering the formal nature of such portmanteau tests. They have been derived without the explicit formulation of an alternative hypothesis, which is an appropriate strategy when there is little information available about the likely nature of any model misspecification. Godfrey and Tremayne (1988) argue that such tests may therefore be regarded as 'pure significance' tests, in the sense of Cox and Hinkley (1974, section 3.2); in which case, the larger the sample size, the stronger the evidence of departure from the null hypothesis is required for rejection. As Cox and Hinkley note (page 81), the marginal significance level of a pure significance test may be very small but the 'amount of departure from H_0' may be of little practical interest and, even if strong evidence against the null is obtained, the test gives no more than 'a guide to the type and magnitude of the departure'.

Such problems with portmanteau tests have prompted the development of alternative test statistics. One particular form of test that has been found to be useful is that based on the Lagrange Multiplier (LM) principle: see, for example, Godfrey (1979) and Poskitt and Tremayne (1980). Unlike the above portmanteau tests, a test based on the LM principle requires an explicit alternative hypothesis, although the model given by this alternative does not need to be estimated. Thus, extending our notation to these circumstances, let us denote the ARMA(p,q) specification under the null hypothesis H_0 as:

$$\phi_p(B)\,w_t = \theta_q(B)\,a_t, \tag{8.21}$$

where, as usual, $w_t = \nabla^d x_t$. We will consider two alternative hypotheses;

H_A: that the w_t are ARMA $(p+r, q)$, and H_B: that the w_t are ARMA $(p, q+s)$. The models corresponding to H_A and H_B may then be written as

$$\phi_{p+r}(B)\, w_t = \theta_q(B)\, a_t \tag{8.22}$$

$$\phi_p(B)\, w_t = \theta_{q+s}(B)\, a_t \tag{8.23}$$

respectively, where $\phi_{p+r}(B) = (1-\phi_1 B - \ldots - \phi_{p+r} B^{p+r})$, etc. The more general alternative hypothesis, H_C: that the w_t are ARMA $(p+r, q+s)$, is well known to lead to a singularity problem which rules out the use of standard large sample test procedures, but Poskitt and Tremayne (1980) show that the LM test, or score test as it is often known, may still be used in this case if certain conditions are met.

Godfrey (1979) shows that the LM test can be calculated as the sample size, T, times the coefficient of determination (R^2) of an auxiliary regression of $\hat{a}_t$, the fitted residuals, on the partial derivatives of a_t with respect to the model parameters evaluated at their ML estimates. These partial derivatives can easily be calculated from w_t and $\hat{a}_t$. Following Godfrey (1979), we define y_t and z_t by $y_t = z_t = 0$ for $t \leqslant 0$, and

$$z_t = -w_t + \hat{\theta}_1 z_{t-1} + \ldots + \hat{\theta}_q z_{t-q}, \quad t = 1, 2, \ldots, T \tag{8.24}$$

$$y_t = \hat{a}_t + \hat{\theta}_1 y_{t-1} + \ldots + \hat{\theta}_q y_{t-q}, \quad t = 1, 2, \ldots, T. \tag{8.25}$$

Providing that relevant starting values are set equal to zero, z_{t-i} is $\partial a_t / \partial \phi_i$ and similarly y_{t-i} is $\partial a_t / \partial \theta_i$, both evaluated at $(\hat{\phi}_1, \ldots, \hat{\phi}_p, \hat{\theta}_1, \ldots, \hat{\theta}_q)$. It follows that in order to test H_0: that the w_t are ARMA (p, q), against H_A: that the w_t are ARMA $(p+k, q)$, it is necessary only to estimate the regression equation

$$\hat{a}_t = \alpha_1 z_{t-1} + \ldots + \alpha_{p+k} z_{t-p-k} + \beta_1 y_{t-1} + \ldots + \beta_q y_{t-q} + u_t \tag{8.26}$$

by ordinary least squares and to calculate the LM statistic as $T \cdot R^2$. If the alternative is H_B: that the w_t are ARMA $(p, q+k)$, then the appropriate regression equation is

$$\hat{a}_t = \alpha_1 z_{t-1} + \ldots + \alpha_p z_{t-p} + \beta_1 y_{t-1} + \ldots + \beta_{q+k} y_{t-q-k} + v_t. \tag{8.27}$$

In both cases the test statistic will be asymptotically distributed as χ^2 with k degrees of freedom when H_0 is true, with significantly large values indicating that (8.21) is an inadequate specification.

Although the LM approach requires the specification of an alternative hypothesis, so that the LM test is asymptotically optimal for a particular alternative, this does not imply that it will have poor power against other alternatives. Thus, if an ARMA model has been misspecified, this may be revealed even if the wrong alternative is selected. Conversely, if the null hypothesis is rejected, the model specified under the alternative should not

Exhibit 8.19 *One degree of freedom LM tests for interbank model portfolio*

H_0	H_A	LM
(2, 0)	(3, 0)	0.03
(2, 0)	(2, 1)	0.03
(0, 1)	(1, 1)	4.56
(0, 1)	(0, 2)	4.56
(0, 2)	(1, 2)	0.93
(0, 2)	(0, 3)	0.93
(1, 1)	(2, 1)	0.02
(1, 1)	(1, 2)	0.02

automatically be adopted without further checking. Another difference of LM tests when compared to portmanteau tests is that k, the number of parameters constrained to be zero by the null hypothesis, is not to be thought of in the same way as m, the number of residual autocorrelations included in the Q statistic: the validity of the LM approach does not require that k be increased as the sample size T increases.

It was mentioned above that McLeod (1978) had provided improved approximations to the standard errors of the residual autocorrelations. If we denote the estimated covariance matrix of $\hat{r}' = (\hat{r}_1, \ldots, \hat{r}_k)$ obtained using McLeod's approach as $\hat{C}$, then a further 'portmanteau type' test statistic is given by

$$S = T\hat{r}'\hat{C}^{-1}\hat{r} \tag{8.28}$$

which will be asymptotically distributed as χ^2 with k degrees of freedom. Newbold (1980) shows that S and the LM statistic from (8.26) for testing the alternative ARMA$(p+k, q)$ specification are equivalent. However, it would seem that the LM approach is the most suitable in terms of computational convenience.

Example 8.19: LM tests for the income models

Consider again the real income models shown in Exhibit 8.13. There is some doubt as to the significance of $\hat{\theta}_2$ in the ARMA(0, 2) model, and while a decision might be made by comparing $\hat{\theta}_2$ with its standard error (the t-ratio being 1.94, this test being of the Wald variety), we may also construct an LM test of H_0: the model is ARMA(0, 1), against the alternative H_A: the model is ARMA(0, 2). This involves constructing the auxiliary variable y_t (no z_t variable is required since $p = 0$) as

$$y_{-1} = y_0 = 0$$

$$y_1 = \hat{a}_1$$

$$y_t = \hat{a}_t - 0.451 y_{t-1}, \quad t = 2, \ldots, 36,$$

estimating the regression

$$\hat{a}_t = \beta_1 y_{t-1} + \beta_2 y_{t-2} + u_t, \quad t = 1, 2, \ldots, 36,$$

and computing the test statistic LM $= 36 \cdot R^2$. This yields a value of 1.36, which is compared with the $\chi^2(1)$ distribution and obviously does not allow rejection of the null hypothesis that the appropriate model is ARMA (0, 1).

Example 8.20: LM tests for the interbank rate models

In Example 8.18, all specifications in the model portfolio for the interbank rate were accompanied by significant Q statistics, although no obvious respecifications were signalled by examination of the residual autocorrelations. Poskitt and Tremayne (1981) argue that, if the model selection process has been carried out carefully, then any misspecification is likely to be captured by models having orders close to, but larger than, those of the selected, i.e. null, specification. This suggests constructing one degree of freedom LM tests by entertaining alternative models in which either p or q, but not both, are increased by one. This was done for the four competing interbank specifications, the results being shown as Exhibit 8.19. We see that the ARMA (0, 1) model is rejected in favour of either the ARMA (1, 1) or ARMA (0, 2) models, while the other specifications are not rejected when tested against higher order alternatives.

That the test statistics shown in Exhibit 8.19 are the same for each alternative to a given null is no chance occurrence. Once $s = \max(f, g)$ has been fixed, the same LM statistic is appropriate for all restricted ARMA $(p+f, q+g)$ models that are identified under the null ARMA (p, q) specification. The case here has $s = f = g = 1$, so that the same LM test is valid when the null is ARMA (p, q) and the alternative is either ARMA $(p+1, q)$ or ARMA $(p, q+1)$. These latter models are, in the terminology of Godfrey (1981), 'locally equivalent alternatives' with respect to the null specification, and are essentially the same for small departures from the null.

8.9 Choosing between alternative models

The use of model selection techniques, supplemented by diagnostic checks, has led to the same three closely competing models for the interbank rate as were identified using the conventional 'Box–Jenkins' procedure. This is reassuring, for the interbank series is extremely long, 740 observations, and the SACF should in these circumstances give a good indication of the underlying process generating the series. The close connection between all the models, when expressed in linear filter form, has already been demonstrated in section 8.5.

The use of selection techniques and diagnostic checking for the much shorter real income and consumption series reveal moving average models that are somewhat superior to the autoregressive models previously identified. When the autoregressive models are expressed in linear filter form we obtain, for real consumption and real income respectively,

$$\nabla x_t = 0.024 + (1 + 0.32B - 0.25B^2 - 0.19B^3 - \ldots)\, a_t$$

and

$$\nabla x_t = 0.024 + (1 + 0.36B - 0.28B^2 - 0.25B^3 - \ldots)\, a_t,$$

both of which are considerably different to the moving average specifications shown in Exhibits 8.14 and 8.15 respectively. A more detailed inspection of the differences between these models is provided by the implied eventual forecast functions. From Chapter 7, the moving average models for ∇x_t imply forecast functions of the form

$$f_{n,h} = b_0 + b_1^{(n)},$$

i.e. straight line projections with $b_1^{(n)}$ being updated as a function of the moving average parameters and the latest forecast error. The ARMA (2, 1, 0) models both have complex roots and hence forecast functions of the form

$$f_{n,h} = b_0 + b_1^{(n)} + b_2^{(n)} \delta^h \sin[2\pi f_0(n+h) + F]$$

where, from Box and Jenkins (1976, page 60), the parameters of the damped sine wave are given by

$$\delta = \sqrt{-\phi_2},$$

where the positive square root is taken,

$$\cos 2\pi f_0 = \frac{|\phi_1|}{2\sqrt{-\phi_2}},$$

and

$$\tan F = \frac{1+\phi_2}{1-\phi_2} \tan 2\pi f_0.$$

For real consumption, these parameters are estimated as $\delta = 0.64$ (the damping factor), a period of $f_0^{-1} = 4.9$ years and a phase angle $F = 83°$. For real income, almost identical estimates of $\delta = 0.64$, $f_0^{-1} = 4.9$ and $F = 83.3°$ are obtained. Thus, if these models are used, forecasts of the series will be of the form of a straight line with a damped, nonoscillatory sine wave exhibiting a five-year period superimposed upon it, and if one wishes models to display 'pseudo cyclical' behaviour, then these ARIMA (2, 1, 0) models would be preferred to the integrated moving average models which do not exhibit such behaviour.

9 Exponential smoothing and its relationship to ARIMA modelling

9.1 Introduction

The approach to time series modelling developed in the previous four chapters is not the only one that may be taken. Exponential smoothing, excellently surveyed recently by Gardner (1985), is a methodology that was originally developed by operations researchers in the 1950s and has since found widespread use in such areas as short-range sales forecasting and inventory control. Indeed, recent empirical studies have found little difference in forecast accuracy between exponential smoothing techniques and ARIMA models identified using the methodology just outlined; see, for example, Makridakis and Hibon (1979) and Makridakis et al. (1982).

As we have remarked earlier, forecasting in itself is not of primary interest to us here. What we are especially concerned with are the links between exponential smoothing and ARIMA models that have been explored by, for example, Gardner (1985) and E. McKenzie (1974, 1984), the interpretation, through these links, of ARIMA models as reduced forms of certain 'structural' models (Harvey (1984)), and the ability of exponential smoothing methods to be extended to capture non-linearities in observed series that may be difficult for ARIMA models to cope with.

It is often argued that exponential smoothing models can be ignored because they are just special cases of ARIMA models. But as McKenzie (1985) points out, this 'special case' argument is flawed, for the two approaches are considerably different in the criteria they use and the model building philosophy they employ. It is our contention that using the two approaches in conjunction with each other can lead to a much better understanding of the processes generating observed time series. This is best seen by the ability of many exponential smoothing models to be represented as 'structural' models, thus enabling, when the equivalencies between these and ARIMA models are taken into account, the unobservable structural parameters to be expressed as functions of the estimable coefficients of fitted ARIMA processes.

9.2 Simple exponential smoothing

In Chapter 6 we assumed that a time series could be decomposed into two unobserved components: a mean level and a random error. Thus, in equation (6.2) we wrote

$$x_t = \mu_t + \varepsilon_t. \tag{9.1}$$

The idea that an observed time series can be thought of as being composed of, perhaps several, unobserved components has a long history in the statistical literature dealing with the analysis of economic time series, as the survey by Nerlove et al. (1979, chapter 1) clearly shows, and it is one that we will continually find useful. We typically assume that the error, or noise, component, ε_t, is indeed white noise, and attention then focuses on the specification of the mean level, often termed the trend component, μ_t. It is also generally assumed that the two components are independent.

The simplest model for μ_t is, of course, to assume that it is constant, so that x_t is itself a white noise process. As we have seen in previous chapters, this is unlikely to be a realistic model for most economic time series, but a popular alternative is to assume that the current level is an exponentially weighted moving average (EWMA) of current and past observations on x_t:

$$\begin{aligned}\mu_t &= \alpha x_t + \alpha(1-\alpha)x_{t-1} + \alpha(1-\alpha)^2 x_{t-2} + \dots \\ &= \alpha \sum_{j=0}^{\infty} (1-\alpha)^j x_{t-j}, \\ &= \alpha[1 + (1-\alpha)B + (1-\alpha)^2 B^2 + \dots + (1-\alpha)^j B^j + \dots] x_t.\end{aligned} \tag{9.2}$$

Since

$$(1 + (1-\alpha)B + (1-\alpha)^2 B^2 + \dots) = [1-(1-\alpha)B]^{-1},$$

equation (9.2) can be written

$$[1-(1-\alpha)B]\mu_t = \alpha x_t$$

or

$$\mu_t = \alpha x_t + (1-\alpha)\mu_{t-1}, \tag{9.3}$$

showing that the current level, μ_t, is a weighted average of the current observation, x_t, and the previous level, μ_{t-1}, the weight being given by the 'smoothing constant' α. Alternatively, (9.3) can be expressed in 'error correction' form as

$$\mu_t = \mu_{t-1} + \alpha(x_t - \mu_{t-1}) = \mu_{t-1} + \alpha e_t,$$

so that the current level is updated from the previous level by a proportion of the current error $e_t = x_t - \mu_{t-1}$, the proportion again being given by the parameter α. Equation (9.3) is the basic algorithm of a forecasting technique known as *simple exponential smoothing*, which has been used for many years in industry, particularly in inventory control, and in economics where, for example, Friedman's (1957) concept of permanent income is simply an EWMA formulation and equation (9.3) can then be interpreted as an adaptive expectations mechanism: see Nerlove et al. (1979, chapter 4).

Detailed analysis of the forecasting aspects of simple exponential smoothing may be found in many textbooks, Abraham and Ledolter (1983, chapter 3) being a good example, and such analysis will not be developed here. However, substituting (9.3) into (9.1) obtains

$$\left(1 - \frac{\alpha}{[1-(1-\alpha)B]}\right) x_t = \varepsilon_t$$

which can be shown to lead to (Muth, 1960)

$$\nabla x_t = (1-\theta B)\, a_t, \tag{9.4}$$

where $\theta = 1-\alpha$ and $a_t = \varepsilon_t/\theta$. Thus, simple exponential smoothing is an optimal method of forecasting if x_t follows an ARIMA (0, 1, 1) process and the smoothing constant is set equal to $(1-\theta)$. Furthermore, Ledolter and Abraham (1984) show that if backcasting is used to obtain an estimate of the initial mean level, μ_0, then simple exponential smoothing leads to the same forecasts as the ARIMA (0, 1, 1) model when estimated by ULS.

It is often suggested, Montgomery and Johnson (1976) for example, that α should be set in the range 0.1 to 0.3, but the equivalence of simple exponential smoothing with the ARIMA (0, 1, 1) model, which is invertible for $-1 < \theta < 1$, suggests that the range for the smoothing parameter is, theoretically, $0 < \alpha < 2$. Indeed, both Chatfield (1978) and Makridakis et al. (1982) found that the most accurate parameters were frequently in the range 0.3 to 1.0 and sometimes above 1.0.

Through its equivalence with the ARIMA (0, 1, 1) model, simple exponential smoothing is also optimal for an unobserved components (UC) model in which the mean level follows a random walk:

$$\mu_t = \mu_{t-1} + \nu_t, \tag{9.5}$$

where ν_t is a white noise process with variance σ_v^2. From Box and Jenkins (1976, Appendix A4.4.4), substituting (9.5) into (9.1) again yields (9.4), with the relationships between the parameters being given by

$$\frac{(1-\theta)^2}{\theta} = \frac{\sigma_v^2}{\sigma_\varepsilon^2} \quad \text{and} \quad \sigma_\varepsilon^2 = \theta \sigma_a^2,$$

σ_a^2 and σ_ε^2 being the variances of a_t and ε_t respectively. It then follows that θ is given by the invertible solution of

$$(2+\eta \pm (\eta^2+4\eta)^{\frac{1}{2}})/2,$$

where $\eta = \sigma_v^2/\sigma_\varepsilon^2$ is the 'signal-to-noise' variance ratio.

Example 9.1: Modelling expected real rates of interest

An important economic example of the unobserved random walk buried in noise is provided by the analysis of expected real rates of interest under the assumption of rational expectations, or equivalently, financial market efficiency; see, for example, Fama (1975) and Nelson and Schwert (1977). In this model, the unobservable expected real rate is assumed to follow the random walk process given by equation (9.5), and it differs from the observed real rate by the amount of unexpected inflation which, under the assumption of market efficiency, will be a white noise process. The observed real rate will then follow the ARIMA (0, 1, 1) process of equation (9.4).

Mills and Stephenson (1985) employ such a model to analyse the UK Treasury bill market over the period 1952 to 1982. They find that observed real Treasury bill returns are adequately modelled by the ARIMA (0, 1, 1) process

$$\nabla x_t = (1-0.678B)\, a_t, \quad \hat{\sigma}_a^2 = 14.5.$$

From the relationships linking σ_v^2 and σ_ε^2 to θ and σ_a^2, it follows that the unobserved variances may be estimated as

$$\hat{\sigma}_v^2 = (1-0.678)^2 \hat{\sigma}_a^2 = 1.51$$

$$\hat{\sigma}_\varepsilon^2 = 0.678\hat{\sigma}_a^2 \qquad = 9.85,$$

yielding a signal-to-noise ratio $\hat{\sigma}_v^2/\hat{\sigma}_\varepsilon^2$ of 0.15, so that variations in the expected real return are small compared to variations in unexpected inflation. Expected real returns can be estimated using the algorithm (9.3) with smoothing parameter $\alpha = 0.322$, and unexpected inflation can then be obtained by residual as $\hat{\varepsilon}_t = x_t - \hat{\mu}_t$. Exhibit 9.1 provides plots of x_t, $\hat{\mu}_t$ and $\hat{\varepsilon}_t$, showing that the expected real return is considerably smoother than the observed real return, as was suggested by the small signal-to-noise ratio. In the early part of the 1950s expected real returns were generally negative, but from 1956 to 1970 they were consistently positive. From the middle of 1970 and for the subsequent decade the expected real return was always negative, reaching a minimum of -13.65% in 1975Q1 after inflation peaked at 36.61 % in the previous quarter as a consequence of the OPEC price rise, and a local minimum of -5.25% in 1979Q2, this

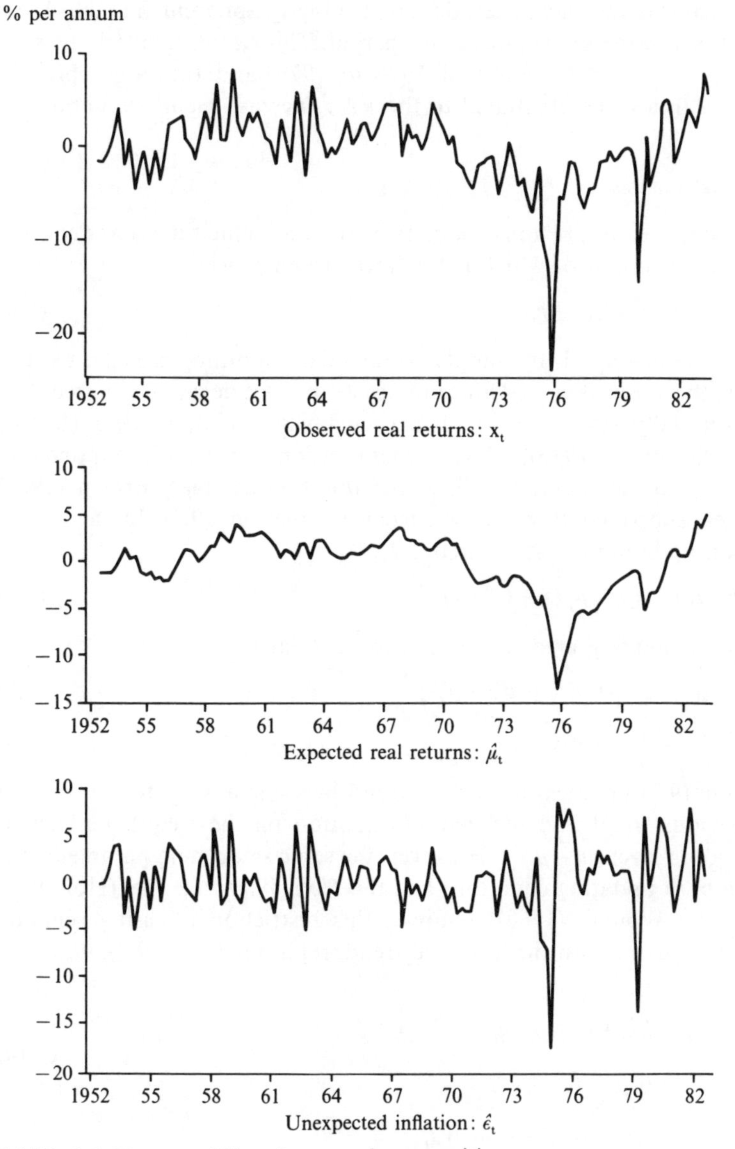

Exhibit 9.1 Treasury bill real return decomposition

being a result of the VAT increase in the Budget of that year. From mid-1980 the series is again consistently positive, standing at the end of the sample period at a maximum value of 5.76%. Until the early 1970s fluctuations in unexpected inflation were relatively small, but the turbulence of the next decade is strikingly apparent in the large fluctuations of the series during this period. This can particularly be seen in the post oil-shock period of 1974 to 1976, and the large spike in 1979 Q2 can also be attributed to the VAT increase mentioned above.

9.3 Trend models

As we have seen in previous chapters, many economic time series contain trends, the simplest of which is the linear trend process

$$x_t = b_0 + b_1 t + \varepsilon_t, \tag{9.6}$$

so that $\mu_t = b_0 + b_1 t$. If the simple exponential smoothing model is used to forecast this series it will produce forecasts that are negatively biased, this bias eventually stabilising at an expected value of $b_1(1-\alpha)/\alpha$ (Brown (1963), Gardner (1985)). Two generalisations of simple exponential smoothing are available to adjust for this forecast bias. Brown (1963) proposes supplementing the recurrence equation (9.3) by a second equation updating the series trend, T_t:

$$T_t = \alpha(\mu_t - \mu_{t-1}) + (1-\alpha) T_{t-1}, \tag{9.7}$$

with corresponding error-correction forms (Gardner (1985))

$$\mu_t = \mu_{t-1} + T_{t-1} + \alpha(2-\alpha) e_t \tag{9.8}$$

$$T_t = T_{t-1} + \alpha^2 e_t. \tag{9.9}$$

Equation (9.7) updates the current trend by using a weighted average of the previous trend and the new information on the trend, the current difference in level $(\mu_t - \mu_{t-1})$. However, the same smoothing parameter α is used in both updating equations and the *Holt–Winters model* (Holt et al. (1960) and Winters (1960)) removes this restriction by using separate parameters to smooth the level and trend, replacing (9.3), (9.7), (9.8) and (9.9) by

$$\begin{aligned} \mu_t &= \beta x_t + (1-\beta)(\mu_{t-1} + T_{t-1}) \\ &= \mu_{t-1} + T_{t-1} + \beta e_t \end{aligned} \tag{9.10}$$

and

$$\begin{aligned} T_t &= \gamma(\mu_t - \mu_{t-1}) + (1-\gamma) T_{t-1} \\ &= T_{t-1} + \beta\gamma e_t \end{aligned} \tag{9.11}$$

respectively. Using the recurrence equations (9.10) and (9.11), it can be shown that the Holt–Winters model is equivalent to the ARIMA (0, 2, 2) process (Harrison (1967), Harvey (1984))

$$\nabla^2 x_t = (1 - (2 - \beta - \beta\gamma) B - (\beta - 1) B^2) a_t, \tag{9.12}$$

so that, in terms of the general process

$$\nabla^2 x_t = (1-\theta_1 B - \theta_2 B^2)\, a_t, \tag{9.13}$$

the smoothing parameters are given by $\beta = 1+\theta_2$ and $\gamma = (1-\theta_1-\theta_2)/(1+\theta_2)$. Note that if these expressions are substituted into equations (9.10) and (9.11), then the error correction forms are identical to the updating equations of the ARIMA (0, 2, 2) model given in Example 7.2(c), on interpreting $b_1^{(n)}$ and $b_2^{(n)}$ as the level and trend of x_t respectively.

Brown's one parameter model, known as *double exponential smoothing*, is equivalent to the restricted ARIMA (0, 2, 2) model

$$\begin{aligned}\nabla^2 x_t &= (1-(1-\alpha)B)^2 a_t\\ &= (1-2(1-\alpha)B+(1-\alpha)^2 B^2)\, a_t,\end{aligned}$$

which places the restriction $\theta_1^2 + 4\theta_2 = 0$ on the parameters of (9.13). If $\beta = \alpha(2-\alpha)$ and $\gamma = \alpha/(2-\alpha)$, then Holt–Winters is equivalent to double exponential smoothing.

The Holt–Winters model is also optimal for an extension of the random walk plus noise model of equation (9.5). Here we may allow for a drift or linear growth term, also following a random walk, and thus replace (9.5) by the two equations:

$$\mu_t = \mu_{t-1} + T_{t-1} + v_t \tag{9.14}$$

$$T_t = T_{t-1} + u_t. \tag{9.15}$$

The relationships between the error variances σ_ε^2, σ_v^2 and σ_u^2 in the 'structural model', comprised of equations (9.1), (9.14) and (9.15), and the parameters σ_a^2, θ_1 and θ_2 of the ARIMA model (9.13) are given by (Harrison (1967), Harvey (1984))

$$\begin{aligned}\sigma_\varepsilon^2 &= (1-\beta\gamma)\,\sigma_a^2\\ \sigma_v^2 &= (\beta^2+\beta^2\gamma-2\gamma\beta)\,\sigma_a^2\\ \sigma_u^2 &= \gamma^2\sigma_a^2,\end{aligned}$$

where $\beta = (1+\theta_2)$ and $\gamma = (1-\theta_1-\theta_2)/(1+\theta_2)$ are the smoothing parameters defined above. Double exponential smoothing then effectively imposes the restriction that $\bar\sigma_u^2 = (\bar\sigma_v^2/2)^2$, where $\bar\sigma_u^2 = \sigma_u^2/\sigma_\varepsilon^2$ and $\bar\sigma_v^2 = \sigma_v^2/\sigma_\varepsilon^2$ (Harvey (1984)). When $\sigma_v^2 = \sigma_u^2 = 0$, then $\beta = \gamma = 0$ and so $\theta_1 = 2$ and $\theta_2 = -1$. Thus we obtain in these circumstances

$$\nabla^2 x_t = (1-2B+B^2)\, a_t = \nabla^2 a_t$$

or, equivalently, the *global linear trend* model

$$x_t = b_0 + b_1 t + \varepsilon_t.$$

Example 9.2: Fitting a local trend model to UK industrial production

An ARIMA (0, 2, 2) model was fitted to Crafts *et al.*'s (1989a) UK industrial production index for the period 1700–1913, obtaining

$$\nabla^2 x_t = (1-1.191B+0.341B^2)\,a_t, \quad \hat{\sigma}_a^2 = 0.00737.$$

Since $\hat{\theta}_1 = 1.191$ and $\hat{\theta}_2 = -0.341$, we obtain estimates of the smoothing parameters as $\hat{\beta} = 0.66$ and $\hat{\gamma} = 0.23$, so that the updating equations for the level and trend are given by

$$\mu_t = 0.66x_t + 0.34(\mu_{t-1} + T_{t-1}) = \mu_{t-1} + T_{t-1} + 0.66e_t$$

and

$$T_t = 0.23(\mu_t - \mu_{t-1}) + 0.77T_{t-1} = T_{t-1} + 0.15e_t$$

respectively. Equivalently, the estimated variances of the structural model given by equations (9.14) and (9.15) are

$$\hat{\sigma}_\varepsilon^2 = 0.85\hat{\sigma}_a^2 = 0.00626$$

$$\hat{\sigma}_v^2 = 0.28\hat{\sigma}_a^2 = 0.00206$$

$$\hat{\sigma}_u^2 = 0.05\hat{\sigma}_a^2 = 0.00039,$$

yielding a signal-to-noise variance ratio $\hat{\sigma}_v^2/\hat{\sigma}_\varepsilon^2$ of 0.33, reflecting a smooth underlying trend contaminated with considerable noise.

Some further special cases are worth mentioning. If $\beta = 1$ then $\theta_2 = 0$ and x_t follows an ARIMA (0, 2, 1) process. In this case $\sigma_\varepsilon^2 = \sigma_v^2$ and the updating and recurrence equations become

$$\mu_t = x_t$$

and

$$T_t = T_{t-1} + \gamma e_t,$$

so that the current level is always the observed value and the trend is updated by a proportion $\gamma = 1-\theta_1$ of the current error. If $\gamma = 1$ as well, then $\theta_1 = 0$ and x_t follows an ARIMA (0, 2, 0) process with $\sigma_\varepsilon^2 = \sigma_v^2 = 0$, $\sigma_u^2 = \sigma_a^2$, $\mu_t = x_t$ and $\nabla T_t = e_t$, the trend therefore being updated by the full amount of the current error. If $\gamma = 0$ then $\theta_1 + \theta_2 = 1$ and we have

$$\nabla^2 x_t = (1-(1-\theta_2)B - \theta_2 B^2)\,a_t$$

$$= (1-B)(1+\theta_2 B)\,a_t$$

i.e.

$$\nabla x_t = \theta_0 + (1+\theta_2 B)\,a_t,$$

an ARIMA (0, 1, 1) process with a constant included. In this case $\sigma_u^2 = 0$ and the trend is therefore constant, with $\theta_0 = T_0$.

Nerlove and Wage (1964) and Theil and Wage (1964) consider

the structural model with $\sigma_v^2 = 0$, thus imposing the restriction $\beta^2 + \beta^2\gamma - 2\gamma\beta = 0$, which can be shown to be equivalent to placing the non-linear restriction $\theta_1 - \theta_1\theta_2 + 4\theta_2 = 0$ on the parameters of (9.13).

Example 9.3: A structural interpretation of an ARIMA model for nominal TFE

Example 8.10 presents an ARIMA (0, 2, 1) model for nominal TFE:

$$\nabla^2 x_t = (1 - 0.437B)\, a_t, \quad \hat{\sigma}_a = 0.0327.$$

This can be interpreted as a structural model of the form

$$\mu_t = x_t = \mu_{t-1} + T_{t-1} + e_t$$

$$T_t = T_{t-1} + \gamma e_t,$$

where

$$\hat{\gamma} = 1 - 0.437 = 0.563,$$

$$\hat{\sigma}_e = \hat{\sigma}_v = (1 - \hat{\gamma})^{\frac{1}{2}} \hat{\sigma}_a = 0.0216,$$

and

$$\hat{\sigma}_u = \hat{\gamma}\hat{\sigma}_a = 0.0184.$$

Example 9.4: A structural interpretation of an ARIMA model for real income

In Example 8.19 an adequate model for the real income series was found to be the ARIMA (0, 1, 1) model:

$$\nabla x_t = 0.025 + (1 + 0.451B)\, a_t, \quad \hat{\sigma}_a = 0.0234.$$

This can be interpreted as a structural model of the form

$$x_t = \mu_t + \varepsilon_t$$

$$\mu_t = \mu_{t-1} + T_{t-1} + v_t$$

$$T_t = T_0 = 0.025,$$

since $\gamma = \sigma_u^2 = 0$. The estimates of the remaining structural parameters are

$$\hat{\beta} = 1 + \hat{\theta} = 1.451$$

$$\hat{\sigma}_\varepsilon = \hat{\sigma}_a = 0.0234$$

$$\hat{\sigma}_v = \hat{\beta}\hat{\sigma}_a = 0.0340,$$

so that the signal-to-noise ratio $\hat{\sigma}_v^2/\hat{\sigma}_\varepsilon^2$ is 2.11 and hence the systematic variation totally dominates the noise component, thus reflecting the extremely smooth evolution of the series.

9.4 Extensions of Holt–Winters type models

A number of extensions to the Holt–Winters class of structural models have recently been proposed. The simple exponential smoothing model can be extended to allow the current level always to decay back to zero (more generally to a nonzero process mean) by replacing the error correction form, for example, by

$$\mu_t = \phi\mu_{t-1} + \alpha e_t,$$

with $0 < \phi < 1$. Thus, since

$$x_t = \mu_t + (1-\alpha)e_t,$$

it follows that x_t must be generated by the ARIMA (1, 0, 1) process

$$(1-\phi B)\, x_t = (1-\theta B)\, e_t,$$

where $\theta = \phi(1-\alpha)$. If $\phi = 1$, we return to the original model (9.4).

Exponential, rather than linear, trends may be modelled by replacing the error correction forms of equations (9.10) and (9.11) by

$$\mu_t = \mu_{t-1} + \phi T_{t-1} + \beta e_t$$

and

$$T_t = \phi T_{t-1} + \beta\gamma e_t,$$

the corresponding updating equations being

$$\mu_t = \beta x_t + (1-\beta)(\mu_{t-1} + \phi T_{t-1})$$

and

$$T_t = \gamma(\mu_t - \mu_{t-1}) + (1-\alpha)\phi T_{t-1}.$$

If $\phi > 1$, growth in x_t is exponential, while if $\phi < 1$ growth has a damped exponential form, declining in both relative and absolute terms each period. The difference between a linear and a damped trend can be substantial over a long time span, even with a relatively large setting of ϕ, say 0.9 or 0.95. Roberts (1982) has shown that this model is optimal for the ARIMA (1, 1, 2) process

$$(1-\phi B)\nabla x_t = (1-\theta_1 B - \theta_2 B^2)\, a_t,$$

where $\theta_1 = 1 + \phi - \beta - \phi\beta\gamma$ and $\theta_2 = -\phi(1-\beta)$. No direct empirical evidence is available on the performance of damped exponential models, although some indirect evidence is available from Lewandowski (1982), in which a damped trend model, the rate of decay of which increases with the noise in the series, was found to produce very accurate forecasts at long horizons.

Higher order polynomial trends can be analysed, although models above the first degree would seem to have little practical importance in economics. The use of local nth-degree polynomial trends can be analysed through *exponential smoothing of order* $n+1$ (Brown (1963)). In general, such exponential smoothing is optimal for the ARIMA $(0, n+1, n+1)$ process (Cogger (1974)). Thus quadratic (or triple exponential smoothing) implies the ARIMA (0, 3, 3) process, which is rarely observed in practice.

10 Modelling seasonal time series

10.1 Introduction

As we saw in Chapter 2, many economic time series exhibit seasonal fluctuation, which we may informally describe as the movement of a component with a period of approximately one year. Typically, such fluctuations have been removed from the observed series by the process of *seasonal adjustment* (for historical surveys of this important area of time series modelling see, for example, Nerlove et al. (1979 chapter 1), Bell and Hillmer (1984), and Hylleberg (1986, chapter 1)). In recent years, however, a great deal of research has been undertaken on developing models capable of analysing seasonal time series and it is now generally felt that seasonal adjustment and modelling should be regarded as being linked together, rather than carried out as distinctly separate forms of analysis (see on this Box et al. (1978) and Hillmer, Bell and Tiao (1983) and, for surveys, Pierce (1980) and Bell and Hillmer (1984)).

With this modern synthesis in mind, this chapter begins by developing ARIMA models for seasonal time series and then investigates the correspondences existing between such models and seasonal versions of the familiar exponential smoothing techniques. The most widely used method of seasonal adjustment (X-11) is then introduced and its relationship with the ARIMA class of models discussed. Finally, model-based seasonal adjustment through the application of signal extraction techniques is developed, with examples of the various procedures being provided.

10.2 Seasonal ARIMA models

Seasonality is often clearly shown in plots of time series, with Exhibits 2.5, 2.6 and 2.10 being obvious examples. How, though, does seasonality manifest itself in the SACFs and PACFs needed for the identification of

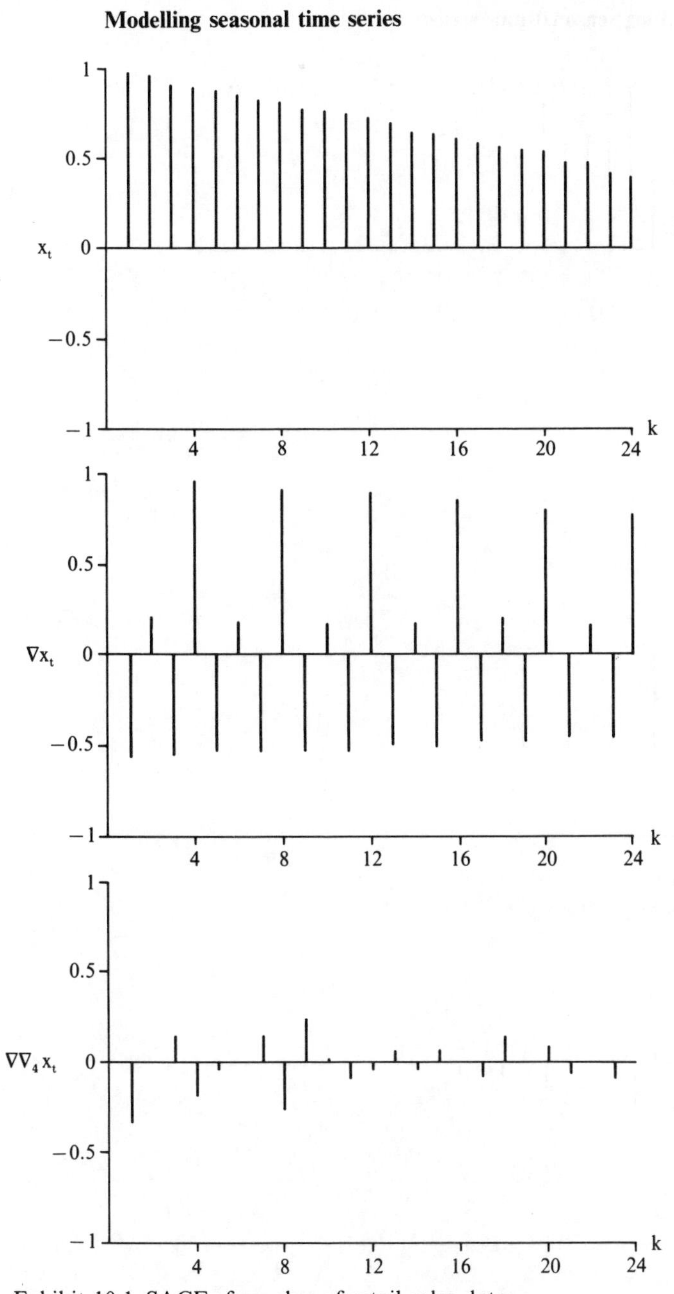

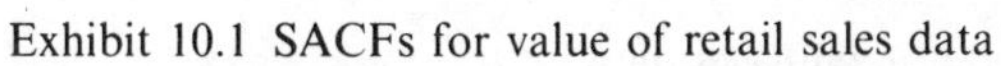
Exhibit 10.1 SACFs for value of retail sales data

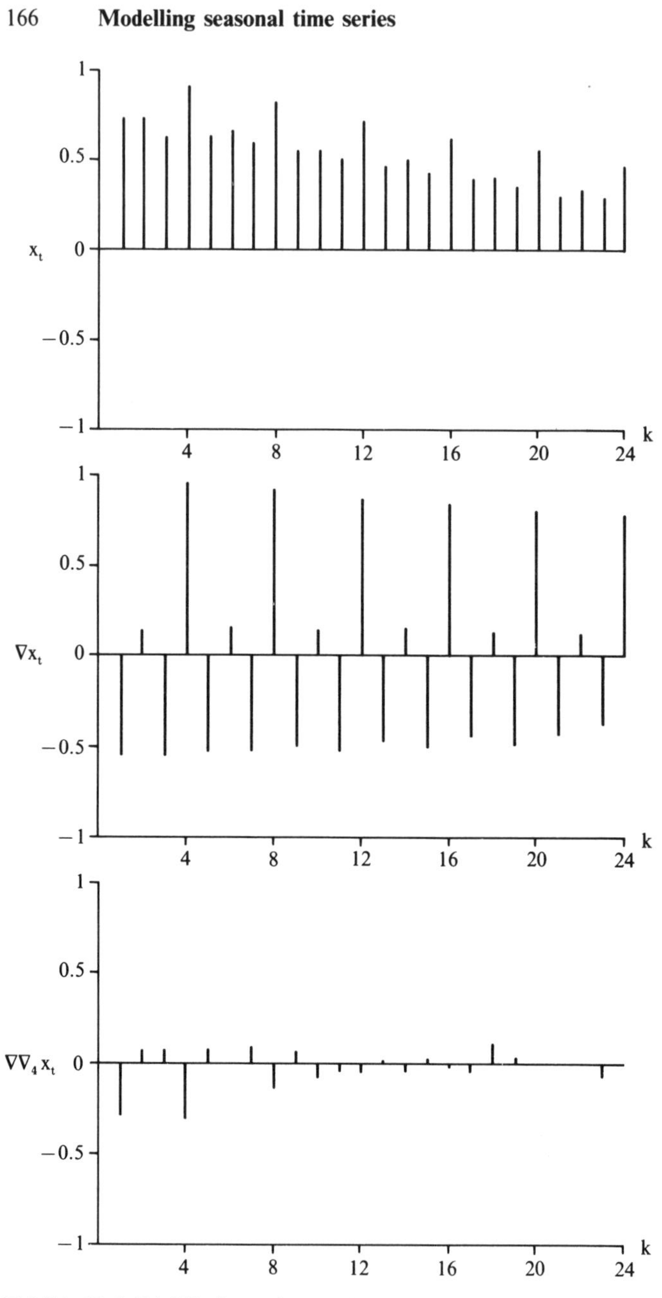

Exhibit 10.2 SACFs for volume of retail sales data

suitable ARIMA models? Exhibits 10.1 and 10.2 show the SACFs of various differences of the logarithms of the value and volume of UK retail sales originally plotted as Exhibit 2.5. From the analysis of the levels data, the value series is clearly nonstationary, but it is difficult to isolate any seasonal pattern as all autocorrelations are dominated by the effect of the nonseasonal unit root. The SACF of the volume series, although still indicating nonstationarity, does display seasonality, as revealed by the large and slowly declining autocorrelations at lags 4, 8, 12, etc. (since the data are quarterly, seasonality should show up at such lags). First differencing produces a very distinct pattern in the SACFs of both series, with very large positive autocorrelations at the seasonal lags ($4k, k \geqslant 1$) being flanked by negative autocorrelations at the 'satellites' ($4(k-1)$, $4(k+1)$). The slow decline of these seasonal autocorrelations is indicative of *seasonal nonstationarity* and, analogous to the analysis of 'nonseasonal nonstationarity', this may be dealt with by *seasonal differencing*; i.e. by using the $\nabla_4 = (1-B^4)$ operator in conjunction with the usual ∇ operator. The SACFs of the two series transformed by the $\nabla\nabla_4$ operator, i.e. by seasonally differencing the first differences, show much more interpretable patterns, both now displaying the characteristics of a stationary time series.

In general, we may denote the seasonal period by s and the seasonal differencing operator as ∇_s; for quarterly data, as above, $s = 4$, and for monthly data, $s = 12$. Noting that nonseasonal and seasonal differencing may be applied, in general, d and D times respectively, a seasonal ARIMA model for a series x_t may then take the general form

$$\nabla^d \nabla_s^D x_t = \frac{\theta(B)}{\phi(B)} a_t, \tag{10.1}$$

where the elements making up the right hand side of the equation are as defined in, for example, equation (6.9). Appropriate forms of the $\theta(B)$ and $\phi(B)$ polynomials can then, at least in principle, be obtained by the usual methods of identification and/or model selection. Unfortunately, two difficulties are encountered. The PACFs of seasonal models are extremely difficult to interpret, although some headway has been made by Hamilton and Watts (1978), so conventional identification is usually based solely on the behaviour of the appropriate SACF. Furthermore, since the $\theta(B)$ and $\phi(B)$ polynomials must account for seasonal autocorrelations, at least one of them must be of minimum order s. This often means that the number of models which need to be considered in model selection procedures can become prohibitively large.

As examples of the identification of models of the form (10.1), consider again the SACFs of the $\nabla\nabla_4$ transformations shown in Exhibits 10.1 and

10.2. For the (logarithm of the) value series, significant autocorrelations are found at lags 1, 4, 8 and 9, and since there is no tendency for the SACF to exhibit any form of sinusoidal or exponential decay, being rather of the appearance of a set of 'spikes', a high order, but sparse, moving average model is therefore suggested. On estimation, the following model was obtained:

$$\nabla\nabla_4 x_t = \begin{pmatrix} 1 - 0.254B - 0.427B^4 - 0.220B^8 + 0.378B^9 \\ \quad (0.081) \quad (0.088) \quad (0.089) \quad (0.082) \end{pmatrix} a_t$$

$$\hat{\sigma}_a = 0.0167; \quad Q_{12}(8) = 7.1.$$

All parameter estimates are significantly different from zero and the Ljung–Box (1978) portmanteau statistic (see Chapter 8) does not indicate any model misspecification. Twelve autocorrelations were included in the calculation of this statistic to enable any residual autocorrelation at seasonal lags to be picked up. Lagrange Multiplier tests could, of course, be constructed in the manner outlined in Chapter 8, but with seasonal models of this type it is more difficult to specify a reasonable alternative hypothesis, thus suggesting that portmanteau tests may be more useful in these circumstances.

For the (logarithm of the) volume series, spikes in the SACF are found at lags 1 and 4, leading to the following estimated model:

$$\nabla\nabla_4 x_t = \begin{pmatrix} 1 - 0.268B - 0.713B^4 \\ \quad (0.055) \quad (0.056) \end{pmatrix} a_t$$

$$\hat{\sigma}_a = 0.0187; \quad Q_{12}(10) = 13.9,$$

and again this would appear to be an adequate model.

10.2.1 *Multiplicative seasonal ARIMA models*

As the value series demonstrates, identification of models of the form (10.1) can lead to a large number of parameters having to be fitted and may result in a model being difficult to interpret. Box and Jenkins (1976, chapter 9) develop an argument for using a restricted version of (10.1) which, they feel, should provide an adequate fit to many seasonal time series. By way of introducing this model, consider Exhibit 10.3, which shows the first ten years' observations on the logarithms of the volume of retail sales arranged in a year-by-quarter format, which emphasises the fact that, in seasonal data, there are not one but two time intervals of importance. In this example, these intervals correspond to quarters and years and we expect, therefore, relationships to occur:

Exhibit 10.3 *Logarithms of volume of retail sales in year-by-quarter format*: 1955 *to* 1964

	Q1	Q2	Q3	Q4
1955	4.047	4.091	4.096	4.244
1956	4.060	4.111	4.133	4.278
1957	4.076	4.126	4.116	4.272
1958	4.074	4.130	4.124	4.300
1959	4.098	4.194	4.177	4.350
1960	4.156	4.234	4.220	4.381
1961	4.200	4.248	4.246	4.393
1962	4.196	4.257	4.260	4.402
1963	4.211	4.282	4.297	4.444
1964	4.258	4.309	4.315	4.472

(a) between observations for successive quarters in a particular year, and

(b) between observations for the same quarter in successive years.

Referring to the retail sales data of Exhibit 10.3, it is clear that the seasonal effect implies that an observation for a particular quarter, say Q4, is related to the observations for previous Q4s. We may be able to link the Q4 observations by a model of the form:

$$\Phi(B^s)\,\nabla_s^D x_t = \Theta(B^s)\,\alpha_t, \tag{10.2}$$

where $s = 4$ in the above example and $\Phi(B^s)$ and $\Theta(B^s)$ are polynomials in B^s of degrees P and Q respectively, i.e.

$$\Phi(B^s) = 1 - \Phi_1 B^s - \Phi_2 B^{2s} - \ldots - \Phi_P B^{Ps}$$

$$\Theta(B^s) = 1 - \Theta_1 B^s - \Theta_2 B^{2s} - \ldots - \Theta_Q B^{Qs},$$

which satisfy standard stationarity and invertibility conditions. Now, suppose that the same model holds for the observations from each quarter. This implies that all errors that correspond to a fixed quarter in different years are uncorrelated. However, the errors corresponding to adjacent quarters need not be uncorrelated: i.e. the error series $\alpha_t, \alpha_{t-1}, \ldots$ may be autocorrelated. For example, retail sales in 1984 Q4, while related to previous Q4 values, will also be related to the values in 1984 Q3, 1984 Q2, etc. These autocorrelations may be represented by a second, nonseasonal, model:

$$\phi(B)\,\nabla^d \alpha_t = \theta(B)\,a_t, \tag{10.3}$$

so that α_t is ARIMA (p, d, q), with a_t being a white noise process. Substituting (10.3) into (10.2) yields a general *multiplicative seasonal* model

$$\phi_p(B)\,\Phi_P(B^s)\,\nabla^d\nabla_s^D x_t = \theta_q(B)\,\Theta_Q(B^s)\,a_t. \tag{10.4}$$

The subscripts p, P, q, Q have been added to emphasise the orders of the various polynomials and the ARIMA process in (10.4) is said to be of order $(p, d, q)(P, D, Q)_s$. A constant θ_0 can be included in (10.4), and this will introduce a deterministic trend component into the model. A comparison with the 'non-multiplicative' model (10.1) reveals that the $\phi(B)$ and $\theta(B)$ polynomials have been factored as

$$\phi_{p+P}(B) = \phi_p(B)\,\Phi_P(B^s)$$

and

$$\theta_{q+Q}(B) = \theta_q(B)\,\Theta_Q(B^s)$$

10.2.2 *The 'airline' model*

Because the general multiplicative model (10.4) is rather complicated, it is difficult to provide explicit expressions for its ACF and PACF, although Hamilton and Watts (1978) provide expressions for the PACFs of some low order models and Peña (1984) obtains the formula of the theoretical ACF for the general form. However, since no new principles of analysis are involved, we will proceed by considering one simple special case of (10.4), only pointing out those aspects of the general multiplicative seasonal model which call for special mention.

To develop this particular model, recall that a simple and widely applicable stochastic model for the analysis of nonstationary, nonseasonal time series is the ARIMA (0, 1, 1) process. Suppose that we employ such a model to link the x_t's one *year* apart:

$$\nabla_s x_t = (1 - \Theta B^s)\,\alpha_t, \tag{10.5}$$

and that a similar model is used to link α_t's one *period* apart:

$$\nabla\alpha_t = (1 - \theta B)\,a_t, \tag{10.6}$$

where, in general, θ and Θ will have different values. On combining (10.5) and (10.6) we obtain the multiplicative ARIMA $(0, 1, 1)(0, 1, 1)_s$ model:

$$\nabla\nabla_s x_t = (1 - \theta B)(1 - \Theta B^s)\,a_t. \tag{10.7}$$

For invertibility, we require the roots of $(1 - \theta B)(1 - \Theta B^s)$ to satisfy the conditions $|\theta|, |\Theta| < 1$. Note that the model may be written as

$$w_t = (1 - B - B^s + B^{s+1})\, x_t = (1 - \theta B - \Theta B^s + \theta\Theta B^{s+1})\, a_t,$$

and, since the autocovariances of $w_t = \nabla\nabla_s x_t$ are

$$\begin{aligned} \gamma_k &= E(w_t w_{t-k}) \\ &= E\{(a_t - \theta a_{t-1} - \Theta a_{t-s} + \theta\Theta a_{t-s-1}) \times \\ &\quad (a_{t-k} - \theta a_{t-k-1} - \Theta a_{t-s-k} + \theta\Theta a_{t-s-1-k})\}, \end{aligned}$$

we have

$$\begin{aligned} \gamma_0 &= (1+\theta^2)(1+\Theta^2)\,\sigma_a^2 \\ \gamma_1 &= -\theta(1+\Theta^2)\,\sigma_a^2 \\ \gamma_{s-1} &= \theta\Theta\sigma_a^2 \\ \gamma_s &= -\Theta(1+\theta^2)\,\sigma_a^2 \\ \gamma_{s+1} &= \theta\Theta\sigma_a^2, \end{aligned}$$

with all other γ_k's equal to zero. Hence the ACF is

$$\begin{aligned} \rho_1 &= -\frac{\theta}{1+\theta^2} \\ \rho_{s-1} &= \frac{\theta\Theta}{(1+\theta^2)(1+\Theta^2)} \\ \rho_s &= -\frac{\Theta}{1+\Theta^2}, \\ \rho_{s+1} &= \rho_{s-1} \end{aligned}$$

and $\rho_k = 0$ otherwise.

On the assumption that the model is of the form (10.7), the variances for the estimated sample autocorrelations at lags higher than $s+1$ are given by

$$V(r_k) = \frac{1 + 2(r_1^2 + r_{s-1}^2 + r_s^2 + r_{s+1}^2)}{n}, k > s+1.$$

Using this result in conjunction with the known form of the ACF will enable the ARIMA $(0,1,1)(0,1,1)_s$ model to be identified. If the SACF follows a more complicated pattern, then other members of the ARIMA $(p,d,q)(P,D,Q)_s$ class must be entertained.

The ARIMA $(0,1,1)(0,1,1)_s$ model is often referred to as the 'airline model', as Box and Jenkins (1976, chapter 9) first illustrated its use on airline travel data previously used by Brown (1963).

The ACFs for various multiplicative seasonal models may be found in, for example, Box and Jenkins (1976, Appendix A9.1). For such low order models, the PACF can be thought of as a repetition of a combination of the auto- and partial correlation functions of the seasonal component about the seasonal partial values (Hamilton and Watts (1978)). What can be said in general is that seasonal and nonseasonal moving average components introduce exponential decays and damped sine waves at the seasonal and nonseasonal lags, whereas with autoregressive processes the PACF cuts off.

10.2.3 *Forecasting from the airline model*

Forecasts for the ARIMA $(0,1,1)(0,1,1)_s$ model are best computed directly from

$$x_{n+h} = x_{n+h-1} + x_{n+h-s} - x_{n+h-s-1} + a_{n+h} - \theta a_{n+h-1} - \Theta a_{n+h-s} + \theta\Theta a_{n+h-s-1}$$

so that

$$f_{n,h} = E\{(x_{n+h-1} + x_{n+h-s} - x_{n+h-s-1} + a_{n+h} - \theta a_{n+h-1} - \Theta a_{n+h-s} + \theta\Theta a_{n+h-s-1} | x_n, x_{n-1}, \ldots)\}$$

and

$$E(x_{n+j}) = \begin{cases} x_{n+j}, & j \leqslant 0 \\ f_{n,j}, & j > 0 \end{cases}$$

$$E(a_{n+j}) = \begin{cases} a_{n+j}, & j \leqslant 0 \\ 0, & j > 0. \end{cases}$$

In general, the eventual forecast function is

$$(1-B)(1-B^s)f_{n,h} = 0, \quad h > s+1.$$

It can be shown that the 'ψ-weights' in the process $x_t = \psi(B)a_t$, where

$$\psi(B) = (1-B)^{-1}(1-B^s)^{-1}(1-\theta B)(1-\Theta B^s),$$

are

$$\psi_0 = 1$$

$$\psi_1 = \psi_2 = \ldots = \psi_{s-1} = 1-\theta$$

$$\psi_s = 2-\theta-\Theta$$

$$\psi_{s+1} = \psi_{s+2} = \ldots = \psi_{2s-1} = (1-\theta)(2-\Theta)$$

$$\psi_{2s} = (1-\theta)(2-\Theta)+(1-\Theta)$$

$$\psi_{2s+1} = \psi_{2s+2} = \ldots = \psi_{3s-1} = (1-\theta)(3-2\Theta)$$

$$\psi_{3s} = (1-\theta)(3-2\Theta)+(1-\Theta),$$

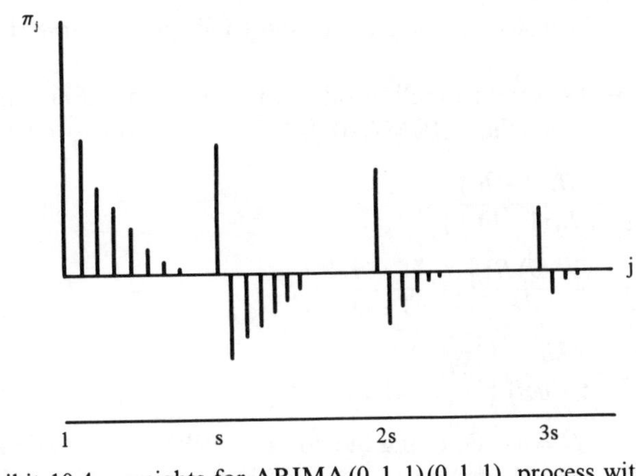

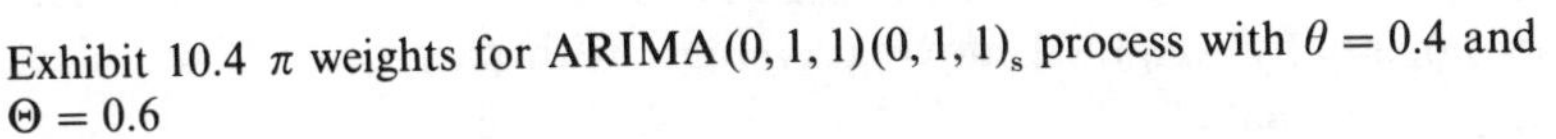

Exhibit 10.4 π weights for ARIMA$(0, 1, 1)(0, 1, 1)_s$ process with $\theta = 0.4$ and $\Theta = 0.6$

and, in general,

$$\psi_{rs+1} = \psi_{rs+2} = \ldots = \psi_{(r+1)s-1} = (1-\theta)(r+1-r\Theta)$$

$$\psi_{(r+1)s} = (1-\theta)(r+1-r\Theta)+(1-\Theta).$$

Using these expressions, the h-step ahead forecast error variance can be calculated from

$$V(e_{n,h}) = \sigma_a^2(1+\psi_1^2+\ldots+\psi_{h-1}^2).$$

It is instructive to consider the one-step ahead forecasts in 'π-weight' form. From

$$\pi(B)\,x_t = a_t,$$

it follows that

$$f_{n,1} = \sum_{j=1}^{\infty} \pi_j\, x_{n+1-j}.$$

The π weights may be obtained by equating coefficients in

$$(1-B)(1-B^s) = (1-\theta B)(1-\Theta B^s)(1-\pi_1 B-\pi_2 B^2-\ldots),$$

yielding

$$\pi_j = \theta^{j-1}(1-\theta) \qquad j = 1, 2, \ldots, s-1$$

$$\pi_s = \theta^{s-1}(1-\theta)+(1-\Theta)$$

$$\pi_{s+1} = \theta^s(1-\theta)-(1-\theta)(1-\Theta)$$

$$(1-\theta B-\Theta B^s+\theta\Theta B^{s+1})\,\pi_j = 0 \quad j \geqslant s+2.$$

These weights, for $\theta = 0.4$ and $\Theta = 0.6$, are shown graphically in Exhibit 10.4.

The reason that the weight function takes this particular form can be understood as follows: the ARIMA $(0,1,1)(0,1,1)_s$ process may be written as

$$a_{t+1} = \frac{(1-B)(1-B^s)}{(1-\theta B)(1-\Theta B^s)} x_{t+1}$$

$$= \left\{1 - \frac{(1-\theta)B}{1-\theta B}\right\}\left\{1 - \frac{(1-\Theta)B^s}{1-\Theta B^s}\right\} x_{t+1}$$

$$= \left\{1 - \frac{\lambda B}{1-\theta B}\right\}\left\{1 - \frac{\Lambda B^s}{1-\Theta B^s}\right\} x_{t+1},$$

where $\lambda = 1-\theta$ and $\Lambda = 1-\Theta$. Using the notation $\text{EWMA}_\lambda(x_t)$ to denote an exponentially weighted moving average, with parameter λ, of values $x_t, x_{t-1}, \ldots$, we have

$$\text{EWMA}_\lambda(x_t) = \frac{\lambda}{1-\theta B} x_t = \lambda x_t + \lambda\theta x_{t-1} + \lambda\theta^2 x_{t-2} + \ldots$$

and, similarly, for the seasonal pattern we have

$$\text{EWMA}_\Lambda(x_t) = \frac{\Lambda}{1-\Theta B^s} x_t = \Lambda x_t + \Lambda\Theta x_{t-s} + \Lambda\Theta^2 x_{t-2s} + \ldots$$

Thus, since $a_{n+1} = f_{n,1} - x_{n+1}$,

$$f_{n,1} = \text{EWMA}_\lambda(x_n) + \text{EWMA}_\Lambda(x_{n-s+1} - \text{EWMA}_\lambda(x_{n-s})).$$

Hence the one-step ahead forecast is an EWMA taken over previous periods, modified by a second EWMA of discrepancies found between similar EWMAs and actual performances in previous years.

Example 10.1: Fitting an airline model to the retail sales volume series

As an example of analysing an 'airline model', consider again the SACF of $\nabla\nabla_4 x_t$, where x_t is the logarithm of the retail sales volume series, shown in Exhibit 10.2. As we remarked earlier, there are large autocorrelations at lags 1 and 4, which led to us fitting a (restricted) fourth order moving average to $\nabla\nabla_4 x_t$. On closer inspection of the SACF, satellite correlations of the appropriate sign (although small) are found at lags 3 and 5, thus suggesting that an ARIMA $(0,1,1)(0,1,1)_4$ model could be a possibility. Estimation obtained

$$\nabla\nabla_4 x_t = \left(\underset{(0.091)}{1 - 0.301B}\right)\left(\underset{(0.065)}{1 - 0.754B^4}\right) a_t$$

$$\hat{\sigma}_a = 0.0187; \quad Q_{12}(10) = 12.4$$

which yields an almost identical fit to the previous model and may, therefore, be preferred as it has a ready interpretation as a model in which between- and within-year autocorrelation are simultaneously modelled in an additive fashion (see Box and Jenkins (1976, pages 322–4) for further interpretation of this 'additivity' concept and analogies with two-way analysis of variance tables). Note that the airline model with $s = 4$ can be written as

$$\nabla\nabla_4 x_t = (1 - \theta_1 B - \theta_4 B^4 + \theta_1 \theta_4 B^5)\, a_t,$$

and hence a test of the adequacy of a multiplicative model vis-à-vis a non-multiplicative model of the form

$$\nabla\nabla_4 x_t = (1 - \theta_1 B - \theta_4 B^4 - \theta_5 B^5)\, a_t$$

is equivalent to testing the nonlinear restriction $\theta_5 + \theta_1 \theta_4 = 0$. For this volume series, a likelihood ratio test of this restriction could not reject the null hypothesis of a multiplicative model at any conventional levels of significance.

Forecasts for the airline model fitted to the volume series can be routinely computed, noting that $n = 120$, as

$$f_{120,1} = x_{120} + x_{117} - x_{116} - 0.30a_{120} - 0.75a_{117} + 0.23a_{116}$$

$$f_{120,2} = f_{120,1} + x_{118} - x_{117} - 0.75a_{118} + 0.23a_{117}$$

$$\vdots$$

$$f_{120,5} = f_{120,4} + f_{120,1} - x_{120} + 0.23a_{120}$$

$$f_{120,6} = f_{120,5} + f_{120,2} - f_{120,1}, \quad \text{etc.}$$

The 'ψ-weights' are then calculated as:

$$\psi_0 = 1;\ \psi_1 = \psi_2 = \psi_3 = 0.70;\ \psi_4 = 0.95;$$

$$\psi_5 = \psi_6 = \psi_7 = 0.87;\ \psi_8 = 1.12;$$

$$\psi_9 = \psi_{10} = \psi_{11} = 1.04;\ \psi_{12} = 1.29;\ \text{etc.}$$

Hence the h-step ahead forecast error variances are

$$V(e_{n,1}) = \sigma_a^2 = 0.00035$$

$$V(e_{n,2}) = \sigma_a^2(1 + \psi_1^2) = 0.0187^2(1 + 0.49) = 0.00052$$

$$V(e_{n,3}) = \sigma_a^2(1 + \psi_1^2 + \psi_2^2) = 0.0187^2(1 + 0.49 + 0.49) = 0.00104 \text{ etc.}$$

The 'π-weights' are calculated as

$$\pi_1 = 0.70;\ \pi_2 = 0.21;\ \pi_3 = 0.06;\ \pi_4 = 0.27;\ \pi_5 = -0.17$$

$$(1 - 0.30B - 0.75B^4 + 0.23B^5)\,\pi_j = 0 \quad \text{for} \quad j \geqslant 6.$$

The equivalent form of $f_{n,1}$ is, therefore,

$$f_{n,1} = \mathrm{EWMA}_{0.70}(x_n) + \mathrm{EWMA}_{0.25}(x_{n-3} - \mathrm{EWMA}_{0.70}(x_{n-4})).$$

Thus, if we were attempting to predict the Q4 value of this retail sales series, then the first term would be an EWMA taken over previous quarters up to Q3. However, we know this will be an underestimate, so we correct it by taking a second EWMA over previous years of the discrepancies between actual Q4 retail sales and the corresponding quarterly EWMAs taken over previous third quarters in those years.

Example 10.2: Fitting a multiplicative seasonal ARIMA model to the retail sales value series

The SACF for the $\nabla\nabla_4$ transformation of the value series shown in Exhibit 10.1 displays, as we have noted above, large autocorrelations at lags 7, 8, and 9 and hence, if a multiplicative model is to be fitted, a more complex one than the airline model is required. The obvious extension is to fit the ARIMA (0, 1, 1)(0, 1, 2)$_4$ model, which yields

$$\nabla\nabla_4 x_t = \begin{pmatrix} 1 - 0.218B \\ \quad (0.094) \end{pmatrix} \begin{pmatrix} 1 - 0.377B^4 - 0.291B^8 \\ \quad (0.093) \qquad (0.093) \end{pmatrix} a_t$$

$$\hat{\sigma}_a = 0.0172; \quad Q_{12}(9) = 14.5.$$

Comparing this to the non-multiplicative model previously fitted shows an increase in $\hat{\sigma}_a$ and a somewhat larger Q value, a consequence, in fact, of a large residual autocorrelation at lag 9. The test of the appropriateness of this multiplicative model vis-à-vis the nonmultiplicative model

$$\nabla\nabla_4 x_t = (1 - \theta_1 B - \theta_4 B^4 - \theta_5 B^5 - \theta_8 B^8 - \theta_9 B^9)\, a_t$$

is equivalent to the joint test of the restrictions $\theta_5 + \theta_1\theta_4 = 0$ and $\theta_9 + \theta_1\theta_8 = 0$ and, indeed, a likelihood ratio test rejects this hypothesis at a low marginal significance level, thus indicating that in this case a nonmultiplicative seasonal model is to be preferred.

Example 10.3: Modelling monthly M0

A further example concerning the analysis of multiplicative models, which illustrates in detail the use of forecast functions and their updating, is provided by Mills (1986). In this example the following model for the logarithm of the monthly M0 money supply series, adjusted for a change in definition in August 1981, was developed for the period January 1976 to December 1984:

$$\nabla\nabla_{12} x_t = -0.0008 + (1 - 0.297B)(1 - 0.674B^{12})\, a_t, \hat{\sigma}_a = 0.0102. \tag{10.8}$$

Such a model has a forecast function (see Chapter 7) satisfying the difference equation

$$\nabla\nabla_{12} f_{n,h} = \theta_0 + (1-\theta B)(1-\Theta B^{12})\,\hat{a}_{n,h},$$

where the lead time h is defined as $h = 12r+m$, $r = 0, 1, 2, \ldots$ being years and $m = 1, 2, \ldots, 12$ being months, and where

$$\hat{a}_{n,h} = E(a_{n,h}|x_n, x_{n-1}, \ldots) = \begin{cases} a_{n+h}, & h \leqslant 0 \\ 0, & h > 0. \end{cases}$$

By developing the results of Box and Jenkins (1976, chapters 5 and 9), Box et al. (1987) and Newbold (1988), the forecast function of (10.8) can be written as

$$f_{n,h} = \beta_{0m}^{(n)} + \beta_*^{(n)} h. \tag{10.9}$$

This shows that the forecast function is of the form of a local linear trend. The slope, β_*, and monthly intercept coefficients, β_{0m}, are updated each period according to:

$$\beta_*^{(n+1)} = \theta_0 + \beta_*^{(n)} + (\lambda\Lambda/12)\,a_{n+1}$$

$$\beta_{0m}^{(n+1)} = \beta_{0,m+1}^{(n)} + \beta_*^{(n)} + \left[\lambda(1-\frac{m}{12}\Lambda) + \delta_{12,m}\Lambda\right] a_{n+1},$$

where $\lambda = 1-\theta$, $\Lambda = 1-\Theta$ and $\delta_{12,m} = 1$ if $m = 12$ and 0 otherwise. Note that the presence of a nonzero intercept θ_0 introduces a deterministic, as well as the usual stochastic, component into the slope coefficient. The monthly seasonal intercepts incorporate both an overall level effect

$$\beta_0^{(n)} = \frac{1}{12}\sum_{m=1}^{12} \beta_{0m}^{(n)},$$

and a seasonal effect specific to the mth month:

$$\beta_m^{(n)} = \beta_{0m}^{(n)} - \beta_0^{(n)}.$$

These may be isolated by rewriting (10.9) as

$$f_{n,h} = \beta_0^{(n)} + \beta_m^{(n)} + \beta_*^{(n)} h, \tag{10.10}$$

in which $\beta_0^{(n)}$ and $\beta_m^{(n)}$ can be regarded as the level and seasonal components of the forecast. Using this decomposition the current projection of the trend, based upon information up to and including time n, can then be defined as

$$T_{n,h} = \beta_0^{(n)} + \beta_*^{(n)} h.$$

The level and seasonal coefficients can also be updated each period using

$$\beta_0^{(n+1)} = \beta_0^{(n)} + \beta_*^{(n)} + [\lambda(1 - \tfrac{13}{24}\Lambda) + \tfrac{1}{12}\Lambda]\, a_{n+1}$$

and

$$\beta_m^{(n+1)} = \beta_{m+1}^{(n)} + \left[\frac{13-2m}{24}\lambda\Lambda - (1 - 12\delta_{12,m})\tfrac{1}{12}\Lambda\right] a_{n+1}.$$

Thus the previously forecasted level for each month, $\beta_{0m}^{(n)}$, is adjusted by the previously forecasted increment to the trend, now incorporated into the level at the new origin $n+1$, and all the coefficients are adjusted by varying fractions of the new information, a_{n+1}. The seasonal and level components, $\beta_m^{(n)}$ and $\beta_0^{(n)}$, are adjusted in a similar way.

The components of (10.9) and (10.10) can be evaluated in terms of the forecasts from origin n through the following expressions (see Box et al. (1987), Mills (1986)):

$$\hat{\beta}_*^{(n)} = \tfrac{1}{12}(f_{n,13} - f_{n,1})$$

$$\hat{\beta}_{0m}^{(n)} = f_{n,m} - \frac{m}{12}(f_{n,13} - f_{n,1})$$

$$\hat{\beta}_0^{(n)} = \frac{1}{12}\sum_{j=1}^{12} f_{n,j} - \tfrac{13}{24}(f_{n,13} - f_{n,1})$$

$$\hat{\beta}_m^{(n)} = f_{n,m} - \frac{1}{12}\sum_{j=1}^{12} f_{n,j} + \frac{13-2m}{24}(f_{n,13} - f_{n,1}).$$

Using the parameter estimates from (10.8), Exhibit 10.5 shows the implied forms of equations (10.9) and (10.10), the coefficient updating equations, the estimates of the components at the end of the sample period (December 1984), and the updating coefficients to be applied to the new information (observation).

The estimated current trend at the end of the sample period is, therefore,

$$T_{n,h} = 9.5597 + 0.0030h$$

and, since the metric of the model is logarithms, the current projected trend growth rate is given (in annualised terms) as

$$f_{n,13} - f_{n,1} = 12(0.0030) = 0.036,$$

i.e. 3.6 per cent. An estimate of the error to be attached to this estimated trend growth is readily available. Since the annualised trend growth rate is

$$12\hat{\beta}_*^{(n)} = f_{n,13} - f_{n,1},$$

Exhibit 10.5 *Forecast function for the logarithm of M*0

$$f_{n,h} = \beta_{0m}^{(n)} + \beta_{*}^{(n)} h = \beta_{0}^{(n)} + \beta_{m}^{(n)} + \beta_{*}^{(n)} h$$

where

$$\beta_{*}^{(n+1)} = -0.0008 + \beta_{*}^{(n)} + 0.002a_{n+1}$$

$$\beta_{0m}^{(n+1)} = \beta_{0,m+1}^{(n)} + \beta_{*}^{(n)} + (0.70 - 0.02m)\, a_{n+1},\ 1 \leqslant m \leqslant 11$$

$$\beta_{0,12}^{(n+1)} = \beta_{01}^{(n)} + \beta_{*}^{(n)} + 0.70a_{n+1}$$

$$\beta_{0}^{(n+1)} = \beta_{0}^{(n)} + \beta_{*}^{(n)} + 0.60a_{n+1}$$

$$\beta_{m}^{(n+1)} = \beta_{m+1}^{(n)} + [0.01(13-2m) - 0.03]\, a_{n+1},\ 1 \leqslant m \leqslant 11$$

$$\beta_{12}^{(n+1)} = \beta_{1}^{(n)} + 0.30a_{n+1}$$

h	$\hat{\beta}_{0m}^{(n)}$	$\hat{\beta}_{m}^{(n)}$	$\beta_{0m}^{(n+1)}$	$\beta_{m}^{(n+1)}$
1	9.5806	0.0209	0.68	−0.08
2	9.5276	−0.0322	0.66	−0.06
3	9.5316	−0.0282	0.64	−0.04
4	9.5494	−0.0104	0.62	−0.02
5	9.5487	−0.0111	0.60	0
6	9.5541	−0.0058	0.58	−0.02
7	9.5637	0.0038	0.56	0.04
8	9.5716	0.0117	0.54	−0.06
9	9.5673	0.0074	0.52	−0.08
10	9.5577	−0.0023	0.50	−0.10
11	9.5573	−0.0027	0.48	−0.12
12	9.5854	0.0254	0.70	0.30
*	0.0030[a]	9.5597[b]	0.02[c]	0.60[d]

(a) $\hat{\beta}_{*}^{(n)}$ (b) $\hat{\beta}_{0}^{(n)}$ (c) $\beta_{*}^{(n+1)}$ (d) $\beta_{0}^{(n+1)}$

the accompanying error, e_n, is

$$e_n = e_{n,h} - e_{n,1},$$

where the $e_{n,h} = x_{n+h} - f_{n,h}$ are the forecast errors from the model. This trend growth error can be written as

$$e_n = a_{n+13} + \psi_1 a_{n+12} + \ldots + \psi_{11} a_{n+2} + (\psi_{12} - 1)\, a_{n+1},$$

with mean square error

$$\sigma^2[e_n] = \sigma^2 \left[\sum_{j=0}^{11} \psi_j^2 + (\psi_{12} - 1)^2 \right],$$

where $\psi_0 = 1, \psi_1 = \psi_2 = \ldots = \psi_{11} = \lambda$, and $\psi_{12} = \lambda + \Lambda$. With $\lambda = 0.70$, $\Lambda = 0.33$ and $\hat{\sigma}^2 = 0.000105$, this yields a mean square error of 0.00067, or a standard error of 0.0259, approximately $2\frac{1}{2}\%$ per annum. Thus, although the forecasted trend growth rate at the end of the sample period is 3.6%, a 95% (say) confidence interval spans the range $-1\frac{1}{2}\%$ to $8\frac{1}{2}\%$.

10.2.4 *ARIMA models with multiple seasonality*

The multiplicative seasonal ARIMA model developed above can be extended to deal with time series that exhibit more than one period of seasonality, as in Exhibit 2.15, which shows UK bank lending to have both an annual (twelve month) and, to a lesser extent, a quarterly (three month) seasonal pattern. Mills and Stephenson (1987) modelled this series by employing a general three-factor class of ARIMA model, obtaining

$$\nabla_{12} x_t = \underset{(18.4)}{149} + \Big(1 + \underset{(0.097)}{0.271} B^3\Big)\Big(1 - \underset{(0.075)}{0.837} B^{12}\Big) a_t$$

$$\hat{\sigma}_a = 548; \quad Q_{24}(21) = 32.1.$$

Although the Q statistic is rather large (the marginal significance level being only 0.06), its size is primarily due to large residual autocorrelations at high lags; using only twelve lags produces a statistic with a marginal significance level of the much more acceptable value of 0.28. As we shall see in Chapter 12, however, an improvement in the fit of this model can be obtained by an extension to cope with certain known, extraneously occurring, events.

The identification of such extended seasonal models has been discussed by Polasek (1980) and similar models have been employed, for example, by Johannes and Rasche (1979) to analyse US money multipliers, and by Thompson and Tiao (1971) to analyse monthly telephone disconnections. For certain time series, however, homogeneous ARMA models are inappropriate for modelling seasonality. These are series in which there are periodic components, and models for such series are discussed in Tiao and Grupe (1980).

10.3 Structural seasonal models

In Example 10.3 we showed that the forecast function of an airline model could be decomposed into current trend and seasonal components. The appropriate decomposition for general seasonal ARIMA models is given in Box et al. (1987) and provides projected components based on all current and past information. An alternative way of decomposing a time series has been introduced in Chapter 9 and can also easily be extended to deal with seasonal fluctuations. The additive decomposition of equation (9.1), used as the basis for the Holt–Winters class of models, can be replaced by

$$x_t = \mu_t + \psi_t + \varepsilon_t, \tag{10.11}$$

where ψ_t is the seasonal component, assumed to be independent of the level and noise components, μ_t and ε_t respectively. The *additive Holt–Winters seasonal model* replaces the updating and error-correction equations for the level by

$$\begin{aligned}\mu_t &= \beta(x_t - \psi_{t-s}) + (1-\beta)(\mu_{t-1} + T_{t-1}) \\ &= \mu_{t-1} + T_{t-1} + \beta e_t, \end{aligned}\tag{10.12}$$

the trend equations remain unchanged;

$$\begin{aligned}T_t &= \gamma(\mu_t - \mu_{t-1}) + (1-\gamma) T_{t-1} \\ &= T_{t-1} + \beta\gamma e_t, \end{aligned}\tag{10.13}$$

and seasonal updating and error-correction equations are incorporated as (see Gardner (1985))

$$\begin{aligned}\psi_t &= \delta(x_t - \mu_t) + (1-\delta)\psi_{t-s} \\ &= \psi_{t-s} + \delta(1-\beta) e_t. \end{aligned}\tag{10.14}$$

Newbold (1988), for example, shows that equations (10.11)–(10.14) are equivalent to the ARIMA model

$$\nabla\nabla_s x_t = \theta_{s+1}(B) a_t, \tag{10.15}$$

where

$$\begin{aligned}\theta_1 &= 1 - \beta - \beta\gamma \\ \theta_2 &= \ldots = \theta_{s-1} = -\beta\gamma \\ \theta_s &= 1 - \beta\gamma - (1-\beta)\delta \\ \theta_{s+1} &= -(1-\beta)(1-\delta). \end{aligned}$$

Note that if $\gamma = 0$, so that the trend is constant, (10.15) becomes

$$\nabla\nabla_s x_t = \theta_{s+1}(B) a_t,$$

where

$$\begin{aligned}\theta_1 &= 1 - \beta \\ \theta_2 &= \ldots = \theta_{s-1} = 0 \\ \theta_s &= 1 - (1-\beta)\delta \\ \theta_{s+1} &= -(1-\beta)(1-\delta). \end{aligned}$$

Thus, if $\theta_1 \theta_s + \theta_{s+1} = 0$, or equivalently, $2 - 2\delta + \beta\delta = 0$, the model reduces to the ARIMA $(0, 1, 1)(0, 1, 1)_s$ airline model. Moreover, the airline model will also result if, in the general model (10.15) above, both $\beta\gamma$ and $\beta\delta$ are negligibly small.

It is often the case, however, that the additive decomposition (10.11) is felt to be inappropriate, for seasonal movements are often thought to be proportional to the level, but the noise component still enters additively. (Note, of course, that if all components enter multiplicatively, a logarithmic transformation of x_t yields the additive decomposition.) In such circumstances, in which the implied decomposition is

$$x_t = \mu_t \psi_t + \varepsilon_t, \tag{10.16}$$

the *multiplicative Holt–Winters* model can be used:

$$\mu_t = \beta(x_t/\psi_{t-s}) + (1-\beta)(\mu_{t-1} + T_{t-1})$$

$$= \mu_{t-1} + T_{t-1} + \beta e_t/\psi_{t-s} \tag{10.17}$$

$$T_t = \gamma(\mu_t - \mu_{t-1}) + (1-\gamma) T_{t-1}$$

$$= T_{t-1} + \beta\gamma e_t/\psi_{t-s} \tag{10.18}$$

$$\psi_t = \delta(x_t/\mu_t) + (1-\delta)\psi_{t-s}$$

$$= \psi_{t-s} + \delta(1-\beta) e_t/\mu_t. \tag{10.19}$$

This model does not appear to have an equivalent ARIMA process. One problem with these models is that the seasonal factor ψ_t is updated only once a year, viz. (10.14) and (10.19), rather than being normalised to sum to zero (for the additive model), or s (for the multiplicative model), in each period. Roberts (1982) introduces a variety of extensions to these models which both allow for normalisation and for the inclusion of exponential growth trends. Furthermore, by focusing attention on structural models such as those of the Holt–Winters class, Roberts and Harrison (1984) provide arguments against the prevalent use of the $\nabla\nabla_s$ operator for inducing stationarity in both the level and seasonal components. They argue that for time series which do not have a growth component, use of the $\nabla\nabla_s$ operator will often lead to multiplicative seasonal ARIMA models which provide a reasonably adequate fit, but which are, in fact, overdifferenced, although this is difficult to detect in practice because diagnostic checks of unit roots in the moving average operator are well known to be rather weak (see, for example, Plosser and Schwert (1977)). Roberts and Harrison (1984) show that the appropriate ARIMA process for a model for nonstationary seasonal variation about a constant level is

$$U(B)\,x_t = \theta_{s-1}(B)\,a_t$$

where

$$U(B) = 1+B+B^2+\ldots+B^{s-1} = (1-B^s)/(1-B) = \nabla_s/\nabla,$$

and recommend that processes of this type should seriously be considered in the identification stage of model building.

As we have discussed in Chapter 9, the nonseasonal Holt–Winters model can be interpreted in terms of the structural equations

$$\mu_t = \mu_{t-1}+T_{t-1}+v_t$$

$$T_t = T_{t-1}+u_t.$$

When dealing with seasonal time series, Harvey (1981) suggests extending these structural equations with

$$\psi_t = \psi_{t-s}+\omega_t,$$

where ω_t is a normally distributed white noise process, so that ψ_t follows a seasonal random walk and the seasonal pattern is therefore allowed to change through time. It can be shown that the implied ARIMA process for x_t is of the form (10.15) and, if $\sigma_u^2 = 0$, so that the trend is constant, the implied ARIMA process is the airline model

$$\nabla\nabla_s x_t = (1-\theta_1 B)(1-\Theta_s B^s)\,a_t,$$

where the relationships between the model parameters θ and Θ and the structural variances σ_v^2, σ_ω^2 and σ_ε^2 can be written in terms of the relative (to the irregular) variances as (see, for example, Harvey (1981, pages 180–1)),

$$\frac{\sigma_v^2}{\sigma_\varepsilon^2} = \frac{(1-\theta)^2}{\theta} \quad \text{and} \quad \frac{\sigma_\omega^2}{\sigma_\varepsilon^2} = \frac{(1-\Theta)^2}{\Theta}.$$

Example 10.4: A structural interpretation of the retail sales volume model

The volume of retail sales series was found to be adequately fitted by the airline model with parameters $\hat{\theta} = 0.301$ and $\hat{\Theta} = 0.754$. The estimated relative variances of the trend and seasonal components are thus 1.62 and 0.08, showing that the seasonal pattern changes quite slowly, while changes in trend dominate changes in the irregular component.

In Harvey (1984), an alternative model for the seasonal component is proposed:

$$\sum_{j=0}^{s-1} \psi_{t-j} = U(B)\,\psi_t = \omega_t.$$

This allows the seasonal pattern to change over time while imposing the condition that the *expectation* of the sum of seasonal effects over s consecutive periods should be zero. The overall model then has a forecast function consisting of a linear trend with a fixed seasonal pattern. If the variances σ_u^2, σ_v^2 and σ_ω^2 are all zero, the regression model

$$x_t = \alpha + \beta t + \sum_{j=1}^{s-1} \psi_j D_{jt} + \varepsilon_t$$

is obtained, where the D_{jt} are a set of 0–1 seasonal dummies. Harvey (1984) also suggests that seasonal effects might be allowed to increase or decrease by adding a slope term to $U(B)\,\psi_t$, and that seasonality could alternatively be modelled by a set of sines and cosines. Note that the use of the operator $U(B)$ to model the seasonal component is consistent with the arguments of Roberts and Harrison (1984).

10.3.1 *General exponential smoothing*

A related approach to modelling seasonal time series is provided by an extension of exponential smoothing, termed by Brown (1963) as *adaptive* or *general exponential smoothing* (GES). A GES model is formulated as a multiple linear regression and has a forecast equation of the form

$$f_{n,h} = \sum_{j=1}^{p} b_j^{(n)} f_j^{(h)} = \boldsymbol{f}'(h)\,\boldsymbol{\beta}^{(n)}, \tag{10.20}$$

$\boldsymbol{\beta}^{(n)}$ being the vector of coefficients at time n and $\boldsymbol{f}(h)$ a vector of 'fitting functions', restricted to be polynomials, sinusoids, exponentials, and their sums or products (cf. the eventual forecast function for an ARIMA model shown as equation (7.7)). Brown (1963) discusses the estimation of $\boldsymbol{\beta}^{(n)}$ in great detail, with estimates being obtained by minimising

$$\sum_{i=0}^{t} w^i [x_{t-j} - \boldsymbol{f}'(h)\,\boldsymbol{\beta}^{(t)}]^2,$$

so that the procedure is, in effect, a weighted, or discounted, least squares estimator, the solution requiring that we smooth (by w) the model parameters, rather than the components of the time series as in the Holt–Winters approach. The error-correction form of the solution is (cf. equation (7.10))

$$\boldsymbol{\beta}^{(t)} = L'\boldsymbol{\beta}^{(t-1)} + \boldsymbol{h}e_t, \tag{10.21}$$

where $\boldsymbol{h} = F^{-1}\boldsymbol{f}(0)$ and, as usual, $e_t = x_t - f_{t-1,1}$. L is a fixed square matrix such that $\boldsymbol{f}(t) = L\boldsymbol{f}(t-1)$ and $\boldsymbol{h}$ depends on both the fitting functions and the discount factor w. For large t, F is given by

$$F_t = F = \sum_{i=0}^{\infty} w^i \boldsymbol{f}(-j)\boldsymbol{f}'(-j).$$

Clearly, use of this approach requires knowledge of the smoothing vector $\boldsymbol{h}$, which depends on both the fitting functions and the discount factor w. Although explicit expressions for $\boldsymbol{h}$ have been derived in some cases, such expressions are usually obtained numerically (see Montgomery and Johnson (1976), Abraham and Ledolter (1983)).

If the data are nonseasonal, there is no reason to use GES, for equivalent and simpler models from the Holt–Winters class can always be formulated. If the data are seasonal, then GES differs considerably from Holt–Winters. The seasonal terms in GES are coefficients of cosine and sine functions, whereas in Holt–Winters the seasonal terms are indices of the typical level of the series each period. However, if the seasonal pattern is multiplicative, it is necessary to multiply each cosine or sine term by the trend, thus making the model extremely complex. This may well be the reason why only additive seasonal patterns are typically used in practice, unlike the Holt–Winters models in which multiplicative seasonality is the standard assumption. For additive seasonal patterns, GES has an advantage over Holt–Winters in that the seasonal terms are revised with each observation, thus making it more responsive to changing seasonal patterns, as well as having only one parameter to determine. For detailed discussion of GES models and their implementation, see Brown (1963), Abraham and Ledolter (1983), Sweet (1981) and E. McKenzie (1976, 1984), and for extensions requiring two discount factors, see Harrison (1965) and Ameen and Harrison (1984).

McKenzie defines an equivalent ARIMA model to any GES model as that process for which the MMSE forecasts are identical to the GES forecasts for all lead times. By examining the structure of GES forecasts, he shows that such forecasts are optimal, in a MMSE sense, for the ARIMA process

$$\phi_p(B)\, x_t = \theta_p(wB)\, a_t, \tag{10.22}$$

in which $\theta(B)$ is the characteristic polynomial of the matrix L in equation (10.21) and $\phi_k = \theta_{p-k}/\theta_p, k = 1, 2, \ldots, p$. In fact, $\phi(B)$ is the characteristic polynomial of the difference equation of order p whose solutions are the fitting functions contained in $\boldsymbol{f}(t)$. We say that (10.22) is the equivalent ARMA process for the GES model with matrix L.

Furthermore, if no real exponential terms appear amongst the fitting functions then $\phi(B) = \theta(B)$, and since such terms are rarely used for modelling economic time series, the equivalent ARMA process of most interest is

$$\phi_p(B)\, x_t = \phi_p(wB)\, a_t. \tag{10.23}$$

As an example, suppose our GES forecast equation is

$$f_{n,h} = b_1^{(n)} + b_2^{(n)}\, h + b_3^{(n)} \sin(\lambda h) + b_4^{(n)} \cos(\lambda h).$$

The fitting functions are, therefore, $f_1^{(h)} = 1$, $f_2^{(h)} = h$, $f_3^{(h)} = \sin(\lambda h)$ and $f_4^{(h)} = \cos(\lambda h)$. Clearly, $(1-B)f_1^{(h)} = 0, (1-B)^2 f_2^{(h)} = 0, \{1-2\cos(\lambda)\, B + B^2\} f_i^{(h)} = 0$, for $i = 3$ and 4, and the equivalent ARMA process is

$$(1-B)^2\{1-2\cos(\lambda)\, B+B^2\}\, x_t$$
$$= (1-wB)^2\{1-2\cos(\lambda)\, wB + w^2B^2\}\, a_t.$$

From this general result, the following equivalencies may be established (E. McKenzie (1984), Abraham and Ledolter (1983)):

(i) For the 12-period sinusoidal model

$$f_{n,h} = b_1^{(n)} + b_2^{(n)} \sin\left(\frac{2\pi h}{12}\right) + b_3^{(n)} \cos\left(\frac{2\pi h}{12}\right)$$

the equivalent ARMA model is

$$(1-B)(1-\surd 3B+B^2)\, x_t = (1-wB)(1-w\surd 3B+w^2B^2)\, a_t.$$

(ii) For the exponential smoothing model

$$f_{n,h} = b_1^{(n)} + b_2^{(n)}\, h + b_3^{(n)} \sin\left(\frac{2\pi h}{12}\right) + b_4^{(n)} \cos\left(\frac{2\pi h}{12}\right)$$
$$+ b_5^{(n)}\, h\sin\left(\frac{2\pi h}{12}\right) + b_6^{(n)}\, h\cos\left(\frac{2\pi h}{12}\right)$$

the equivalent ARMA model is

$$(1-B)^2(1-\surd 3B+B^2)^2 x_t = (1-wB)^2(1-w\surd 3B+w^2B^2)^2\, a_t.$$

(iii) For the trend free seasonal model

$$f_{n,h} = b_1^{(n)} + \sum_{i=2}^{s} b_i^{(n)}\, D_i,$$

where the D_i are $s-1$ seasonal indicators or dummies, the equivalent ARMA model is

$$\nabla_s x_t = (1-w^s B^s)\, a_t.$$

(iv) For the additive seasonal trend model

$$f_{n,h} = b_1^{(n)} + \sum_{i=2}^{s} b_i^{(n)} D_i + b_{s+1}^{(n)} h,$$

the equivalent ARMA model is

$$\nabla\nabla_s x_t = (1 - wB)(1 - w^s B^s) a_t.$$

10.4 Seasonal adjustment

In the previous section we have emphasised the decomposition of a time series into trend, seasonal and irregular components and considered specifying 'structural' models for them. Although ideas of decomposition have played a major role in the analysis of economic time series for many years, until recently most attention focused on the removal of the seasonal component, so that the 'seasonally adjusted' series could then be analysed. Given our standard decomposition

$$x_t = \mu_t + \psi_t + \varepsilon_t,$$

the seasonally adjusted series can be defined as

$$x_t^a = x_t - \hat{\psi}_t = \hat{\mu}_t + \hat{\varepsilon}_t. \tag{10.24}$$

Why do we wish to remove the seasonal component? Pierce (1980) states that it is because

> ...our ability to recognise, interpret, or react to important nonseasonal movements in a series (such as turning points and other cyclical events, emerging patterns, or unexpected occurrences for which possible causes are sought) is hindered by the presence of seasonal movements. (Pierce (1980, page 125)).

Bell and Hillmer (1984) expand on this theme, offering the following *possible* justification for seasonal adjustment:

> Seasonal adjustment is done to simplify data so that they may be more easily interpreted by statistically unsophisticated users without a significant loss of information. (Bell and Hillmer (1984, page 301)).

They regard this as only a possible justification because the validity of the phrase 'without a significant loss of information' cannot be established unless appropriate adjustment methods for a given time series can be found. Maravall (1984), apart from offering a personal financial motive, offers another justification for seasonal adjustment: in markets where supply is controlled, the removal of seasonality from prices requires knowledge of the seasonal variation that has to be induced in supply.

Seasonal adjustment procedures that have been found to be useful in practice can be categorised as either empirically-based or model-based. In either case, we see from (10.24) that seasonally adjusting a time series implicitly requires the estimation of all three unobserved components, or at least ψ_t and the nonseasonal component $n_t = \mu_t + \varepsilon_t$. We also note that seasonal adjustment is inherently different from forecasting. The latter, for which various structural models have been proposed in previous sections, only utilises past and present data. Seasonal adjustment, on the other hand, is aimed at obtaining good estimates of the underlying components using all the data: past, present and future (although, as we shall see, when future data are unavailable, optimal forecasts may be used in their place).

10.4.1 *Empirically-based methods*: *X*-11

As mentioned above, any seasonal adjustment procedure requires estimation of the trend, seasonal and irregular components. It was quickly discovered that, to estimate the seasonal component, the underlying trend movement of the series needed to be estimated and removed. Although a number of approaches to trend estimation have been considered, (see Bell and Hillmer (1984)), the method that gradually gained ascendancy was the use of moving averages whose weights were determined by actuarial graduation formulae to provide smooth, but not deterministic, trend curves. These moving averages are symmetric, taking the general form

$$\delta_m(B) = \delta_o + \sum_{j=1}^{m} \delta_j(B^j + B^{-j}), \tag{10.25}$$

and their application to the observed series x_t leads to the trend estimator

$$\hat{\mu}_t = \delta_m(B)\, x_t.$$

Usually we impose the restriction that the δ-weights sum to unity, i.e.

$$\delta_o + 2\sum_{j=1}^{m} \delta_j = 1.$$

An important example of $\delta_m(B)$ is the Henderson moving average (HMA), which is defined by the requirement that the moving average should follow a cubic polynomial trend without distortion (see Kenny and Durbin (1982, pages 26–8) for a derivation of the HMA). HMAs are available for $m = 2$, 4, 6 and 11, and the δ_j weights associated with each of these settings are given in, for example, Abraham and Ledolter (1983, page 176). The HMA is important because of the central role it plays in the most widely used empirically-based seasonal adjustment procedure, the Bureau of the Census Method II–X–11 variant, known simply as

X–11 (see Shiskin et al. (1967) and, for detailed textbook discussions, Abraham and Ledolter (1983, chapter 4), Levenbach and Cleary (1981, chapter 18) and Makridakis et al. (1983, chapter 4)). This procedure consists largely (other features include possible adjustment for trading day variation and methods for dealing with outliers) of a set of symmetric moving averages applied to the series to estimate its seasonal component and hence, through (10.24), obtain x_t^a. The net effect of these moving average operations can be represented by a single, composite, moving average (see Young (1968) or Wallis (1974)). For a time period sufficiently far in the past then, following Burridge and Wallis (1984), the *final* or *historical* adjusted value $x_t^{(m)}$ is obtained by application of the symmetric filter

$$\alpha_m(B) = \alpha_o + \sum_{j=1}^{m} \alpha_j(B^j + B^{-j}),$$

yielding

$$x_t^{(m)} = \alpha_m(B)\, x_t = \sum_{j=-m}^{m} \alpha_{mj}\, x_{t-j}.$$

For current and recent observations this filter cannot be applied, and truncated asymmetric filters must be used:

$$x_t^{(k)} = \alpha_k(B)\, x_t = \sum_{j=-k}^{m} \alpha_{kj}\, x_{t-j}, k = 0, 1, \ldots, m. \tag{10.26}$$

For the filter $\alpha_k(B)$, the subscript k indicates the number of future values of x entering the moving average. Equivalently, $x_t^{(k)}$ is the adjusted value of x_t calculated from $x_{t-m}, x_{t-m+1}, \ldots, x_t, \ldots, x_{t+k}$, so that, for example, $x_t^{(o)}$ is the *first announced* or *preliminary* seasonally adjusted figure for x_t. For the standard X–11 filters, if a 13-term HMA is selected to estimate the trend component (which is the usual selection for monthly data), then $m = 84$. The α_{mj} weights for three X–11 filters of particular interest, $\alpha_0(B)$, $\alpha_{12}(B)$ and $\alpha_{84}(B)$, have been obtained by Wallis (1982) and are presented as Exhibit 10.6.

An important question concerning the X–11 procedure is, for what underlying models are the filters in (10.26) in some sense optimal? Although an analytical answer to this question has not, as yet, been found, Cleveland and Tiao (1976) and Burridge and Wallis (1984) have used search procedures based on signal extraction techniques (to be discussed below) to find models yielding filters that closely approximate $\alpha_k(B)$. We will concentrate attention on models that approximate $\alpha_{84}(B)$; Burridge and Wallis (1984) also provide models for the other filters shown in Exhibit 10.6. Thus, let us consider the decomposition of x_t into nonseasonal and seasonal components:

$$x_t = n_t + \psi_t, \tag{10.27}$$

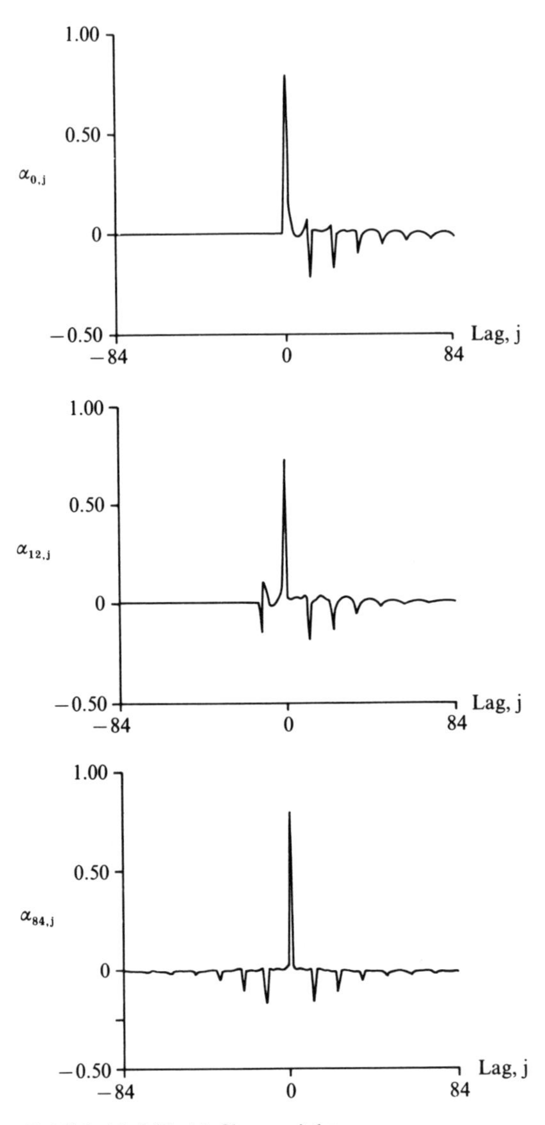

Exhibit 10.6 X–11 filter weights

and assume that the component models are of the following ARMA forms:

$$\Phi^n(B)\, n_t = \theta^n(B)\, u_t, \phi^\psi(B)\, \psi_t = \theta^\psi(B)\, v_t, \tag{10.28}$$

where u_t and v_t are uncorrelated normally distributed white noise series. Cleveland and Tiao (1976) found that the following models for n_t and ψ_t

Exhibit 10.7 *Moving average coefficients of X–11 filter approximations*

Lag	Equation (10.30)	Equation (10.32)
1	−0.34	−0.67
2	0.14	0.29
3	0.14	0.23
4	0.14	0.22
5	0.14	0.21
6	0.13	0.20
7	0.13	0.20
8	0.12	0.19
9	0.11	0.19
10	0.09	0.18
11	0.08	0.12
12	−0.42	−0.33
13	0.23	0.45
14	*	0.02
15	*	*
16	*	*
17	*	*
18	*	*
19	*	*
20	*	*
21	*	*
22	*	*
23	−0.01	−0.01
24	0.04	0.04
25	−0.02	−0.04
26		0.01

* coefficient less than 0.01 in absolute value.

provide a close approximation to the X–11 filter when monthly data are being analysed (see also Bell and Hillmer (1984)):

$$\nabla^2 n_t = (1 - 1.252B + 0.4385B^2)\, u_t$$

$$\nabla_{12}\, \psi_t = (1 + 0.64B^{12} + 0.83B^{24})\, v_t$$

$$\sigma_u^2/\sigma_v^2 = 24.5. \tag{10.29}$$

The models in (10.29) lead to a model for x_t of the form:

$$\nabla\nabla_{12}\, x_t = \theta_{25}(B)\, a_t, \sigma_a^2/\sigma_v^2 = 43.1, \tag{10.30}$$

where the coefficients of $\theta_{25}(B)$ are shown in Exhibit 10.7. The model

(10.30) for x_t has the conventional $\nabla\nabla_{12}$ operator used extensively in seasonal modelling, and its presence is undoubtedly related to the widespread success of the X–11 procedure. However, there is one disadvantage of the component models of (10.29). The seasonal and nonseasonal autoregressive operators $\phi^n(B) = \nabla^2 = (1-B)^2$ and $\phi^{\psi}(B) = \nabla_{12} = (1-B)(1+B+\ldots+B^{11})$ have a common factor of $(1-B)$, and Pierce (1979a) shows that, under such circumstances, the mean square error of estimating ψ_t is infinite.

Burridge and Wallis (1984) prefer to use the seasonal operator $U(B) = (1+B+\ldots+B^{11})$ in specifying their seasonal model (cf. the arguments of Roberts and Harrison (1984) and Harvey (1984) in favour of this operator), leading to their preferred model:

$$\nabla^2 n_t = (1-1.59B+0.86B^2)\,u_t$$

$$U(B)\,\psi_t = (1+0.71B^{12}+1.00B^{24})\,v_t$$

$$\sigma_u^2/\sigma_v^2 = 58.8. \qquad (10.31)$$

Again, the component models of (10.31) lead to a composite model for x_t of the form

$$\nabla\nabla_{12}\,x_t = \theta_{26}(B)\,a_t, \sigma_a^2/\sigma_v^2 = 90.9, \qquad (10.32)$$

the coefficients of $\theta_{26}(B)$ also being shown in Exhibit 10.7. The composite models (10.30) and (10.32) are similar in a number of respects. The coefficients at lags greater than 13 are all very small, but the intermediate coefficients at lags 2–11 are not, particularly for the Burridge and Wallis model (10.32). Thus, the common multiplicative moving average specification $(1-\theta B)(1-\Theta B^{12})$ can only be regarded as an approximation to $\theta_{25}(B)$, and an even less appropriate approximation to $\theta_{26}(B)$, particularly when it is noted that for this model θ_{13} is *greater* in absolute value than θ_{12}. As Burridge and Wallis (1984) emphasise, simple specifications for the component models typically do not yield simple composite models.

Burridge and Wallis (1984) also examine the robustness of the X–11 filters. Their findings may be summarised as follows:

(i) Increasing (decreasing) the variance ratio σ_u^2/σ_v^2 both lengthens (shortens) the filter and increases (decreases) the weight at lag 0, thus reflecting two effects: first, if the seasonal pattern changes rapidly, then observations at remote lags contain less information relevant to its estimation, and, second, as the relative contribution of n_t to the overall variation of the series is reduced, so α_{om} is driven towards zero. This suggests that X–11 will perform relatively poorly for composite models with moving average specifications of an order higher than that of $(1-\theta B)(1-\Theta B^{12})$.

(ii) Variations in the nonseasonal moving average parameters have the greater effect on the composite moving average $\theta_{26}(B)$, suggesting that the appropriateness of X–11 for a series adequately represented by the ARIMA $(0, 1, 1)(0, 1, 1)_{12}$ specification would depend on the similarity of its key parameters to those given in Exhibit 10.7.

Example 10.5: Seasonal adjustment of retail sales volume series by X–11

The volume of retail sales series was found to be adequately modelled by an airline ARIMA $(0, 1, 1)(0, 1, 1)_4$ model. From the above discussion, the quarterly version of X–11 should provide an adequate decomposition of the series into trend, seasonal and irregular components. The resultant decomposition is shown in Exhibit 10.8. The trend component reveals a fairly smooth upward trend throughout the sample period, although cyclical fluctuations are apparent and the recessions of the mid-1970s and early-1980s are clearly marked. The seasonal component, although fairly stable, does show an evolving pattern consistent with the seasonal moving average parameter being large, but not too close, to unity. The distinctive feature of the irregular component is the 'inverted-V' shaped triple in 1979, reflecting the increase in VAT in the March 1979 Budget, which produced lower than expected values in the first and third quarters. The seasonally adjusted series, not shown, is very similar to the trend component since the irregular component is very small relative to the level of the series.

Example 10.6: Seasonal adjustment of bank lending by X–11

The bank lending series discussed earlier required a model considerably different to that needed for optimal adjustment by X–11, needing only a ∇_{12} operator to induce stationarity but a moving average polynomial of order 15. The X–11 decomposition of this series produced an irregular component having significant autocorrelations at the first four lags and a corresponding portmanteau statistic of $Q(24) = 49.4$, the marginal significance level being 0.002. Since, by assumption, X–11 should produce a white noise irregular component if its decomposition is optimal, this example illustrates the potential disadvantages of using X–11 to seasonally adjust series having ARIMA representations markedly different to that of (10.30) or (10.32).

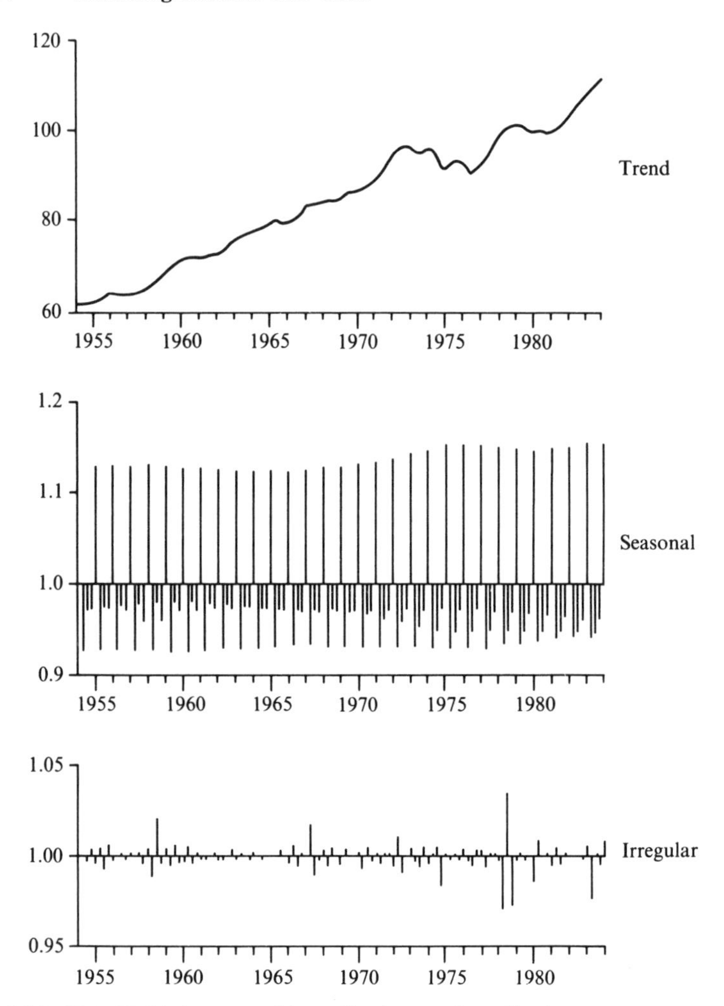

Exhibit 10.8 X–11 decomposition of volume of retail sales

10.4.2 *Extensions of X–11*

Apart from producing optimal seasonal adjustments for only a relatively small class of ARIMA models, an important drawback of X–11 is its use of asymmetric moving averages at the end of the time series. The consequence of this is that sometimes large revisions of the recently adjusted data are necessary when new observations become available. The main proposal to correct this drawback has been to replace $\alpha_0(B)$, for example, by $\alpha_{84}(B)$ applied to the *extended* time series, this being the available series augmented by a set of forecasts. Typically, these forecasts have been conditional on the observed series, assumed to follow an ARIMA process. Such a procedure provides a MMSE forecast of the final seasonal component and hence minimises the mean square of the revision (Pierce (1980)). This extended procedure is called X–11–ARIMA by Dagum (1978), who develops a small family of ARIMA specifications that are automatically selected from. Details of this method and of further extensions to X–11 may be found in, for example, Moore et al. (1981), Dagum and Morry (1984) and S. K. McKenzie (1984). A number of other empirically based seasonal adjustment procedures are in use in various countries: Burman (1979) provides a survey. An important new procedure, discussed in detail in Levenbach and Cleary (1981, chapter 19), is SABL (Cleveland et al., 1978).

10.4.3 *Model-based methods: signal extraction*

In the discussion of the optimality of X–11, we stated that Cleveland and Tiao (1976) and Burridge and Wallis (1984) employed the technique of signal extraction to obtain the various ARIMA models for the observed series and its components. Signal extraction lies at the heart of model-based methods of seasonal adjustment, and is developed in Burman (1980), Box et al. (1978), Hillmer and Tiao (1982), Hillmer, Bell and Tiao (1983), Pierce (1978) and Tiao and Hillmer (1978), all of which use the theoretical foundations provided by Whittle (1983) and extended by Pierce (1979a) and Bell (1984).

Thus, consider again the ARIMA models for the components n_t and ψ_t:

$$\phi^n(B)\, n_t = \theta^n(B)\, u_t \quad \text{and} \quad \phi^\psi(B)\, \psi_t = \theta^\psi(B)\, v_t,$$

where we now explicitly assume that the pairs $\{\phi^n(B), \theta^n(B)\}$, $\{\phi^\psi(B), \theta^\psi(B)\}$ and $\{\phi^n(B), \phi^\psi(B)\}$ have no common factors. The model for x_t is then given by

$$\phi(B)\, x_t = \theta(B)\, a_t \tag{10.33}$$

where

$$\phi(B) = \phi^n(B)\, \phi^\psi(B),$$

and where $\theta(B)$ and σ_a^2 are determined from the equation

$$\sigma_a^2 \frac{\theta(B)\, \theta(B^{-1})}{\phi(B)\, \phi(B^{-1})} = \sigma_u^2 \frac{\theta^n(B)\, \theta^n(B^{-1})}{\phi^n(B)\, \phi^n(B^{-1})} + \sigma_v^2 \frac{\theta^\psi(B)\, \theta^\psi(B^{-1})}{\phi^\psi(B)\, \phi^\psi(B^{-1})}.$$

If all roots of $\phi^n(B) = 0$ and $\phi^\psi(B) = 0$ lie on or outside the unit circle, then the estimated components are given by

$$\hat{n}_t = \pi^n(B)\, x_t \quad \text{and} \quad \hat{\psi}_t = \pi^s(B)\, x_t,$$

where

$$\pi^n(B) = \frac{\sigma_u^2 \theta^n(B)\theta^n(B^{-1})}{\sigma_a^2 \theta(B)\theta(B^{-1})} \phi^\psi(B)\phi^\psi(B^{-1})$$

and

$$\pi^s(B) = \frac{\sigma_v^2 \theta^\psi(B)\theta^\psi(B^{-1})}{\sigma_a^2 \theta(B)\theta(B^{-1})} \phi^n(B)\phi^n(B^{-1}).$$

In practice, the n_t and ψ_t series are unobservable, so that without additional assumptions, the weight functions $\pi^n(B)$ and $\pi^s(B)$, and hence $\hat{n}_t$ and $\hat{s}_t$, cannot be computed. An estimate of the model for the observed series, (10.33), can be obtained though, and it is of interest to investigate to what extent the component models can be obtained from the observable x_t. In fact, the components n_t and s_t can be estimated uniquely if some prior restrictions are imposed. These restrictions are discussed in detail in Bell and Hillmer (1984).

The first of these is that the seasonal operator is taken as $\phi^\psi(B) = U(B)$, so that summing ψ_t over 12 consecutive months, for example, produces a zero mean stationary series (cf. Harvey (1984)). Since most seasonal adjustment procedures, including X–11, have this property built in to them, this seems to be a reasonable assumption to make. This particular choice of $\phi^\psi(B)$ enables us to obtain $\phi^n(B)$ as

$$\phi^n(B) = \phi(B)/\phi^\psi(B) = \phi(B)(1-B)/(1-B^s).$$

Typically, $\phi(B)$ will be of the form

$$\phi(B) = \phi^*(B)(1-B)^d(1-B^s)$$

so that

$$\phi^n(B) = \phi^*(B)(1-B)^{d+1}.$$

The second assumption is that the order of $\theta^\psi(B)$ is at most $s-1$. This implies that the forecast function of the model for ψ_t follows a fixed annual pattern that sums to zero over s consecutive months. If the order of $\theta^\psi(B)$ exceeds $s-1$, then the forecast function will change its annual pattern, but in such circumstances the forecastable change in the seasonal pattern should be part of the trend and hence be included in n_t.

Given these assumptions, Hillmer and Tiao (1982) show that σ_v^2 must lie in some known range $[\bar{\sigma}_v^2, \tilde{\sigma}_v^2]$ and that the models for ψ_t and n_t are uniquely determined once the choice of σ_v^2 is made. A decomposition corresponding to a σ_v^2 in this range is called an *admissable* decomposition, with corresponding admissable seasonal and nonseasonal components. If σ_v^2 is chosen so that $\sigma_v^2 = \bar{\sigma}_v^2$, then this is the *canonical* decomposition and the corresponding seasonal component, $\bar{\psi}_t$, is then the canonical seasonal component. Hillmer and Tiao (1982) then demonstrate that choosing $\sigma_v^2 = \bar{\sigma}_v^2$ minimises the variance of $U(B)\psi_t$, thus making the seasonal

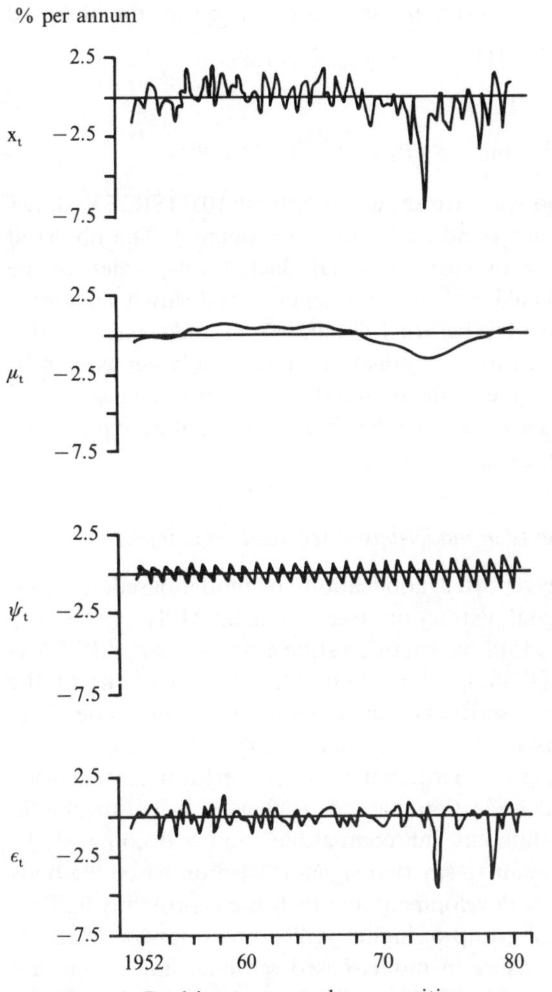

Exhibit 10.9 Real interest rate decomposition

pattern as stable as possible. Any other choice of σ_v^2 leads to a seasonal component that is the sum of $\bar{\psi}_t$ and white noise.

Example 10.7: An example of seasonal adjustment by signal extraction

Burman (1980) has developed a signal extraction seasonal adjustment procedure, based on the above assumptions, which is known as SIGEX (or MSX). Mills (1982) uses the SIGEX methodology to decompose the ex post real return on Treasury bills for the data period

1952 Q1 to 1981 Q4. The following model was estimated for the observed series

$$\nabla\nabla_4 x_t = (1-0.70B)(1-0.88B^4)\, a_t, \hat{\sigma}_a^2 = 1.08.$$

Thus, with $d = 1$ and $s = 4$,

$$\phi^n(B) = (1-B)^2 \quad \text{and} \quad \phi^\psi(B) = (1+B+B^2+B^3),$$

and the estimated components are shown in Exhibit 10.9 (SIGEX allows n_t to be separated out into trend and noise components). The observed series is rather erratic with clear seasonal fluctuations, whereas the extracted trend, the expected real return, is smooth and slowly changing, being predominantly positive before 1970 and from 1980 and negative during the 1970s. The seasonal component is slowly evolving, as is to be expected from the large value of the seasonal moving average parameter. The seasonally adjusted series can then be obtained by subtracting $\hat{\psi}_t$ from x_t or, equivalently, as $\hat{\mu}_t + \hat{\varepsilon}_t$.

10.4.4 *Recent developments in model-based seasonal adjustment*

There are relatively few reported applications of model-based seasonal adjustment through signal extraction (see, though, Mills (1983) and Bilongo and Carbone (1985)), but an interesting example using SIGEX is provided by Maravall (1986a), who investigates the behaviour of the Spanish import and export series. He finds that models which differ little in terms of fit or forecasts may provide remarkably different decompositions, so that, when signal extraction is to be performed, the usual criteria for ARIMA model selection may be insufficient: attention should also be paid to the way different, but compatible, models decompose the observed time series. It is fair to say that signal extraction based methods are still in an early stage of development and their use is probably limited, at the moment, to a small group of highly skilled practitioners. Many of the features and issues arising in model-based seasonal adjustment are addressed by Maravall (1985, 1986b) and Maravall and Pierce (1987), while recent developments in the still prevalent empirically based methods are discussed, along with a survey of the little used regression based methods, in Moore et al. (1981). The seasonal adjustment of weekly time series is discussed in Pierce et al. (1984) and problems associated with series exhibiting calendar variation, e.g. moving festivals such as Easter, are analysed in Bell and Hillmer (1983), Cleveland and Devlin (1982) and Liu (1986). Finally, the effect of measurement error on seasonal adjustment is analysed in Hausman and Watson (1985).

11 Further topics in univariate time series modelling

11.1 Introduction

In this chapter we introduce some extensions of the techniques discussed in previous chapters which we feel are of importance, or potential importance, to modellers of economic time series. The choice of topics is necessarily selective, and reflects our experience in modelling such time series, but is sufficiently broad as to consider a wide range of phenomena that economists are often confronted with.

The next section discusses the concepts of trend stationary and difference stationary processes, the importance of distinguishing between the two, and various methods of discrimination. The examples chosen are designed to illustrate the crucial role these processes play in business cycle modelling, both for currently observed values of time series and in analysing historical data. These ideas lead naturally on to unobserved component models and the decomposition of time series, ideas that we have met earlier but which are analysed further, in section 11.3, in the context of trend and cycle models. The analysis of such models is best carried out through their state space representation and the associated Kalman filter. These are discussed in section 11.4 and many of the models introduced in earlier chapters are shown to have natural state space representations.

Economic time series are often published in time-aggregated form and section 11.5 analyses the implications for ARIMA model specification of dealing with such data. Many series are also aggregates of component series, and section 11.6 examines this type of aggregation of ARIMA models. Section 11.7 considers the question of long run dependence in time series, and related concepts, while section 11.8 develops summary measures for fitted time series models analogous to the R^2 statistic so commonly used for assessing the goodness-of-fit of regression models. The final section briefly discusses the computer routines available on the major statistical packages for applying the techniques introduced in this part of the book.

11.2 Trend stationary and difference stationary processes

In previous chapters, we have normally used differencing to render a nonstationary time series stationary. Typically, for a nonseasonal economic time series, this leads to the model

$$\nabla x_t = \beta + \varepsilon_t, \tag{11.1}$$

where ε_t is a stationary, zero mean series admitting to an ARMA representation. Such a model is termed by Nelson and Plosser (1982) as being a member of the class of *difference stationary* (DS) processes.

An alternative method of modelling nonstationarity, also introduced in Chapter 6, is to express x_t as a deterministic function of time, known as the trend, plus a zero mean stationary series. For the linear trend case, this yields the class of *trend stationary* (TS) processes

$$x_t = \alpha + \beta t + u_t. \tag{11.2}$$

The two models are superficially similar, since accumulating changes in x from any initial value, x_0 say, yields

$$x_t = x_0 + \beta t + \sum_{i=1}^{t} \varepsilon_i. \tag{11.3}$$

This equation is fundamentally different from (11.2), however, for two reasons. Recalling the discussion in Chapter 6.2, the intercept in (11.3) is not a fixed parameter but depends upon the initial value x_0, and the disturbance is not stationary; rather, the variance and autocovariances depend on time. For example, if ε_t is white noise, so that x_t evolves as a random walk with drift, the variance is $t\sigma_\varepsilon^2$, where σ_ε^2 is the variance of ε_t. Thus, whereas the forecast errors of x_t from equation (11.2) are bounded no matter how far into the future forecasts are made because u_t has finite variance, forecasts made from (11.3) have an error variance that will increase without bound. Moreover, while autocorrelation in u_t can be exploited in making short-term forecasts, over long horizons the only information about a future x is its mean, $\alpha + \beta t$, so that neither current nor past events will alter long-term expectations. The long-term forecasts from a DS process, on the other hand, will always be influenced by historical events through the accumulation of the shocks, ε_t.

TS models of the form (11.2) have been a popular means of 'detrending' time series by regression on time (or perhaps a polynomial in time), the residuals then being interpreted as the cyclical component to be explained by business cycle theory. Nelson and Plosser (1982, page 140) provide references to studies that, implicitly or explicitly, regard residuals from fitted time trends as the appropriate data for business cycle analysis, while Crafts et al. (1989a, b) discuss the use of this process in estimating trend growth rates in historical contexts. In these applications, x_t is typically taken as the logarithm of the series in question, so that β can be regarded directly as the trend rate of growth.

Given this prevalence for using the TS model, are there any undesirable effects from using a member of this class of processes when, in fact, the time series we are examining belongs to the DS class? Building on the earlier work of Chan et al. (1977), Nelson and Kang (1981, 1984) consider the effects of such a misspecification in some detail. Their conclusions may be summarised as follows:

(1) Regressing the levels of a series, generated by a random walk, on time by OLS will produce R^2 values of around 0.44 regardless of sample size when the series has zero drift ($\beta = 0$ in equation (11.1)). For random walks with drift, R^2 will increase with sample size, reaching unity in the limit regardless of the actual rate of drift of the series or its variability.

(2) Residuals obtained from regressing a random walk on time will have a variance that is, on average, only about 14% of σ_ε^2, the true stochastic variance of the series. This result holds regardless of sample size or the rate of drift of the series.

(3) The mean values of the sample autocorrelations of a 'detrended' random walk are a function of the sample size, n, being roughly $(1-10/n)$ at lag one, for example, and they are therefore purely artifactual. The SACF oscillates with a period of about $(2/3)n$, so the detrended series exhibits a completely spurious long cycle, this result being quite robust to ε_t exhibiting serial correlation.

(4) A conventional t statistic for the least squares coefficient of time is a very poor test for the presence of trend. Such tests lead to a rejection of the null hypothesis $\beta = 0$ in (11.2) in 87% of the cases for a sample size of 100 at a nominal 5% level when, in fact, the null hypothesis is true. Attempts to correct for residual serial correlation only partially alleviate the problem, for after a first order AR correction, the true null hypothesis would still be rejected at a nominal 5% level with 73% probability. The correct procedure is to take first differences, in which case the size of the test would be correct.

Given that modelling a DS process as a TS process is fraught with potential pitfalls, what dangers are attached to the converse misspecification? Very few it would seem, given the analysis of Plosser and Schwert (1977, 1978). The least squares estimator of β will still be unbiased and will have approximately a normal (or Student t) distribution. Although the efficiency of the estimator is reduced, this is not serious if the induced serial correlation in u_t is simultaneously modelled, even though there are some statistical difficulties in estimating a moving average error structure with one root equal to unity, as is implied by the overdifferencing required to move from (11.2) to (11.1).

It is obviously important, therefore, to ascertain whether an observed time series is a member of the TS or DS class of processes. Pierce (1975a) considers general models of the form

$$x_t = \sum_{j=0}^{m} \beta_t t^j + \frac{\theta(B)}{\nabla^d \phi(B)} a_t, \tag{11.4}$$

where the degree of the polynomial time trend, m, exceeds the degree of differencing, d, since polynomial trends of order d or less are subsumed within the usual class of ARIMA models (see the analysis of forecast functions in Chapter 7). If β_{kj} denotes the coefficient of t^{j-k} in the equation of the kth difference $\nabla^k x_t$ then, with $m-d=h>0$, we have, on dth differencing,

$$\nabla^d x_t = \beta_{dd} + \beta_{d,d+1} t + \ldots + \beta_{dm} t^h + \frac{\theta(B)}{\phi(B)} a_t \tag{11.5}$$

and, on mth differencing,

$$\nabla^m x_t = \beta_{mm} + \frac{\nabla^h \theta(B)}{\phi(B)} a_t. \tag{11.6}$$

It is therefore evident that sufficient differencing can eliminate both trend and homogeneous nonstationarity, but the result for $m > d$ is a stationary series which is *noninvertible*, since $h = m-d$ moving average roots equal unity. If the series is only differenced d times, so that (11.5) results, an invertible stochastic component is obtained, but the series is nonstationary because of its deterministic trend component of order h. Pierce (1975a) considers methods for identifying d and h based on an extension of the Box and Jenkins (1976) approach of considering the SACFs of x_t and its differences.

For our purpose we shall concentrate on the cases when $m=1$ and $d=0$, so that the TS form (11.2) is obtained with $u_t = (\theta(B)/\phi(B))\, a_t$, and $m=d=1$, yielding the DS form (11.1) with $\varepsilon_t = (\theta(B)/\phi(B))\, a_t$ and $\beta = \beta_1$. Pierce's (1975a) procedure involves computing the SACF of $\hat{u}_t = x_t - (\hat{\beta}_0 + \hat{\beta}_1 t)$, where $\hat{\beta}_0$ and $\hat{\beta}_1$ are OLS estimates of β_0 and β_1. If $\hat{u}_t$ is found to be nonstationary then x_t requires differencing, thus eliminating the deterministic linear trend and obtaining the DS process (11.1). On the other hand, if $\hat{u}_t$ is stationary, then x_t must be a TS process of the form

(11.2). Although straightforward to carry out, this procedure is subject to the usual caveats concerning the problem of how to decide whether a series is stationary based on the SACF alone. In conventional ARIMA model building, this has led to the development of the Dickey–Fuller (1979) set of unit root tests, and Nelson and Plosser (1982) have extended this procedure to test whether a series is a TS or DS process. In its simplest form, this extension takes the DS model as the null hypothesis, embodied in the regression

$$x_t = \alpha + \rho x_{t-1} + \beta t + v_t \tag{11.7}$$

as H_0: $\rho = 1, \beta = 0$. As in Chapter 8, under this null hypothesis the usual t-ratios are not distributed as Student's t, but the Dickey–Fuller τ statistic can again be used to test the null hypothesis $\rho = 1$ against the alternative $\rho < 1$. A joint test on ρ and β would seem to be required, but as Nelson and Plosser (1982) note, a test on ρ alone is sufficient given that a process with $\rho = 1$ and $\beta \neq 0$ is excluded from consideration, for it would be one in which ∇x_t follows a deterministic path, implying ever increasing ($\beta >$ 0), or ever decreasing ($\beta < 0$), changes in x_t and, indeed, it is for this reason that quadratic or higher order trends are excluded from both (11.4) and (11.7). Standard t-ratio testing procedures are, in fact, strongly biased towards finding stationarity around a trend, tending to reject the hypothesis $\rho = 1$ when it is true in favour of $\rho < 1$, and tending to reject the hypothesis $\beta = 0$ when it is true.

In practice, the simulation evidence of Schwert (1987) again suggests that the best test to use is the τ statistic computed from the extended regression

$$x_t = \alpha + \rho x_{t-1} + \beta t + \sum_{i=1}^{p} \phi_i \nabla x_{t-i} + v_t, \tag{11.8}$$

where p is chosen by either of the rules given in Chapter 8. Indeed, Schwert (1987) suggests that (11.8) should be used when just testing for unit roots in a time series, since the inclusion of a time trend, even when $\beta = 0$, makes the distribution of $\hat{\rho}$ independent of α.

Numerous examples of this testing procedure using US macroeconomic data may be found in Nelson and Plosser (1982) and Schwert (1987), where the overall impression is that the vast majority of macroeconomic series are members of the DS class of processes, the notable exception being the unemployment rate. This has important implications for macroeconomic modelling for, if an observed series is DS, then its growth component must itself be a nonstationary stochastic process rather than the more generally assumed deterministic trend. Instead of attributing all variation in a series to changes in the cyclical component, the DS model allows for contributions from the variations in both components. Thus,

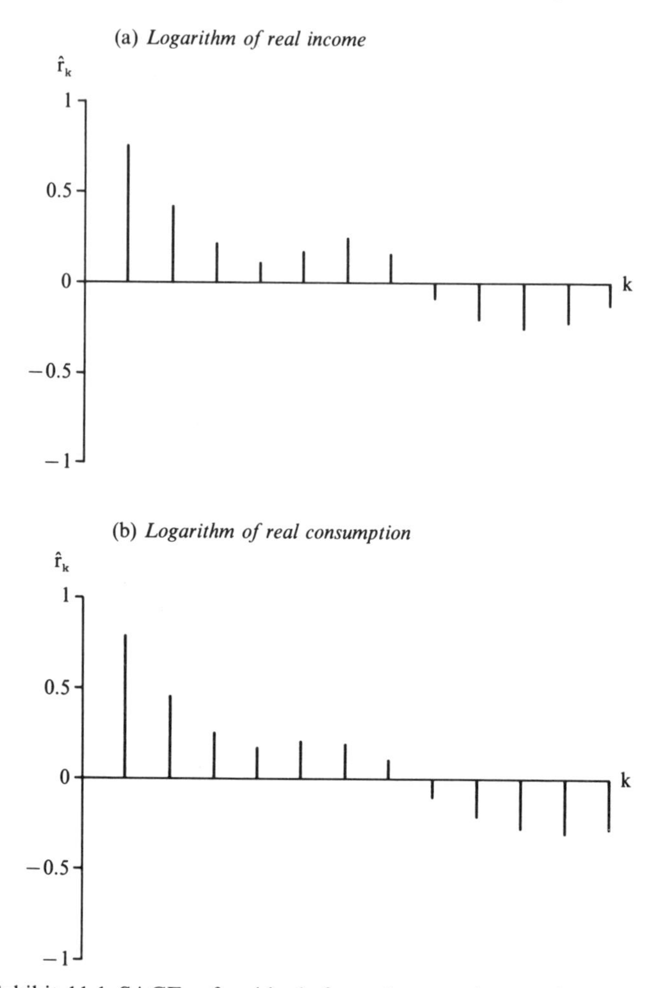

Exhibit 11.1 SACFs of residuals from time trend regressions

empirical analyses of business cycles based on residuals from fitted trend lines are likely to confound the two sources of variation, in general overstating the magnitude and duration of the cyclical component and understating the importance of the growth component.

Example 11.1: Examples of testing whether series are TS or DS processes

The real income and consumption series analysed in previous chapters were examined to ascertain whether they were members of the TS or DS class of processes. Regressing the logarithms of both series on time

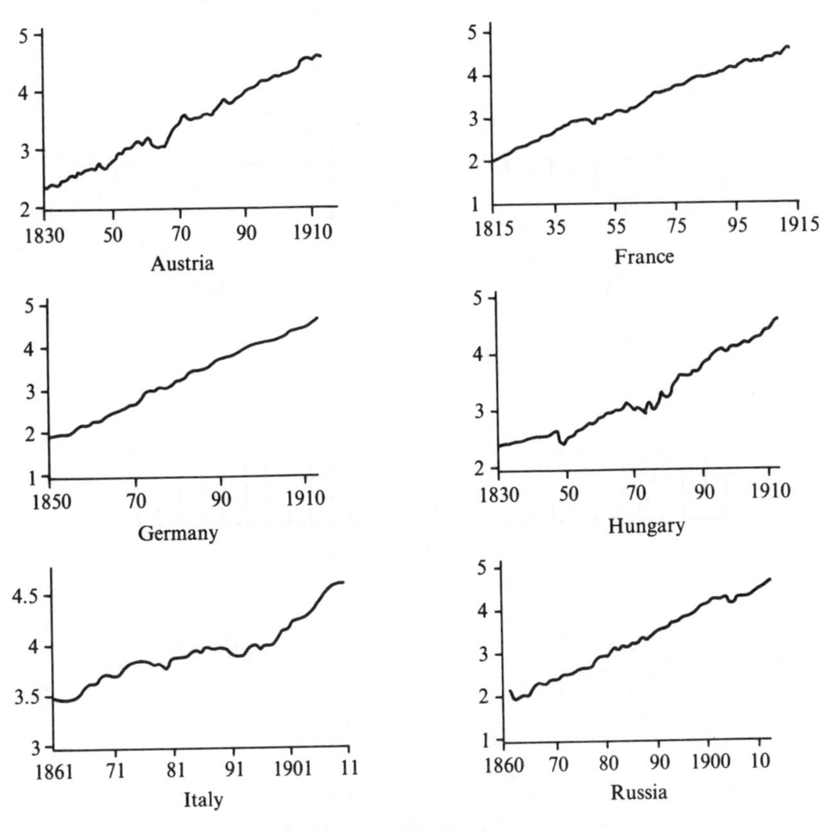

See Crafts et al. (1989b) for data sources

Exhibit 11.2 Logarithms of industrial production for nineteenth-century Europe

yielded the residual SACFs shown in Exhibit 11.1. Neither SACF reveals obvious nonstationarity, but bearing in mind the rather small sample size of 36, the regression (11.8) was estimated for both series. The lag order p was set at two, both because of the limited sample size and since ARIMA (2, 1, 0) models have been selected for each series. The estimated regressions were, for real income,

$$\underset{}{x_t} = \underset{(1.602)}{1.823} + \underset{(0.146)}{0.837}x_{t-1} + \underset{(0.004)}{0.004}t + \underset{(0.184)}{0.389}\nabla x_{t-1} - \underset{(0.205)}{0.240}\nabla x_{t-2},$$

and, for real consumption,

$$x_t = \underset{(1.455)}{1.726} + \underset{(0.133)}{0.845}x_{t-1} + \underset{(0.003)}{0.003}t + \underset{(0.173)}{0.410}\nabla x_{t-1} - \underset{(0.201)}{0.294}\nabla x_{t-2}.$$

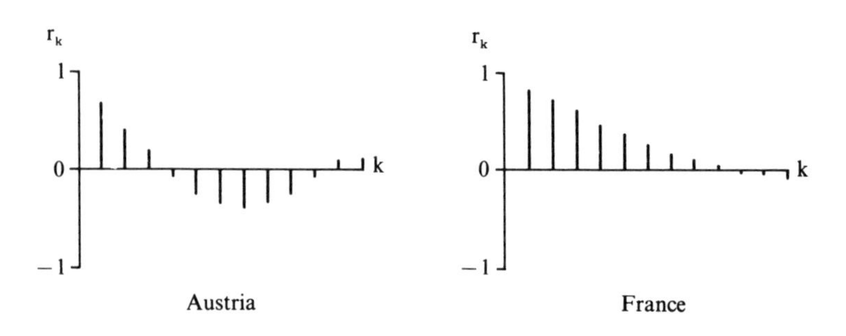

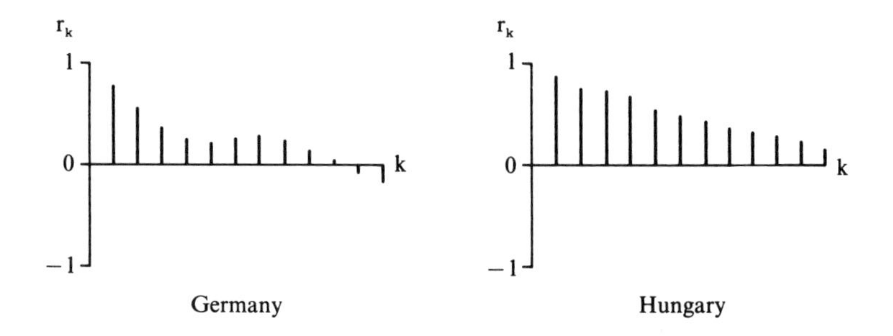

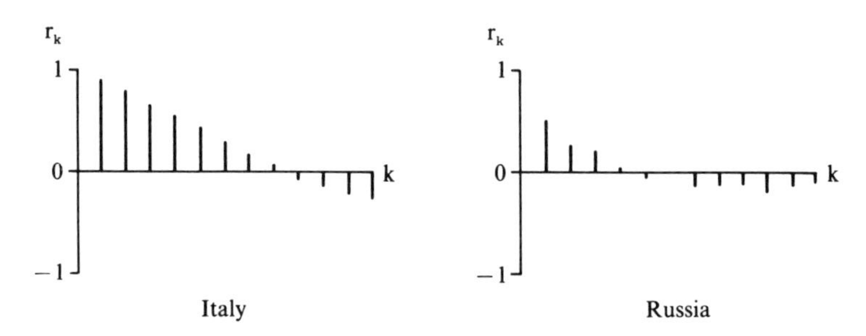

Exhibit 11.3 SACFs of residuals from time trend regressions

Both τ statistics, calculated to be -1.12 and -1.17 respectively, are clearly insignificant, implying that the two series are indeed DS processes, notwithstanding the ambiguous evidence from the above SACFs.

Example 11.2: Nineteenth-century European industrial production

Crafts et al. (1989b) examine the properties of a number of industrial production series for nineteenth-century Europe. Plots of the

Exhibit 11.4 *TS v DS tests for nineteenth-century European industrial production series*

Series	Sample period	τ_3
Austria	1830–1913	−4.63*
France	1815–1913	−3.32
Germany	1850–1913	−2.42
Hungary	1830–1913	−1.77
Italy	1861–1913	−1.31
Russia	1860–1913	−2.50

* Indicates significance at 0.05 level given the critical value = −3.44 (see Schwert (1987)).

logarithms of some of these series are shown in Exhibit 11.2, and all show a distinct upward trend. The important question is, however, whether they are DS or TS processes, for the appropriate separation of the observed series into secular growth and cyclical components is of great importance in answering questions concerning the onset of the industrial revolution and the origin of the trade cycle in the different countries.

Exhibit 11.3 shows the SACFs of the residuals obtained by regressing each series on a time trend. For France, Hungary and Italy the SACFs are suggestive of nonstationarity, implying that these series are DS processes, while for Austria, Germany and Russia the SACFs suggest stationary residuals and series that are therefore TS processes. Nonetheless, such conclusions are to some extent subjective and, as we have seen in Example 11.1, it is important to supplement them by formal tests. Exhibit 11.4 lists the τ_3 statistics obtained by estimating the regression (11.8), with p set equal to 3, for each series. These statistics confirm that Hungary, France and Italy are indeed DS processes, and that Austria is a TS process, but fail to reject the hypothesis that Germany and Russia are DS processes, notwithstanding the visual evidence provided by the SACFs.

11.3 Unobserved component models and decompositions of time series

If a time series is a member of the DS class of processes then, as stated earlier, its trend or secular growth component will also be of the DS class. This is most easily shown by considering an unobserved component decomposition of x_t as

$$x_t = \bar{x}_t + c_t, \tag{11.9}$$

where $\bar{x}_t$ is a secular component and c_t is a cyclical component. If this cyclical component is assumed to be transitory, i.e. stationary, then any underlying nonstationarity in x_t must be attributable to the secular component. Hence, if x_t is a member of the DS class, then so must $\bar{x}_t$. Thus, if $\nabla\bar{x}_t = \theta(B)\, v_t$ and $c_t = \lambda(B)\, u_t$, we have

$$x_t = \nabla^{-1}\theta(B)\, v_t + \lambda(B)\, u_t, \tag{11.10}$$

so that nonstationarity in x_t is assigned to $\bar{x}_t$ though the factor $\nabla^{-1} = (1+B+B^2+\dots)$. Separation of the secular component from the observed data may thus be thought of as a signal extraction problem when only information on the observed series is used.

In the set up of equation (11.10), specific models for the components of x_t are assumed. Thus, prior to the adoption of a particular UC model, such a formulation should be investigated for consistency with the data or, alternatively, a class of models should be identified from observed sample autocorrelations. The classic example of signal extraction in economics is the permanent income model of Friedman (1957), whose formal statistical foundations were investigated by Muth (1960). The Friedman–Muth model may be written as a special case of (11.10) with $\theta(B) = \lambda(B) = 1$:

$$x_t = \bar{x}_t + u_t, \bar{x}_t = \mu + \bar{x}_{t-1} + v_t, \tag{11.11}$$

where the secular component is generated by a random walk with drift μ, having innovations v_t, and the cyclical component is a purely random series, assumed to be independent of v_t. (Note the similarity of this model to the updating equations of simple exponential smoothing.) From (11.11) it follows that ∇x_t is a stationary process

$$\nabla x_t = \mu + v_t + u_t - u_{t-1}, \tag{11.12}$$

illustrating the fact that differencing does not 'remove the trend', since part of ∇x_t is the innovation in the secular component, v_t. ∇x_t has an ACF that cuts off at lag one with coefficient

$$\rho_1 = -\frac{\sigma_u^2}{\sigma_u^2 + 2\sigma_v^2}, \tag{11.13}$$

where σ_u^2 and σ_v^2 are the variances of u_t and v_t respectively. It is clear that $-0.5 \leqslant \rho_1 \leqslant 0$, the exact value depending on the relative sizes of these variances.

In general, if a UC version of (11.10) is restricted a priori by assuming that $\bar{x}_t$ is a random walk and that v_t and u_t are independent, then the parameters of the UC model will be identified. This is clearly the case for the model (11.11) since σ_u^2 can be estimated by the lag one autocovariance of ∇x_t (the numerator of (11.13)) and σ_v^2 from the variance of ∇x_t (the denominator of (11.13)) and the estimated value of σ_u^2. This example illustrates, however, that a decomposition satisfying both of the above restrictions may not always be feasible, for it is unable to account for a positive first order autocorrelation in ∇x_t. To do so requires relaxing either the assumption that $\bar{x}_t$ is a random walk, so that the secular component can contain both permanent and transitory movements, or the assumption that v_t and u_t are independent. If either of these assumptions is relaxed, the parameters of the UC model (11.11) will not be identified.

Nelson and Plosser (1982) find that many of the series they investigate have just a positive lag one autocorrelation in their first differences, but they go on to show that the assumption that the cyclical component is stationary, combined with positive lag one autocorrelations in ∇x_t (all others being zero), is sufficient to imply that the variation in ∇x_t is dominated by changes in the secular component $\bar{x}_t$ rather than the cyclical component c_t, i.e. that if $\rho_1 > 0$ then $\sigma_v^2 > \sigma_u^2$.

The assumption that the secular component is a random walk is, in fact, not as restrictive as it may first seem. Recall from Chapter 6 that Wold's decomposition implies that $w_t = \nabla x_t$ can be expressed as

$$w_t = \mu + a_t + \psi_1 a_{t-1} + \ldots, \tag{11.14}$$

where μ is the mean of w_t and $\{a_t\}$ is a white noise series with zero mean and constant variance σ_a^2. If we now denote the expectation of x_{t+k}, conditional on data for x up to and including time t, as $\hat{x}_t(k)$, then

$$\begin{aligned} \hat{x}_t(k) &= E(x_{t+k} \mid x_t, x_{t-1}, \ldots) \\ &= x_t + E(w_{t+1} + \ldots + w_{t+k} \mid w_t, w_{t-1}, \ldots) \\ &= x_t + \hat{w}_t(1) + \ldots + \hat{w}_t(k), \end{aligned} \tag{11.15}$$

since x_t can be expressed as an accumulation of past w's. From (11.14), the conditional expectation $\hat{w}_t(i)$ is given by

$$\begin{aligned} \hat{w}_t(i) &= \mu + \psi_i a_t + \psi_{i+1} a_{t-1} + \ldots \\ &= \mu + \sum_{j=0}^{\infty} \psi_{i+j} a_{t-j}, \end{aligned} \tag{11.16}$$

since future innovations a_{t+j}, $j = 1, \ldots, i$, are unknown but have zero expectation. Substituting (11.16) into (11.15) and gathering terms in each a_{t-j} yields

$$\hat{x}_t(k) = k\mu + x_t + \left(\sum_{j=1}^{k} \psi_j\right) a_t + \left(\sum_{j=2}^{k+1} \psi_j\right) a_{t-1} + \ldots. \tag{11.17}$$

Since w_t is stationary, each of the summations $\sum \psi_j$ converge (Box and Jenkins (1976, pages 49–50)), so that for large k we have approximately

$$\hat{x}_t(k) \simeq k\mu + x_t + \left(\sum_{j=1}^{\infty} \psi_j\right) a_t + \left(\sum_{j=2}^{\infty} \psi_j\right) a_{t-1} + \ldots.$$

Thus the forecast function is asymptotic to a linear function of the forecast horizon k, with slope equal to μ and a 'level' which is itself a stochastic process. Beveridge and Nelson (1981) interpret this level as the secular component, i.e.

$$\bar{x}_t = x_t + \left(\sum_{j=1}^{\infty} \psi_j\right) a_t + \left(\sum_{j=2}^{\infty} \psi_j\right) a_{t-1} + \ldots. \quad (11.18)$$

Taking first differences of (11.18) yields

$$\nabla \bar{x}_t = w_t + \left(\sum_{j=1}^{\infty} \psi_j\right) a_t - (\psi_1 a_{t-1} + \psi_2 a_{t-2} + \ldots),$$

which, given (11.14), reduces to

$$\nabla \bar{x}_t = \mu + \left(\sum_{j=0}^{\infty} \psi_j\right) a_t, \quad \psi_0 = 1. \quad (11.19)$$

Since a_t is white noise, the secular component is therefore a random walk with rate of drift equal to μ and an innovation equal to $(\sum_0^{\infty} \psi_j) a_t$. The variance of this innovation is $(\sum_0^{\infty} \psi_j)^2 \sigma_a^2$, which may be larger or smaller than σ_a^2, depending on the signs and pattern of the ψ's. In particular, these innovations in the secular component will be 'noisier', in this sense, than those of x_t if the ψ's are positive, which would typically be the case if the changes in x, the w's, are positively autocorrelated. For example, in the discussion of the Friedman–Muth model above, $\psi_1 > 0$, with all higher ψ's zero, so that the variation of the secular component is $(1+\psi_1)^2 \sigma_a^2 > \sigma_1^2$.

Equation (11.18) can equivalently be written as

$$\bar{x}_t = x_t + \lim_{k \to \infty} \{[\hat{w}_t(1) + \hat{w}_t(2) + \ldots + \hat{w}_t(k)] - k\mu\}, \quad (11.20)$$

so that the secular component can be interpreted as the current observed value of x plus all forecastable future changes in the series beyond the mean rate of drift. Bearing in mind the decomposition (11.9), Beveridge and Nelson (1981) regard the cyclical component of x_t as

$$c_t = \bar{x}_t - x_t = \lim_{k \to \infty} \{[\hat{w}_t(1) + \ldots + \hat{w}_t(k)] - k\mu\}$$

$$= \left(\sum_{j=1}^{\infty} \psi_j\right) a_t + \left(\sum_{j=2}^{\infty} \psi_j\right) a_{t-1} + \ldots, \quad (11.21)$$

the equivalence being apparent from (11.18). It is clearly the case that c_t will be a stationary process. However, the decomposition used by Beveridge and Nelson does force the secular and cyclical components to be *perfectly* correlated, which certainly does not correspond with conventional views about the behaviour of these components.

Nevertheless, we have shown that any time series whose first differences are stationary can be decomposed into a secular component that is a random walk, with the same rate of drift as the observed series and an innovation which is proportional to that of the observed series, and a stationary cyclical component representing the forecastable 'momentum' present at each time period but which is expected to be dissipated as the series tends to its 'permanent' level.

As an illustration, consider the ARIMA(0, 1, 1) model found adequately to characterise many macroeconomic time series:

$$w_t = \nabla x_t = \mu + a_t + \psi a_{t-1}. \tag{11.22}$$

From (11.19) the secular component is given by

$$\nabla \bar{x}_t = \mu + [(1+\psi)\, a_t], \tag{11.23}$$

while from (11.21) the cyclical component becomes

$$c_t = \psi a_t, \tag{11.24}$$

which is white noise and simply proportional to both the current innovations in x and the secular component of x. If $\psi = 0$, so that x itself evolves as a random walk, then $\bar{x}_t = x_t$ and $c_t = 0$, thus reflecting the fact that a random walk contains no forecastable momentum and no meaningful cycles.

The unobserved secular and cyclical components can be estimated in the following way. We assume that the Wold decomposition of (11.14) can be approximated by an ARMA(p, q) model:

$$w_t = \mu + \frac{(1-\theta_1 B - \ldots - \theta_q B^q)}{(1-\phi_1 B - \ldots - \phi_p B^p)} a_t. \tag{11.25}$$

As Cuddington and Winters (1987) point out, this allows (11.19) to be written as

$$\nabla \bar{x}_t = \mu + \frac{(1-\theta_1 - \ldots - \theta_q)}{(1-\phi_1 - \ldots - \phi_p)} a_t = \mu + \Omega a_t,$$

where $\Omega = (1-\sum \theta_i)/(1-\sum \phi_i)$. Following Miller (1988), (11.25) can be written as

$$\frac{(1-\phi_1 B - \ldots - \phi_p B^p)}{(1-\theta_1 B - \ldots - \theta_q B^q)} \Omega \nabla x_t = \mu + \Omega a_t,$$

so that

$$\bar{x}_t = \Omega \frac{(1-\phi_1 B - \ldots - \phi_p B^p)}{(1-\theta_1 B - \ldots - \theta_q B^q)} x_t.$$

Exhibit 11.5 *ARIMA models for nineteenth-century European industrial production series*

Series	Model
France	$\nabla x_t = 0.027 + \dfrac{(1+0.754B)}{(1+1.023B+0.357B^2)} a_t, \hat{\sigma}_a = 0.0356, Q_{12}(8) = 6.4$
Germany	$\nabla x_t = 0.043 + a_t, \hat{\sigma}_a = 0.0354, Q_{12}(11) = 11.6$
Hungary	$\nabla x_t = 0.027 + \dfrac{a_t}{(1+0.391B^2)}, \hat{\sigma}_a = 0.0648, Q_{12}(10) = 10.8$
Italy	$\nabla x_t = 0.022 + a_t, \hat{\sigma}_a = 0.0444, Q_{12}(11) = 14.0$
Russia	$\nabla x_t = 0.046 + (1-0.266B)\, a_t, \hat{\sigma}_a = 0.0789, Q_{12}(10) = 10.9$

Thus $\bar{x}_t$ is a weighted average of current and past values of x_t, the presence of Ω acting as a normalisation factor. Given $\bar{x}_t$, the cyclical component can then be obtained by residual as

$$c_t = \bar{x}_t - x_t.$$

Example 11.3: Decompositions for European industrial production series

Exhibit 11.5 presents the ARIMA models fitted to the five series found to be DS processes in Example 11.2 (i.e. all but Austria). Germany and Italy are both found to be random walks, so that there are no cyclical components, and secular growth is 4.3% and 2.2% per annum respectively. Russia is adequately modelled by an ARIMA (0, 1, 1) process, so that the cyclical component is white noise, given by $c_t = -0.266a_t$. The secular component ($\nabla \bar{x}_t$) is a random walk with a drift of 4.6% per annum and an innovation, given by $0.73a_t$, whose variance is $0.54\hat{\sigma}_a^2$. France and Hungary are modelled by ARIMA (2, 1, 1) and ARIMA (2, 1, 0) processes respectively. Given the parameter estimates, the innovations in the secular components are, therefore, $0.74a_t$ and $0.72a_t$ respectively, with variances $0.55\hat{\sigma}_a^2$ and $0.52\hat{\sigma}_a^2$. The similarity of the innovations to secular growth in Russia, France and Hungary is quite striking, given that, on the face of it, they are generated by rather different ARIMA processes.

The secular growth and cyclical components for France and Hungary, calculated using the procedure discussed above, are plotted in Exhibit 11.6. The growth series are white noise processes fluctuating about their mean rates of drift, whereas the cyclical components display some serial correlation while fluctuating around zero.

The above UC decomposition, comprising the 'structural' equations (11.11) and their 'updating' equations (11.19) and (11.21), may be

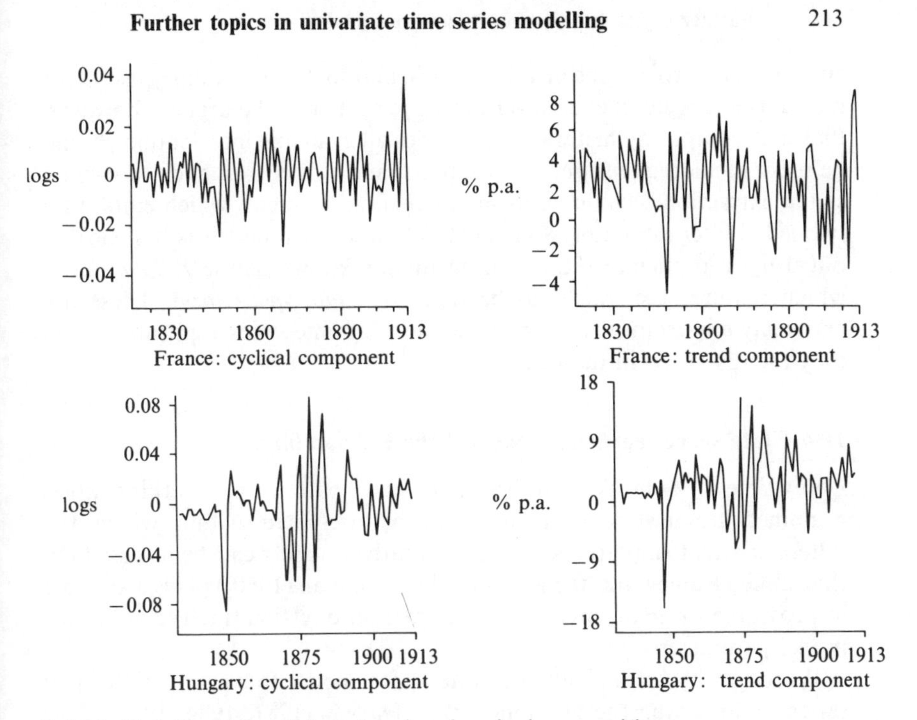

Exhibit 11.6 Beveridge–Nelson trend and cycle decompositions

regarded as an extension of the (constant trend) Holt–Winters procedure discussed in Chapter 9, and a similar model has been analysed by Watson (1986).

Harvey (1985) considers a more general formulation in which the drift term is not constant, but also evolves as a random walk (cf. the standard Holt–Winters model of Chapter 9). As we found in that chapter, such a model corresponds to requiring that $\nabla^2 x_t$ is stationary, thus ruling out the Beveridge–Nelson type decompositions developed above. Harvey (1985) also assumes explicit models for the cyclical component, either an AR (2) process, which is, of course, capable of generating pseudo-cyclical behaviour for certain parameter values, or an explicit model using cosine and sine functions (this model will be discussed in greater detail in the subsequent section).

As Watson (1986, page 55) shows, the Beveridge and Nelson (1981) procedure for decomposing a time series actually corresponds to an optimal one-sided (i.e. using only present and past data) estimator for the secular or trend component, given the random walk model for this component assumed in (11.11). As Beveridge and Nelson (1981) point out, the restriction to a one-sided estimator is very convenient for computing current trend projections in a time series, and so it is particularly useful for monitoring developments in business cycles. It may be argued, however, that in decomposing historical time series all available information should be used, including 'future' observations, and this corresponds to using a signal extraction technique, as in seasonal adjustment, which employs a two-sided filter estimator. Signal extraction in such models is best carried out using a particular estimation technique known as the *Kalman filter*, which requires the model to be recast in *state space form*. These are extremely important tools for a wide range of time series applications and they are discussed in the next section.

11.4 State space representations and the Kalman filter

Many time series models can be cast in state space form, and this enables a unified framework of analysis to be presented within which the differences and similarities of the alternative models can be assessed. In this unified framework, the Kalman filter (Kalman (1960)) plays a key role in providing optimal forecasts and a method of estimating the unknown model parameters.

We begin by setting out the state space model for a univariate time series x_t, following the development in Harvey (1981, 1984, 1987). This model consists of a *measurement equation*

$$x_t = z_t' \boldsymbol{\alpha}_t + \xi_t, \quad t = 1, 2, \ldots, T, \tag{11.26a}$$

and a *transition equation*

$$\boldsymbol{\alpha}_t = T_t \boldsymbol{\alpha}_{t-1} + R_t \boldsymbol{\eta}_t, \quad t = 1, 2, \ldots, T, \tag{11.26b}$$

in which $\boldsymbol{\alpha}_t$ is an $(m \times 1)$ *state* vector, z_t is an $(m \times 1)$ fixed vector, T_t and R_t are fixed matrices of orders $(m \times m)$ and $(m \times s)$, and ξ_t and $\boldsymbol{\eta}_t$ are, respectively, a scalar disturbance and an $(s \times 1)$ vector, which are distributed independently of each other. It will be assumed that $\xi_t \sim$ NID $(0, \sigma^2 g_t)$ and $\boldsymbol{\eta}_t \sim \text{NID}(0, \sigma^2 Q_t)$, where g_t is a fixed scalar, Q_t is a fixed $(m \times m)$ matrix, and σ^2 is a scalar.

Assuming that z_t, T_t, R_t, g_t and Q_t are known, let $\boldsymbol{a}_{t-1}$ be the optimal MMSE estimator of $\boldsymbol{\alpha}_{t-1}$ based on all the information up to and including time $t-1$, and let $\sigma^2 P_{t-1}$ be the MSE matrix of $\boldsymbol{a}_{t-1}$, i.e. the covariance matrix of $\boldsymbol{a}_{t-1} - \boldsymbol{\alpha}_{t-1}$. Given $\boldsymbol{a}_{t-1}$ and P_{t-1} at $t-1$, the MMSE of $\boldsymbol{\alpha}_t$ is given by the *prediction equations*

$$\boldsymbol{a}_{t/t-1} = T_t \boldsymbol{a}_{t-1} \tag{11.27a}$$

with associated covariance matrix

$$P_{t/t-1} = T_t P_{t-1} T_t' + R_t Q_t R_t'. \tag{11.27b}$$

This estimator can be updated when x_t becomes available. The appropriate *updating equations* are

$$\boldsymbol{a}_t = \boldsymbol{a}_{t/t-1} + P_{t/t-1} z_t (x_t - z_t' \boldsymbol{a}_{t/t-1})/f_t \tag{11.28a}$$

$$P_t = P_{t/t-1} - P_{t/t-1} z_t z_t' P_{t/t-1}/f_t \tag{11.28b}$$

with

$$f_t = z_t' P_{t/t-1} z_t + g_t. \tag{11.28c}$$

Together, the prediction (11.27) and updating (11.28) equations make up the Kalman filter.

Prior information on the state vector can be used to provide starting values $\boldsymbol{a}_0$ and P_0 for the Kalman filter, but in its absence the recursions can be started off at $t = 0$ by setting $\boldsymbol{a}_0 = 0$ and $P_0 = \kappa I$, where κ is a large, but finite, number. If, however, T_t, R_t and Q_t are time invariant and the roots of T_t are less than unity, the state vector is stationary and appropriate starting values are given by the unconditional mean and covariance matrix of $\boldsymbol{\alpha}_t$, the latter being obtained by solving, for $T_t = T$, $R_t = R$ and $Q_t = Q$,

$$P_0 = TP_0 T' + RQR'.$$

If h-step ahead forecasts are required, the prediction equations can be used repeatedly without the updating equations. Thus the MMSE of $\boldsymbol{\alpha}_{n+h}$, made at time n, is given by

$$\boldsymbol{a}_{n+h/n} = T_{n+h} \boldsymbol{a}_{n+h-1/n}, \quad h = 1, 2, \ldots, \tag{11.29}$$

with $\boldsymbol{a}_{n/n} = \boldsymbol{a}_n$. Similarly, $P_{n+h/n}$ may be computed as

$$P_{n+h/n} = T_{n+h} P_{n+h-1/n} T'_{n+h} + R_{n+h} Q_{n+h} R'_{n+h},$$

so that the MMSE forecast of x_{n+h} is

$$f_{n,h} = \hat{x}_{n+h/n} = z'_{n+h} \boldsymbol{a}_{n+h/n},$$

with associated MMSE

$$\text{MMSE}(\hat{x}_{n+h/n}) = (z'_{n+h} P_{n+h/n} z_{n+h} + g_{n+h}) \sigma^2.$$

The Kalman filter thus yields the MMSE of the state vector, $\boldsymbol{\alpha}_t$, given the information available at time t. Once all the observations (T) are available, a better estimator can usually be obtained by taking account of the observations obtained after time t. The technique for computing such 'two-sided' estimators is known as *smoothing*, and the usual method is to use a *fixed interval* smoother. This consists of a set of recursions which start with the final Kalman filter estimates, $\boldsymbol{a}_T$ and P_T, and work backwards through time. Denoting the smoothed estimator and its covariance matrix at time t by $\boldsymbol{a}_{t/T}$ and $P_{t/T}$, the smoothing equations are

$$\boldsymbol{a}_{t/T} = \boldsymbol{a}_t + P_t^* (\boldsymbol{a}_{t+1/T} - T_{t+1} \boldsymbol{a}_t) \tag{11.30a}$$

and
$$P_{t/T} = P_t + P_t^*(P_{t+1/T} - P_{t+1/t})\, P_t^{*\prime}, \tag{11.30b}$$
where
$$P_t^* = P_t\, T_{t+1}'\, P_{t+1/t}^{-1}, \quad t = T-1, T-2, \ldots, 1, \tag{11.30c}$$

with $\boldsymbol{a}_{T/T} = \boldsymbol{a}_T$ and $P_{T/T} = P_T$.

If there are unknown parameters within T_t, R_t, z_t, g_t and Q_t, then these can be estimated in the following way. The one-step ahead prediction errors obtained from running the Kalman filter are

$$v_t = x_t - z_t'\, \boldsymbol{a}_{t/t-1}, \quad t = 1, 2, \ldots, T. \tag{11.31}$$

The likelihood function of $(x_1, x_2, \ldots, x_T)$ can then be defined in terms of the *prediction error decomposition*

$$\Lambda = -\frac{T}{2}\ln(2\pi) - \tfrac{1}{2}\ln\sigma^2 - \frac{1}{2}\sum_{t=1}^{T}\ln f_t - \frac{1}{2\sigma^2}\sum_{t=1}^{T}\frac{v_t^2}{f_t}, \tag{11.32}$$

where Λ is the log-likelihood (see Harvey (1981, 1984)). The ML estimator of σ^2, conditional on the unknown parameters, is given by

$$\hat{\sigma}^2 = T^{-1}\sum_{t=1}^{T} v_t^2/f_t. \tag{11.33}$$

Substituting (11.33) into (11.32) yields the concentrated likelihood function, Λ_c, which may be written as

$$\Lambda_c = \sum_{t=1}^{T}\ln f_t + T\ln\hat{\sigma}^2. \tag{11.34}$$

(If k initial observations are used to calculate starting values for the Kalman filter, rather than setting $\boldsymbol{a}_0 = 0$ and $P_0 = \kappa I$, then all summations in equations (11.31)–(11.34) will run from k to T, and T will be replaced by $T-k$.) ML estimates of the unknown parameters can then be obtained by maximising Λ_c non-linearly with respect to them.

On first sight, the state space model, and associated Kalman filter, seems to have little connection with the range of time series models introduced in previous chapters. However, this lack of connection is illusory, as the following examples reveal:

(i) Consider the general ARMA(p, q) model, written in the form

$$x_t = \phi_1 x_{t-1} + \ldots + \phi_m x_{t-m} + \eta_t + \theta_1 \eta_{t-1} + \ldots + \theta_{m-1}\eta_{t-m+1} \tag{11.35}$$

where $m = \max(p, q+1)$. This can be written in the state space form (11.26) with transition equation

$$\boldsymbol{\alpha}_t = \left[\begin{array}{c|c} \phi_1 & \\ \phi_2 & \\ \vdots & I_{m-1} \\ \phi_{m-1} & \\ \hline \phi_m & O \end{array}\right] \boldsymbol{a}_{t-1} + \begin{bmatrix} 1 \\ \theta_1 \\ \vdots \\ \cdot \\ \theta_{m-1} \end{bmatrix} \eta_t, \tag{11.36}$$

so that $s = 1$, T_t and R_t are constant and $Q_t = \sigma^2$. Since $\boldsymbol{\alpha}_t = (\alpha_{1t}, \ldots, \alpha_{mt})'$, the original ARMA model (11.35) may be recovered from (11.36) by repeated substitution, starting at the bottom row of the system. The first element of $\boldsymbol{\alpha}_t$ is then identically equal to x_t, leading to the measurement equation

$$x_t = z_t' \boldsymbol{a}_t$$

with $z_t' = (1, 0_{m-1}')$ and $\xi_t = 0$.

(ii) The simple exponential smoothing model was shown in Chapter 9 to be equivalent, with appropriate changes in notation, to the structural, or UC, model

$$x_t = \alpha_t + \xi_t, \quad \xi_t \sim \text{NID}(0, \sigma^2)$$

$$\alpha_t = \alpha_{t-1} + \eta_t, \eta_t \sim \text{NID}(0, \sigma^2 q), \quad 0 \leqslant q \leqslant \infty,$$

for $t = 1, 2, \ldots, T$, which is in state space form with $T_t = R_t = m = s = 1$, q being the 'signal-to-noise' ratio. Harvey (1984) shows that the standard recursion for the EWMA forecasting procedure (equation (9.3)) is obtained as $T \to \infty$, since the Kalman filter then tends to a steady state in the sense that P_t becomes time invariant.

(iii) The Holt–Winters local linear trend model is equivalent to the structural model

$$x_t = \alpha_{1t} \quad + \xi_t$$

$$\alpha_{1t} = \alpha_{1t-1} + \alpha_{2t-1} + \eta_{1t}$$

$$\alpha_{2t} = \qquad\quad \alpha_{2t-1} + \eta_{2t}$$

and thus is in state space form with $z_t = (1, 0)$, $\alpha_t = (\alpha_{1t}, \alpha_{2t})'$, $R_t = I_2$, $m = s = 2$, and

$$T_t = \begin{bmatrix} 1 & 1 \\ 0 & 1 \end{bmatrix}.$$

Again, if a steady-state is reached as $T \to \infty$, the state updating and prediction equations, when combined, become equivalent to the updating formulae (9.10) and (9.11).

(iv) The addition to the Holt–Winters structural model of a seasonal component of the form (Harvey (1984))

$$\sum_{j=0}^{s-1} \alpha_{3t-j} = \eta_{3t}$$

leads to the state space model with measurement equation

$$x_t = (1, 0, 1, 0, 0)\, \boldsymbol{\alpha}_t + \xi_t \tag{11.37}$$

and transition equation, for quarterly observations,

$$\boldsymbol{\alpha}_t = \begin{bmatrix} \alpha_{1t} \\ \alpha_{2t} \\ \alpha_{3t} \\ \alpha_{3t-1} \\ \alpha_{3t-2} \end{bmatrix} = \left[\begin{array}{cc:ccc} 1 & 1 & & 0 & \\ 0 & 1 & & & \\ \hdashline & & -1 & -1 & -1 \\ & 0 & 1 & 0 & 0 \\ & & 0 & 1 & 0 \end{array}\right] \begin{bmatrix} \alpha_{1,t-1} \\ \alpha_{2,t-1} \\ \alpha_{3,t-1} \\ \alpha_{3,t-2} \\ \alpha_{3,t-3} \end{bmatrix} + \begin{bmatrix} \eta_{1t} \\ \eta_{2t} \\ \eta_{3t} \\ 0 \\ 0 \end{bmatrix}. \tag{11.38}$$

(v) The addition to the Holt–Winters structural model of a cyclical component of the form (Harvey (1985), Clark (1987), Crafts et al. (1989a))

$$\alpha_{3t} = \rho_1 \alpha_{3t-1} + \rho_2 \alpha_{3t-2} + \eta_{3t}$$

leads to measurement and transition equations of the same form as (11.37) and (11.38) except that the first row of the lower right hand partition of T_t contains the elements ρ_1, ρ_2 and 0.

Example 11.4: Structural models for French and Hungarian industrial production

The Holt–Winters structural model with an AR(2) cycle was fitted to the industrial production series for France and Hungary. For France, the structural parameters are estimated to be $\sigma^2(\eta_1) = 8.00 \times 10^{-4}$, $\sigma^2(\eta_2) = 0$, $\sigma^2(\eta_3) = 4.66 \times 10^{-5}$, $\rho_1 = -1.09$ and $\rho_2 = -0.80$. Since $\sigma^2(\eta_2)$ is zero, the trend growth rate (α_{2t}) is constant (equal to 2.71 % per annum) and the ρ_1 and ρ_2 parameters yield a cycle having a period of 6.9 years.

For Hungary, the estimates of the structural parameters are $\sigma^2(\eta_1) = 2.32 \times 10^{-3}$, $\sigma^2(\eta_2) = 4.68 \times 10^{-6}$, $\sigma^2(\eta_3) = 4.63 \times 10^{-4}$, $\rho_1 = -0.31$, $\rho_2 = -0.75$. Even though $\sigma^2(\eta_2)$ is small, trend growth is certainly stochastic, and is plotted as Exhibit 11.7. From this plot, the 'take-off' is seen to be at about 1848, with a levelling off of trend growth at 1890. The ρ_1 and ρ_2

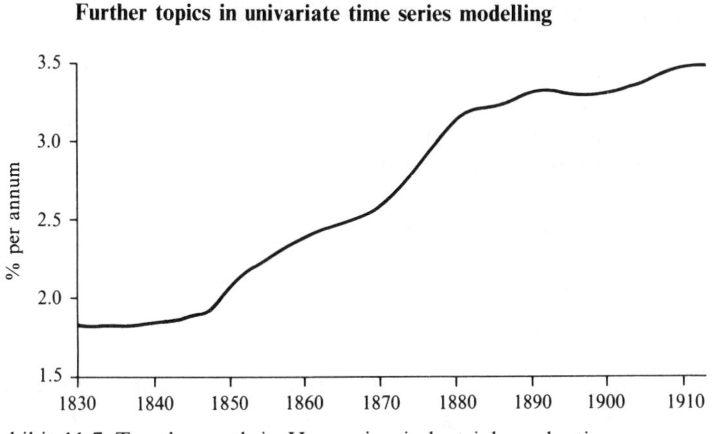

Exhibit 11.7 Trend growth in Hungarian industrial production

parameters again imply pseudo-cyclical behaviour, this time with a cycle having a 4.5 year period.

These results may be compared with those of the 'one-sided' Beveridge–Nelson decompositions of Example 11.3. The Kalman filter smoothing algorithm produces vastly different estimates of the components, and this is particularly apparent in the trend growth estimates. The smooth evolution of such components (indeed, a constant in the case of France) reflects the type of behaviour that economic historians would expect, and such decompositions thus provide invaluable empirical evidence for the analysis of historical time series.

Numerous other problems encountered when dealing with economic time series, such as missing and/or irregularly spaced observations, data revisions and benchmarks, may be dealt with within the state space-Kalman filter set up and are discussed in Harvey (1984). Of particular importance is the problem of temporal aggregation and this is discussed in some detail in the next section, although a more standard analytical approach is taken.

11.5 Temporal aggregation

In our analysis of economic time series, we have assumed so far that the sampling interval of the observed data, be it daily, weekly, monthly, quarterly, annual, etc., is the 'natural' interval to observe. As Brewer (1973) argues, such intervals seldom correspond to any 'causative agency' which might justify their use, and the choice of sampling interval is usually based on the availability of the data. Intuitively, it would seem preferable to have data in as disaggregated form as possible, for example monthly rather than quarterly, because this will yield more observations and, possibly, more information; although we should recognise that the 'natural' sampling interval is probably many times shorter than the actually observed interval between observations.

The question we consider in this section is, what effect does temporal aggregation have on the specification of ARIMA models? We begin by assuming that x_t is the time series observed at 'natural' times $t = 1, 2, \ldots, n$, and that it can be represented by our usual ARIMA (p, d, q) model

$$\phi_p(B)\,\nabla^d x_t = \theta_q(B)\,a_t. \tag{11.39}$$

Let us now suppose that, instead of these basic observations, data is only available in temporally aggregated form. By letting $t = mT$, where $m \geqslant 2$ is an integer, we may define a temporal aggregate as

$$z_T = x_{mT} + x_{mT-1} + \ldots + x_{m(T-1)+1} = \left(\sum_{i=0}^{m-1} B^i\right) x_{mT}. \tag{11.40}$$

Tiao (1972) restricts attention to IMA models ($\phi_p(B) = 1$ in (11.39)) and shows that, in these circumstances, z_T will follow an IMA (d, q^*) model, where

$$q^* = \text{Int}\,[d+1-(d+1-q)/m].$$

As $m \to \infty$, i.e. as the natural sampling interval becomes smaller, the limiting model for the aggregate series is the IMA (d, d) model, the parameters of which can be computed using the formulae given by Tiao (1972, page 528). If the IMA (d, q^*) model for x_T is written

$$(1-L)^d x_T = w_{q^*}(L)\,b_t,$$

where L operates on T and b_t is white noise with variance σ_m^2, then the efficiency of the forecast for the h-step ahead aggregate relative to the forecast obtained from the basic series may be measured by the ratio (Tiao (1972), Wei (1978))

$$\xi_h(m) = \frac{\sigma_m^2 \sum\limits_{j=0}^{m-1} \psi_{j,m}^2}{\sigma_a^2 \sum\limits_{j=0}^{hm-1} \left(\sum\limits_{i=0}^{m-1} \psi_{j-1}\right)^2},$$

where ψ_j is the coefficient of B^j in the expansion of $(1-B)^{-d}\theta_q(B)$ and $\psi_{j,m}$ is the coefficient of L^j in the expansion of $(1-L)^{-d}w_{q^*}(L)$. Tiao (1972) shows that, for $d = 0$, $\xi_h(m) = 1$ for all h when m is large, so that, if x_t follows a stationary moving average model, there is no gain by using it to forecast future z_T when the sampling interval is large relative to the natural interval. However, when the basic series is nonstationary ($d > 0$), substantial gains in efficiency can be achieved by calculating the forecast of z_{T+h} from the data on x_t when h is small. As h increases then, because $\xi_h(m)$ is a decreasing function of h, the advantage of using the basic series x_t becomes less and less. Indeed, as $h \to \infty$, $\xi_\infty(m) = 1$ for every d.

Example 11.5: Temporal aggregation with an IMA (1, 1) process

Consider the important special case when x_t is IMA (1, 1), $\nabla x_t = (1-\theta B)\,a_t$. In this case, for any $m \geqslant 2$, $q^* = 1$, and the temporally aggregated series z_T has the IMA (1, 1) form

$$(1-L)\,z_T = (1-\omega L)\,b_T.$$

Since ρ_1, the lag one autocorrelation of ∇x_t, must lie between -0.5 and $+0.5$, Tiao (1972) shows that the corresponding autocorrelation of $(1-L)\,z_T$, $\rho_1(m)$, must lie between ρ_1 and 0.25, and as $m \to \infty$, $\rho_1^{(m)} \to 0.25$ whatever the value of θ, and therefore ρ_1, in the basic model. This is a generalisation of Working's (1960) result that, for the random walk model for x_t, with $\theta = 0$, $\rho_1^{(m)} \to 0.25$ for large m. In terms of the moving average parameter, this corresponds to $\omega \to 0.268$. It is interesting to note that the ARIMA (0, 1, 1) model fitted to the Russian industrial production index in Exhibit 11.5 has an estimated coefficient of 0.266, and since the observed sampling interval is annual, m can be considered to be large, so that the model for the temporally disaggregated series (if it could have been observed) would be a random walk.

Tiao (1972) then shows that the relative efficiency of forecasts for this model is

$$\xi_h(\infty) = \left(\frac{2}{1+\omega^2}\right)\left\{\frac{1+(h-1)(1-\omega)^2}{1+3(h-1)}\right\}.$$

Since $\omega = 0.268$, $\xi_1(\infty) \simeq 1.866$, so that a substantial gain in efficiency can be achieved by calculating the forecast of z_{T+1} from the data on x_t.

Wei (1978) has extended these results to general multiplicative seasonal ARIMA models. Thus, if the basic series x_t follows an ARIMA (p, d, q) $(P, D, Q)_s$ process, the model for z_T will be ARIMA $(p, d, r)(P, D, Q)_S$, where $s = mS$ and

$$r = \mathrm{Int}\,[(p+d+1)+(q-p-d-1)/m].$$

This result is obtained assuming $m < S$, in which case aggregation affects the model structure through the nonseasonal component. When $m \geqslant S$, aggregation reduces a seasonal model to an ordinary ARIMA model. Exhibit 11.8, based on Abraham (1984), presents the models for aggregate time series corresponding to some common and simple models and further consequences of time aggregation may be found in Engle (1984) and Stram and Wei (1986).

Abraham (1984) investigates this methodology by using some actually

Exhibit 11.8 *Model structures for aggregated series*

	Basic series Nonseasonal		Aggregated series ($m = 3$) Nonseasonal
1	$(1-\phi B)x_t = a_t$	1	$(1-\lambda L)z_T = b_T$
2	$x_t = (1-\theta B)a_t$	2	$z_T = (1-\omega L)b_T$
3	$(1-\phi B)x_t = (1-\theta B)a_t$	3	$(1-\lambda L)z_T = (1-\omega L)b_T$
4	$\nabla x_t = (1-\theta B)a_t$	4	$\nabla z_T = (1-\omega L)b_T$
5	$(1-\phi B)\nabla x_t = a_t$	5	$(1-\lambda L)\nabla z_T = (1-\omega_1 L-\omega_2 L^2)b_T$
	Seasonal $s = 12$		Seasonal $S = 4$ ($m = 3$)
1	$(1-\Phi B^{12})x_t = a_t$	1	$(1-\Phi^* L^4)z_T = b_T$
2	$x_t = (1-\Theta B^{12})a_t$	2	$z_T = (1-\Theta^* L^4)b_T$
3	$\nabla\nabla_{12}x_t = (1-\theta B)(1-\Theta B^{12})a_t$	3	$\nabla\nabla_4 z_T = (1-\omega L)(1-\Theta^* L^4)b_T$
4	$(1-\phi B)\nabla_{12}x_t = (1-\Theta B^{12})a_t$	4	$(1-\lambda L)\nabla_4 z_T = (1-\omega L)(1-\Theta^* L^4)b_t$
5	$(1-\phi B)\nabla\nabla_{12}x_t = (1-\Theta B^{12})a_t$	5	$(1-\lambda L)\nabla\nabla_4 z_T = (1-\omega_1 L-\omega_2 L^2)(1-\Theta^* L^4)b_T$

observed aggregated and disaggregated time series. He finds that there is a tendency for empirical models of aggregates to be somewhat simpler than those dictated by theoretical specifications, and that the loss in forecast efficiency by using these empirically determined aggregate specifications is not large.

11.6 Component aggregation

As well as considering temporal aggregation, we may also consider aggregating component series to obtain an aggregate series. Following Granger and Morris (1976) (see also Anderson (1976, chapter 14) and Engel (1984)), if x_t follows an ARMA (p, q) process, we will denote this as $x_t \sim$ ARMA (p, q). Granger and Morris (1976) prove that if $x_t \sim$ ARMA (p_1, q_1) and $y_t \sim$ ARMA (p_2, q_2), x_t and y_t being independent processes, then $z_t = x_t + y_t \sim$ ARMA (p, q), where

$$p \leqslant p_1 + p_2 \quad \text{and} \quad q \leqslant \max(p_1+q_2, q_1+p_2), \tag{11.41}$$

a result which uses *Granger's lemma*: if $x_t \sim$ MA (q_1) and $y_t \sim$ MA (q_2), then $z_t = x_t + y_t \sim$ MA (q^*), where $q^* = \max(q_1, q_2)$. This result can be extended in a number of ways. The assumption of independence may be weakened to allow for contemporaneous correlation between the innovations of x_t and y_t. If x_t and y_t have k autoregressive roots in common, then the inequalities in (11.41) become

$$p \leqslant p_1 + p_2 - k \quad \text{and} \quad q \leqslant \max(p_1+q_2-k, q_1+p_2-k).$$

If we are dealing with integrated processes, then if $x_t \sim \text{ARIMA}(p_1, d_1, q_1)$ and $y_t \sim \text{ARIMA}(p_2, d_2, q_2)$, with $d_1 \geqslant d_2$, to use an obvious extended notation, then $z_t = x_t + y_t \sim \text{ARIMA}(p, d, q)$, where

$$p \leqslant p_1 + p_2 - k, \quad d = d_1,$$

and

$$q \leqslant \max(p_1 + q_2 - k + d_1 - d_2, q_1 + p_2 - k).$$

With these results, the following special cases emerge:

(i) If $x_t \sim \text{AR}(p)$ and y_t is white noise, then $z_t \sim \text{ARMA}(p, p)$.

(ii) If $x_t \sim \text{AR}(p)$ and $y_t \sim \text{AR}(q)$, then $z_t \sim \text{ARMA}(p+q, \max(p, q))$; in particular, if $x_t \sim \text{AR}(1)$ and $y_t \sim \text{AR}(1)$, then $z_t \sim \text{ARMA}(2, 1)$.

(iii) If $x_t \sim \text{ARMA}(p, q)$ and y_t is white noise, then $z_t \sim \text{ARMA}(p, p)$ if $p > q$, and $z_t \sim \text{ARMA}(p, q)$ if $p < q$.

(iv) If $x_t \sim \text{AR}(p)$ and $y_t \sim \text{MA}(q)$, then $z_t \sim \text{ARMA}(p, p+q)$.

Granger and Morris (1976) argue that if it is believed that component time series are generated by either pure AR or MA processes, then aggregate series will very likely be ARMA processes, given the above cases. Anderson (1976, chapter 14) provides an extension of the basic theorem to the sum of any number of independent processes, and the results can be extended straightforwardly to seasonal models. Engel (1984) provides a unified discussion of this material and also extends the results to include products of ARMA processes.

These authors also consider the *realisability* of simple models: given that a series is fitted by an ARMA(p, q) process, could this have arisen from a simple process or processes? For example, suppose that z_t is fitted by the ARMA(1, 1) model

$$(1 - \phi B) z_t = (1 - \theta B) a_t.$$

What conditions must be satisfied for z_t to be decomposed into $x_t + y_t$, where

$$(1 - \alpha B) x_t = b_t$$

$$y_t = c_t,$$

with b_t and c_t independent? Anderson (1976, chapter 14) shows that the realisability condition is simply that

$$|\theta| < |\phi|,$$

if z_t is stationary and invertible. A number of other realisability conditions for particular decompositions of ARMA(p, q) processes are given in Granger and Morris (1976) and Anderson (1976) and the reader is referred to these sources for further details.

11.7 Long memory models, persistence and fractional differencing

In our analysis of stationary ARMA models in Chapter 6, we found that the ACF of such a model will exhibit an exponential decay as the lag increases, i.e. that $\rho_k \sim \phi^k$, $|\phi| < 1$. Thus, observations separated by a long time span may be assumed to be independent, or at least nearly so. However, many empirically observed time series, although satisfying the assumption of stationarity (perhaps after some differencing transformation), are found to exhibit a dependence between distant observations that, although small, is by no means negligible. Such series are particularly found in hydrology, where this 'persistence' is known as the Hurst phenomenon (see, for example, Mandelbrot and Wallis (1969)), but also many economic, especially financial, time series exhibit characteristics of extremely long range persistence. This may be characterised as a tendency for large values to be followed by large values of the same sign in such a way that the series seem to go through a succession of 'cycles', including long cycles whose length is comparable to the total sample size.

This viewpoint has been persuasively argued by Mandelbrot (1969, 1972), extending his work on non-Gaussian (marginal) distributions in economics (Mandelbrot (1963)) to an exploration of the structure of serial dependence in economic time series. The processes considered by Mandelbrot to model long-term persistence were of the form of discrete time fractional Gaussian noise, the discrete analogue of fractional Brownian motion (see Mandelbrot and Van Ness (1968)). More recently, attention has focused on an alternative approach to modelling long-term persistence, one that is an extension of the more familiar ARIMA class of processes.

We begin by noting that the discrete time analogue of Brownian motion is the random walk, or ARIMA (0, 1, 0) process

$$\nabla x_t = (1-B)\, x_t = a_t, \tag{11.42}$$

where the first difference of x_t is the discrete time white noise process a_t. Fractionally differenced white noise, with parameter d, is then defined to be the dth *fractional* difference of discrete time white noise. This requires the use of the fractional difference operator ∇^d, which is naturally defined by the binomial series expansion

$$\begin{aligned}\nabla^d = (1-B)^d &= \sum_{k=0}^{\infty} \begin{bmatrix} d \\ k \end{bmatrix} (-B)^k \\ &= 1 - dB - \tfrac{1}{2}d(1-d)\,B^2 - \tfrac{1}{6}d(1-d)(2-d)\,B^3 - \ldots. \end{aligned} \tag{11.43}$$

Fractionally differenced white noise is then

$$\nabla^d x_t = a_t, \tag{11.44}$$

so that x_t is an ARIMA $(0, d, 0)$ process, where d is nonintegral. The notion of fractional differencing seems to have been proposed independently by Hosking (1981a) and Granger and Joyeux (1980), and further references include Hosking (1982), Granger (1980, 1981) and Geweke and Porter-Hudak (1983), with ARIMA processes with nonintegral d often being referred to as *long memory* models in this literature.

The reason for this terminology may be found in the properties exhibited by the simplest member of the class, the ARIMA $(0, d, 0)$ process defined above. From Hosking (1981a), it is known that, if $|d| < \frac{1}{2}$,

(i) x_t is stationary and invertible,

(ii) the ACF is generated as

$$\rho_k = \frac{(-d)!\,(k+d-1)!}{(d-1)!\,(k-d)!}, \quad k = 1, 2, \ldots.$$

Thus $\rho_1 = d/(1-d)$ and, as $k \to \infty$,

$$\rho_k \sim \frac{(-d)!}{(d-1)!} k^{2d-1},$$

so that for $d > 0$, $\rho_k > 0$ and the ACF declines monotonically and *hyperbolically* to zero as the lag increases, i.e. at a much slower rate than the exponential decay of an ARMA $(d = 0)$ process,

(iii) the PACF is given by

$$\phi_{kk} = \frac{d}{k-d}, \quad k = 1, 2, \ldots$$

and hence decays as k^{-1}, which is independent of d.

Because $\sum_k \rho_k$ is infinite, when $0 < d < \frac{1}{2}$ the ARIMA $(0, d, 0)$ process is said to have a long memory, and as such should be useful in modelling long-term persistence. When $-\frac{1}{2} < d < 0$, all auto and partial correlations are negative, and even though the autocorrelations have an infinite sum, such negative correlations do not allow for long-term persistence. The ARIMA $(0, d, 0)$ process then has a short memory, also being referred to as 'antipersistent'. When $|d| \geqslant \frac{1}{2}$, the variance of x_t is infinite, and hence the process is nonstationary. When $d = \frac{1}{2}$, x_t will be invertible but not stationary, whereas the converse applies when $d = -\frac{1}{2}$.

The fractional white noise process can naturally be extended by combining fractional differencing with the established family of ARMA (p, q) processes, thereby obtaining the ARIMA (p, d, q) family of processes for $|d| < \frac{1}{2}$. The properties of this class of models have been analysed and discussed by Hosking (1981a) and Granger and Joyeux (1980) and need not detain us here. The important feature for modelling purposes is that,

Exhibit 11.9 *ACFs and PACFs of (a) an ARIMA (1, d, 0) process with $\phi = 0.5$ and $d = 0.2$, and (b) an ARIMA (1, 0, 0) process with $\phi = 0.711$*

	(a) ARIMA (1, d, 0)		(b) ARIMA (1, 0, 0)	
k	ρ_k	ϕ_{kk}	ρ_k	ϕ_{kk}
1	0.711	0.711	0.711	0.711
2	0.507	0.004	0.505	0
3	0.378	0.032	0.359	0
4	0.296	0.031	0.255	0
5	0.243	0.028	0.181	0
10	0.141	0.017	0.033	0
20	0.091	0.009	0.000	0

since the effect of the d parameter on distant observations decays hyperbolically as the lag increases, whereas the effects of the AR and MA parameters decay exponentially, d should be chosen to describe the high-lag correlation structure of a time series while the other parameters will describe the low-lag correlation structure. Indeed, the long-term behaviour of an ARIMA (p, d, q) process may be expected to be similar to that of an ARIMA (0, d, 0) process with the same value of d, since the effects of the AR and MA parameters will be negligible.

Exhibit 11.9 displays the ACF and PACF of the ARIMA (1, d, 0) process

$$(1-0.5B)\nabla^{0.2}x_t = a_t,$$

with those from the ARIMA (1, 0, 0) process

$$(1-0.711B)x_t = a_t,$$

so designed to ensure that ρ_1 (and ϕ_{11}) are identical in both models (see Hosking (1981a, page 173)). Although the PACF for the fractional differenced series has small values after ϕ_{11}, they are all positive, and the ACFs of the two processes differ markedly after the first few lags.

Since it is the fractional difference parameter d that enables long-term persistence to be modelled, the value chosen for d is obviously crucial in any empirical application. Typically, this value will be unknown and must therefore be estimated. To our knowledge, four suggestions have been made in the literature:

(i) Granger and Joyeux (1980) use a grid search of d values, using a measure of h-step ahead forecastibility to determine the chosen value. They emphasise that the method is clearly arbitrary and sub-optimal, but argue that it appears to work quite well in their, admittedly limited, empirical experience with the method.

(ii) Hosking (1981a) suggests that a maximum likelihood estimate of d may be obtained by the methods of McLeod and Hipel (1978), although no details are given.

(iii) Hosking (1981a) also suggests that d could be estimated by the rescaled range, or R/S, exponent. This statistic has been explicitly developed as a measure of long-term persistence, and its use in the analysis of economic time series has been forcibly argued by Mandelbrot (1972). A more recent survey of R/S analysis and applications may be found in Mandelbrot and Taqqu (1979).

(iv) Geweke and Porter-Hudak (1983) propose an estimator of d based on the simple linear regression of the logarithm of the periodogram of the time series on an associated deterministic regressor. The estimator is the OLS estimator of the slope parameter in this regression, formed using only the lowest frequency ordinates of the log periodogram. An alternative frequency domain estimator has recently been proposed by Kashyap and Eom (1988).

Experience with each of these methods seems, at the present time, to be too limited to offer any firm guidance as to which is the most useful. Nevertheless, the empirical experience of forecasting economic time series using fractionally differenced series is quite encouraging. Both Granger and Joyeux (1980) and Geweke and Porter-Hudak (1983) find that such models provide more accurate out-of-sample forecasts, for long forecast horizons, than do conventional ARMA models, and Granger (1980a) shows that fractionally differenced models arise naturally in economics through component aggregation.

Hosking (1981a, 1982) has suggested two further processes involving fractional differencing which may prove useful. He defines the fractional equal-root integrated moving average process as

$$\nabla^q x_t = (1-\theta B)^q a_t, |q| < \tfrac{1}{2}, |\theta| < 1, \tag{11.45}$$

and, noting the conformity of this process with the equivalent ARIMA form of general exponential smoothing (see Chapter 9), as a forecasting procedure it corresponds to 'fractional qth-order general exponential smoothing'. A simple model combining both persistence and seasonality may be written as

$$(1-2\phi B+B^2)^d x_t = a_t, |d| < \tfrac{1}{2}, |\phi| < 1. \tag{11.46}$$

Despite having only two parameters, this model can describe a wide range of seasonal behaviour. Its ACF is characteristic of a hyperbolically damped sine wave, in which the correlations eventually die away to zero, unlike the nonstationary seasonal process $\nabla_s x_t = a_t$, but the rate of decay is less rapid than for the stationary process $(1-\phi B^s) x_t = a_t$. Setting $\phi = 0.866$ corresponds to a seasonal period $s = 12$, negative values of ϕ correspond to short periodicities, and values near 1 yield ACFs reminiscent of the SACFs of ARIMA (0, 1, 0) or nearly nonstationary AR (1) processes.

Fractional differencing can naturally be used in conjunction with conventional integer differencing, so that if $w_t = \nabla x_t$, say, and $\nabla^d w_t = a_t$, where $0 < d < \frac{1}{2}$, then

$$\nabla^d \nabla x_t = \nabla^{1+d} x_t = a_t.$$

In digression, we may note finally that the phrase 'long memory model' has also been used by Parzen (1982) to refer to ARMA-type models in which integer differencing is replaced by an AR polynomial having roots greater than unity, thus leading to the acronym ARARMA; for example, the process

$$(1-\phi^* B)(1-\phi B) x_t = a_t,$$

where $\phi^* > 1$, models the long memory characteristics of x_t through the nonstationary filter $(1-\phi^* B)$, with the short memory properties being modelled by the second, stationary, filter $(1-\phi B)$.

11.8 R^2 statistics for ARIMA models

The R^2 statistic, measuring goodness-of-fit, is one of the most familiar 'outputs' associated with multiple regression analysis. Such a statistic can naturally be defined in the context of stationary ARMA models as one minus the ratio of the residual variance to the total variance of the time series, thus measuring the relative predictability of a time series given its past history. Thus, if x_t is modelled by the ARMA (p, q) process

$$\phi(B) x_t = \theta(B) a_t$$

or

$$x_t = \phi^{-1}(B) \theta(B) a_t = \psi(B) a_t,$$

then the (population) R^2 is defined as

$$R^2 = 1 - \frac{\sigma_a^2}{V(x)}$$

$$= \frac{\sum_{i=1}^{\infty} \psi_i^2}{1 + \sum_{i=1}^{\infty} \psi_i^2}, \qquad (11.47)$$

using the result that (see Chapter 5.2)

$$V(x) = \sigma_a^2\left(1+\sum_{i=1}^{\infty}\psi_i^2\right).$$

The value of R^2 thus depends only on the ψ weights and not on σ_a^2; the more structure there is to the time series (larger ψ weights), the higher will be R^2.

Since the ψ weights determine the ACF of x_t, we should be able to relate R^2 to the ρ_k. This is easy to do for a pure AR process, and Nelson (1976) shows that for an AR(1) process

$$x_t = \phi_1 x_{t-1} + a_t,$$

$$R_1^2 = \rho_1^2 = \phi_1^2,$$

where R_1^2 is the R^2 for an AR(1) process. We can therefore obtain a rough idea of the predictability of the series from the sample estimate of ρ_1: for example, if $r_1 \approx 0.5$, we can anticipate an R^2 of about 0.25. Nelson (1976) goes on to show that if R_j^2 is the R^2 statistic of an AR(j) process, then such values can be obtained recursively as

$$R_j^2 = \phi_{jj}^2(1-R_{j-1}^2)+R_{j-1}^2, \quad j = 1,2,3,\ldots$$

where ϕ_{jj} is the jth partial autocorrelation. Thus the addition of further lags will make little contribution to R^2 if the sample partial autocorrelations for those lags are small.

For pure MA and mixed ARMA models the analysis of R^2 is much less straightforward, and Nelson suggests using a high order AR approximation in such cases. The extension to integrated models also poses difficulties, for when x_t is nonstationary, $V(x)$ is infinite and the R^2 statistic given by (11.47) becomes meaningless. This may also be seen by the following argument. For the stationary case above, the yardstick against which R^2 is being compared is, implicitly, simply the fitting of a mean to the data. When the series is nonstationary, fitting a mean is such a poor alternative that any model which is able to pick up a trend reasonably well will have an R^2 close to unity.

Harvey (1984, appendix 1) suggests replacing the empirical counterpart of (11.47),

$$R^2 = 1-\frac{\sum_{t=k+1}^{n}\hat{a}_t^2}{\sum_{t=1}^{n}(x_t-\bar{x})^2},$$

by the measure

$$R_D^2 = 1-\frac{\sum_{t=k+1}^{n}\hat{a}_t^2}{\sum_{t=2}^{n}(\nabla x_t-\bar{\nabla}x)^2}, \qquad (11.48)$$

where $\hat{a}_t$ are the residuals obtained from fitting any type of model to ∇x_t, k is the number of initial observations lost in estimation and $\overline{\nabla x}$ is the mean of the first differences of x_t. The yardstick being adopted in R_D^2 is the random walk with drift model

$$\nabla x_t = \theta + a_t, \quad t = 2, \ldots, n. \tag{11.49}$$

If $\theta = 0$, so that x_t is a pure random walk, $R_D^2 = 1 - U^2$, where U is the statistic for measuring forecasting performance proposed by Theil (1966). The random walk with drift is thus a more stringent model to use as a yardstick for assessing nonstationary models but, since it is still relatively simple, any model which gives a worse fit, i.e. has $R_D^2 < 0$, should not be seriously entertained.

If the data is seasonal, equation (11.49) can be extended to present a sensible yardstick model by including $s-1$ seasonal dummies on the right hand side. Equivalently, θ can be dropped and s dummies added:

$$\nabla x_t = \sum_{j=1}^{s} \gamma_j z_{tj} + a_t, \tag{11.50}$$

where the z_{tj}'s are dummy variables taking the value unity in season j and zero elsewhere. OLS estimates of $\gamma_1, \ldots, \gamma_s$ are simply the means of ∇x_t in each season. Harvey (1984, appendix 1) suggests using the criterion

$$R_s^2 = 1 - \frac{\text{SSE}}{\text{SSE}_0},$$

where SSE is the residual sum of squares from fitting the model currently under consideration and SSE_0 is the residual sum of squares from fitting the yardstick model (11.50), which assumes random walk behaviour with deterministic seasonality. An obvious model to fit to a seasonal time series is one having the differencing operator $\nabla^d \nabla_s^D$ (e.g. a seasonal multiplicative ARIMA model). In this case $d + sD$ fewer observations are available for estimation than for the yardstick model (11.50). To account for this discrepancy, in a manner analogous to using $\bar{R}^2$ rather than R^2 in regression models, Harvey (1984, appendix 1) defines the adjusted R_s^2 measure as

$$\bar{R}_s^2 = 1 - \frac{\text{SSE}/(T-d-sD-m)}{\text{SSE}_0/(T-1-s)},$$

where m is the number of deterministic components contained in the alternative model. In this form the numerator and denominator of the ratio are the unbiased estimators of the error variances of the two models being compared.

Example 11.6: R^2 statistics for European industrial production series

R_D^2 statistics were computed for the three European industrial production series that were not identified as random walk with drift models in Example 11.3. The statistics were, for Hungary 0.13, for France 0.09 and for Russia 0.03. Thus, even though significant ARMA forms are obtained for each series, they add relatively little to the explanatory power of the random walk with drift model as an explanation of movements in industrial production.

Example 11.7: R^2 statistics for retail sales series

The random walk plus deterministic seasonal model (11.50) was fitted to the (logarithms of the) value and volume of retail sales series analysed in Chapter 10. A comparison of this model to those selected in that chapter yields $\bar{R}_S^2$ values of 0.33 for the value series and 0.18 for the volume series, and in both cases the yardstick model leaves considerable residual autocorrelation, primarily at the seasonal lags. Thus, while a random walk trend model is not too bad an assumption for these series, deterministic seasonality is certainly inadequate, thus providing further confirmation of the finding of evolving, stochastic seasonality shown previously.

11.9 Computing software

Computing software for the analysis of time series is readily available. Routines for the identification, estimation and diagnostic checking of ARIMA models may be found in SAS/ETS as PROC ARIMA (SAS (1985c, chapter 8)) with more specialised routines being found in SAS/IML (SAS (1985e)). SPSS^X has such routines in its BOX–JENKINS procedure (SPSS^X (1985, chapter 38)) and MINITAB (Ryan et al. (1985)) also contains routines for ARIMA modelling. SAS/ETS also has exponential smoothing routines contained within PROC FORECAST (SAS (1985b, chapter 12)). There are also a plethora of microcomputer software in this area; see, for example, Beaumont et al. (1985) for a, by now dated, survey.

Part III

The modelling of multivariate economic time series

A distinguishing feature of economic time series is their interaction, and this part of the book extends the statistical models for univariate time series developed in Part II to the situation where a set of time series are analysed jointly. Chapter 12 begins by considering the case when a single series is known to be affected by certain exogenous events occurring at particular points in time, leading to the analysis of 'intervention models'. These are then generalised to incorporate the detection of outliers, these being characterised as exogenous events whose timings are not known with complete certainty.

Both interventions and outliers can be analysed by extending univariate ARIMA models to include input, or exogenous, variables which take the form of dummy (0–1) variables. Chapter 13 generalises this model to allow for the inclusion of input variables that are themselves stochastic processes, thus leading to the 'transfer function–noise' model. The role of differencing to ensure stationarity of all the variables in this model is emphasised, and it is contrasted with the typical econometric approach of building dynamic models using the levels of time series. This leads on to the problems associated with inference in regressions with integrated processes, and related to this is the concept of 'cointegration'. Cointegration is intimately bound up with the notion of equilibrium, and this connection is discussed in detail. Finally in Chapter 13, we discuss the relationships existing between the transfer function–noise model and some of the more commonly used dynamic models encountered in econometric analysis.

This development assumes that there is no feedback from the output variable to the input variables, but this is clearly untenable in many economic applications. Chapter 14 thus extends the range of models to allow for such feedback between variables by introducing the class of vector ARMA models. The relationship between this class and the dynamic simultaneous equation model is then developed, and the role of exogeneity restrictions is shown to be the crucial factor in making distinctions between the two types of model. Alternative definitions of exogeneity are then discussed, these being linked with an operational definition of causality. Measures of feedback between vectors of time series and associated tests of causal orderings are then developed. Notwithstanding the above discussion, time series analysis can be conducted without making any endogenous/exogenous variable classification. This can be done by using 'innovation accounting' techniques, which are then discussed along with critiques thereof. The consequences of various forms of misspecification in vector ARMA models are then analysed, before estimation and model identification and checking techniques are introduced. The chapter ends by considering forecasting with vector ARMA models.

12 Intervention analysis and the detection of outliers

12.1 Introduction

In the analysis of univariate time series models developed in Part II of the book, little attention was paid to the behaviour of specific observations of the series, or to the residuals associated with them. Rather, attention was devoted to overall patterns of serial correlation in the residual series. It is well known, however, that economic time series are frequently affected by policy changes and other events that are known to have occurred at a particular point of time; for example, the alternative fuel consumption series shown as Exhibit 2.10 reveal markedly different responses to the miners' strike of 1984, while the definitional changes to the monetary aggregates of Exhibit 2.11 produce once-and-for-all upward shifts in the series. It is obvious that ignoring these factors can lead to an inadequate model being fitted and poor forecasts being made.

Events of this type, whose timing are known, have been termed *interventions* by Box and Tiao (1975), and they can be incorporated into a univariate model by extending it to include deterministic (or dummy) input variables. Intervention models, with a number of examples, are discussed in Section 12.2. Often, however, the exact timing of exogenous interventions are unknown and their effects manifest themselves as aberrant observations or *outliers*. Model specification in the presence of outliers can, in principle, be carried out using 'robust' methods (see Martin and Yohai (1985) for a recent survey), but an extension of intervention modelling provides a natural means of dealing with outlying observations within the conventional framework used in this book. The detection of outliers, and the estimation of the ensuing intervention-type models, are discussed, again with examples, in Section 12.3. The final section briefly discusses extensions of this detection procedure to incorporate level shifts and variance changes.

12.2 Intervention models

Interventions can affect a time series in several ways. They can change the level, either abruptly or after some delay, change the trend, or lead to other, more complicated, response patterns. The Student t distribution is traditionally used to test for a change in mean level, but such a test may not be appropriate for time series, both because successive observations will often be correlated, and also because the effect may not just be a step change as postulated by the t-test. A formal demonstration of the inadequacy of the t-test in the presence of serial correlation is given in Abraham (1987).

We begin by considering a single intervention known to occur at time T. If x_t is generated by an ARMA (p, q) process, then an intervention model may be postulated as

$$x_t = v(B) I_t + N_t, \tag{12.1}$$

where

$$N_t = \frac{\theta(B)}{\phi(B)} a_t \tag{12.2}$$

is the 'noise' model, $v(B)$ is a (possibly infinite) polynomial which may admit a 'rational' form, such as

$$v(B) = \frac{\omega(B)}{\delta(B)} B^b, \tag{12.3}$$

where

$$\omega(B) = \omega_0 - \omega_1 B - \ldots - \omega_s B^m,$$

$$\delta(B) = 1 - \delta_1 B - \ldots - \delta_r B^r,$$

and b measures the 'delay' in effect (or 'dead time'). Further discussion of this rational form for $v(B)$ is provided in Chapter 13.

I_t is the intervention variable, and is a 'dummy' or 'indicator' sequence taking the values 1 and 0 to denote the occurrence or nonoccurrence of the exogenous intervention. The following dummy variables have been found to be useful for representing various forms of interventions:

(i) A *pulse* variable, which models an intervention lasting only for the observation T,

$$I_t = \varepsilon_t^{(T)}, \quad \text{where} \quad \varepsilon_t^{(T)} = \begin{cases} 1, t = T \\ 0, t \neq T. \end{cases}$$

(ii) A *step* variable, which models a step change in x_t beginning at observation T,

$$I_t = \xi_t^{(T)}, \quad \text{where} \quad \xi_t^{(T)} = \begin{cases} 0, t < T \\ 1, t \geqslant T. \end{cases}$$

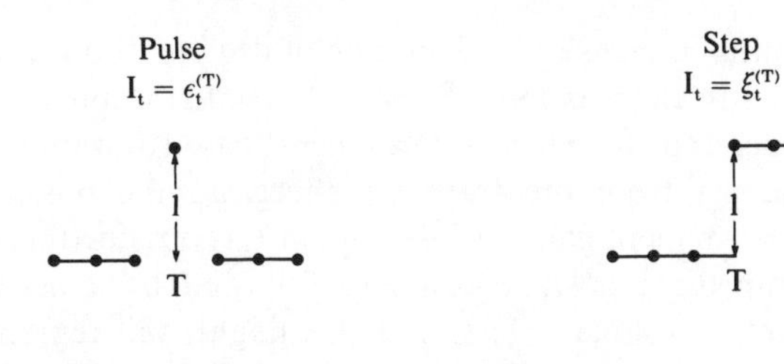

(a) $\nu(\mathrm{B}) = \dfrac{\omega}{1-\delta\mathrm{B}}$

(d) $\nu(\mathrm{B}) = \omega$

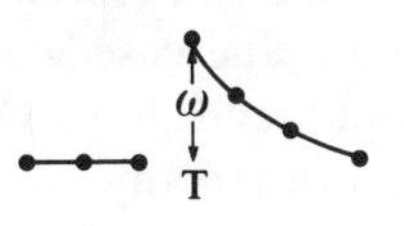

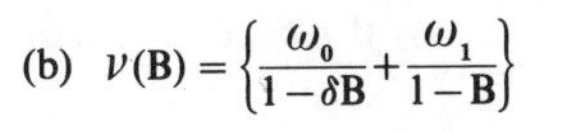

(e) $\nu(\mathrm{B}) = \dfrac{\omega}{1-\delta\mathrm{B}}$

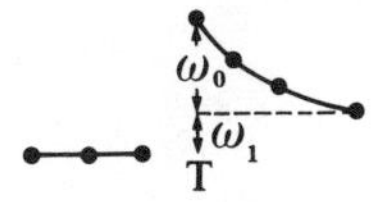

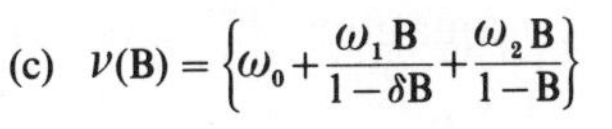

(f) $\nu(\mathrm{B}) = \dfrac{\omega}{1-\mathrm{B}}$

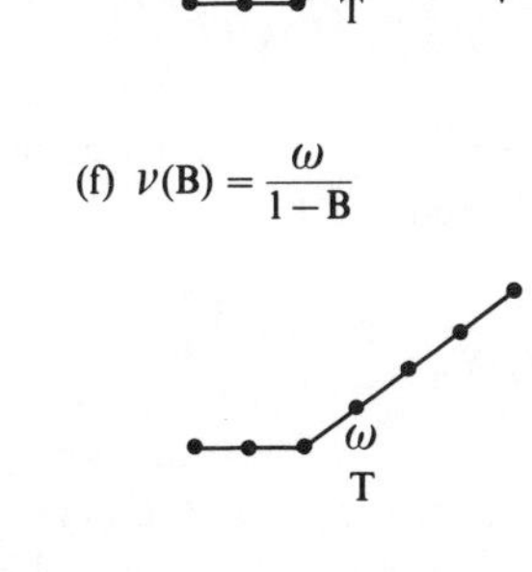

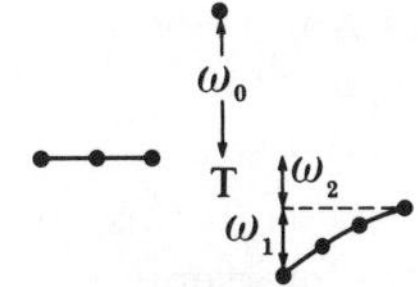

Exhibit 12.1 Examples of dynamic effects which can be simulated using 'pulse' and 'step' interventions

(iii) It is also possible to define an *extended pulse* variable, useful for modelling 'policy on-policy off' interventions, as

$$I_t = \eta_t^{(T_1, T_2)}, \quad \text{where} \quad \eta_t^{(T_1, T_2)} = \begin{cases} 1, T_1 \leqslant t \leqslant T_2 \\ 0, \text{otherwise,} \end{cases}$$

noting that $\eta_t^{(T_1, T_2)} = \sum_{j=0}^{T_2-T_1} \varepsilon(T_1+j)_t = (1+B+\ldots+B^{T_2-T_1})\,\varepsilon_t^{(T_2)}$.

Exhibit 12.1 shows the responses to a pulse and a step change 'input' for various specifications of $v(B)$ that are of practical interest. For a pulse input, (a) shows the case in which I_t has only a transient effect on x_t, with

ω measuring the initial increase and δ the rate of decline. If δ is zero then only an instantaneous effect is felt, whereas if $\delta = 1$, the pulse input is really a step change, and the effect is permanent; case (b) represents the situation where, apart from the transient effect ω_0, the possibility is entertained that a permanent gain (or loss), ω_1, in x_t is obtained; (c) shows the case of an immediate positive response followed by a decay and, possibly, a permanent residual effect, and this might well represent the dynamic response of sales to a price increase.

For a step input, (d) shows an immediate step response of ω; (e) shows the situation of a 'first-order' dynamic response and an eventual, or long-run response, of $\omega/(1-\delta)$; finally, (f) represents the case when $\delta = 1$, in which the step change produces a 'ramp' or trend in x_t. Obviously, these models can be readily extended to represent many situations of potential interest; in general, if the denominator polynomial $\delta(B)$ is unity ($r = 0$), finite responses of length m are obtained, whereas if r exceeds zero, responses of infinite length are obtained.

If the noise model is of the multiplicative form

$$N_t = \frac{\theta(B)\,\Theta(B^s)}{\nabla^d \nabla_s^D \phi(B)\,\Theta(B^s)} a_t, \tag{12.4}$$

and if there are J interventions, the model given by equations (12.1)–(12.3) can then be extended to

$$\nabla^d \nabla_s^D x_t = \sum_{j=1}^{J} \frac{\omega_j(B)}{\delta_j(B)} B^{b_j} \nabla^d \nabla_s^D I_{jt} + \frac{\theta(B)\,\Theta(B^s)}{\phi(B)\,\Phi(B^s)} a_t. \tag{12.5}$$

In building models of this type, parsimonious forms are initially postulated to represent the expected effects of the interventions, with more complex forms only being considered if knowledge of the dynamics of the intervention, or subsequent empirical evidence, suggests so. Identification of the noise model N_t may be carried out in various ways. The identification procedures outlined in Chapter 8 may be applied to data prior to the occurrence of the interventions if a sufficiently large number of such observations is available. If the effects of the interventions are expected to be transient, however, then these identification procedures may be applied to the entire data set. Alternatively, the response polynomials $v_j(B) \simeq \delta_j^{-1}(B)\,\omega_j(B)\,B^{b_j}$ may be estimated by OLS, for suitably large maximum orders, and the identification procedures applied to the residuals $x_t - \sum \hat{v}_j(B)\,I_{jt}$.

Once a model of the form (12.5) has been tentatively specified, the intervention and noise parameters can be estimated simultaneously by maximum likelihood via nonlinear least squares. Diagnostic checks can then be performed on the residuals in the usual manner so as to assess the adequacy of the fitted model and to search for directions of improvement, if needed.

Example 12.1: Modelling interventions in M0

In Example 10.3 we analysed the (logarithms of the) UK monthly M0 series, and stated that the series was adjusted, prior to modelling, for a change in definition in August 1981. This adjustment was effected through an intervention model designed to measure the decrease in the series brought about by redefining the composition of banks' deposits at the Bank of England, a constituent of M0 (for further details, see Mills (1986)). Since the redefinition produced a step change in M0 at a known point in time, a step intervention was defined to take the value 0 up to and including August 1981 and 1 afterwards, being denoted as ξ. The following model was identified and estimated:

$$\nabla\nabla_{12}x_t = \underset{(0.0003)}{-0.0008} + \left(1 - \underset{(0.103)}{0.297}B\right)\left(1 - \underset{(0.085)}{0.674}B^{12}\right)a_t - \underset{(0.009)}{0.030}\nabla\nabla_{12}\xi_t$$

$$\hat{\sigma}_a = 0.0102; Q_{24}(21) = 13.3.$$

The intervention appears very significantly, with the coefficient estimate implying that the definitional change reduces the level of M0 instantaneously, and without any dynamic response, by approximately three per cent. The analysis of Example 10.3 has then been developed using the 'adjusted' series $(x_t + 0.030\xi_t)$, although the original analysis reported in Mills (1986) incorporates the intervention explicitly.

Example 12.2: Modelling interventions in bank lending

In Chapter 10, a model for monthly UK bank lending was presented as

$$\nabla_{12}x_t = \underset{(18)}{149} + \left[1 + \underset{(0.097)}{0.271}B^3\right]\left[1 - \underset{(0.075)}{0.837}B^{12}\right]a_t$$

$$\hat{\sigma}_a = 548; Q_{12}(9) = 11.0.$$

The model eventually selected by Mills and Stephenson (1987) contains two pulse interventions. The first, ε_1, is in July 1980 and corresponds to the removal of the 'corset', which enabled banks to regain much of the business they had sought to keep away from their balance sheets by issuing bank acceptances, and which immediately swelled bank lending. The second, ε_2, attempts to measure the effect on bank lending of the British

Telecom privatisation in November 1984. The fitted model is

$$\nabla_{12} x_t = \underset{(20)}{145} + \left[1 + \underset{(0.101)}{0.360} B^3\right]\left[1 - \underset{(0.075)}{0.816} B^{12}\right] a_t + \underset{(463)}{1569} \nabla_{12} \varepsilon_{1t} + \underset{(499)}{1072} \nabla_{12} \varepsilon_{2t}$$

$$\hat{\sigma}_a = 516; Q_{12}(9) = 9.0.$$

The fit of this model is substantially better, $\hat{\sigma}_a$ being decreased by almost 10 per cent, and with the portmanteau statistic being rather smaller. Both interventions produce instantaneous responses and are highly significant.

Example 12.3: The stock conversion of 1932

Our third example of intervention models is an historical example containing both a step change, with a lagged response, and an 'extended pulse' intervention. Capie et al. (1986) develop a model of the yield on three month bank bills for the period January 1927 to December 1937. Conventional univariate analysis produced

$$\nabla x_t = (1 - \underset{(0.09)}{0.32} B^2) a_t, \hat{\sigma}_a = 0.405.$$

The authors were particularly interested in assessing whether the famous Stock Conversion of 1932 had any quantitative effect on the level of interest rates. To do this, an intervention model was developed, being estimated as

$$\nabla x_t = \underset{(0.23)}{1.14} \nabla \eta_t^{(T_1, T_2)} - \left(\underset{(0.33)}{1.80} + \underset{(0.32)}{1.00} B\right) \nabla \xi_t^{(T_3)} + \left(1 - \underset{(0.09)}{0.22} B^2\right) a_t, \hat{\sigma}_a = 0.318.$$

The previous ARIMA (0, 1, 2) model was found to need modification by a step intervention at T_3(= February 1932), which reduced the bill rate by 2.8 percentage points, spread over two months in an approximately 2:1 ratio. This intervention anticipated the conversion, which took place at the end of June 1932, by four months, and some conjectures for this finding are made in Capie et al. (1986). The second intervention is of the extended pulse variety between T_1(= July 1931) and T_2(= November 1931), during which period the bill rate increased by 1.14 percentage points. This may readily be explained as being a result of exchange rate problems which led to the abandonment of the Gold Standard and the setting up of the Exchange Equalisation Account.

Numerous applications of intervention analysis have appeared in the literature since the original Box and Tiao (1975) analyses of consumer prices and atmospheric pollution. Jenkins and McLeod (1982) present a number of case studies that employ intervention effects, analysing such diverse topics as changing price structures in the US telephone industry, competition between rail and air on London to Scotland passenger routes, employment forecasting in the EEC, and the influence of promotions on product sales. An area that has been a favourite with analysts is the effect of seat belt legislation on traffic accidents. Bhattacharyaa and Layton (1979) analysed the data for Queensland, Australia, and Abraham (1987) the data for Ontario, Canada, both employing the 'ARIMA-augmented' approach developed above. Harvey and Durbin (1986) analysed British data, using both this approach and an intervention augmented version of Harvey's (1984) structural model discussed in Chapters 9 and 10.

12.3 Detecting outliers in time series

In the building of intervention models, it was supposed that the timing of the interventions were known. We now discuss a variant of this methodology for handling situations in which the exact timings of these exogenous events are unknown and whose effects lead to what are called aberrant observations or outliers.

Following the initial work of Fox (1972) and more recent extensions by Hillmer, Bell and Tiao (1983), Tsay (1986a) and Chang et al. (1988), we will concentrate on two types of outliers, additive and innovational. An *additive outlier* (AO) is defined as

$$x_t = N_t + \omega \varepsilon_t^{(T)}, \tag{12.6}$$

while an *innovational outlier* (IO) is defined as

$$x_t = N_t + \frac{\theta(B)}{\phi(B)} \omega \varepsilon_t^{(T)}, \tag{12.7}$$

where, in both cases, the noise model is given by equation (12.2) and ε_t is a pulse intervention as defined above. In terms of the innovations, a_t, of the noise (or outlier-free) model, we have that

$$\text{(AO)} \quad x_t = \frac{\theta(B)}{\phi(B)} a_t + \omega \varepsilon_t^{(T)} \tag{12.8}$$

and

$$\text{(IO)} \quad x_t = \frac{\theta(B)}{\phi(B)} (a_t + \omega \varepsilon_t^{(T)}). \tag{12.9}$$

Thus the AO case may be called a 'gross error' model, since only the level of the Tth observation is affected. On the other hand, an IO represents an extraordinary shock at T influencing $x_T, x_{T+1}, \ldots$ through the memory of the model, given by $\theta(B)/\phi(B)$.

Before proceeding with the outlier detection technique, it is useful to consider the question of what are the effects of outliers on the usual statistics of identification, the SACF and SPACF? Tsay (1986a) summarises the theoretical results pertaining to this, stating that the existence of outliers may cause serious biases in these sample statistics, thus jeopardising their usefulness as tools for model identification. The biases depend upon the number, type, magnitude and relative position of outliers, and for typically sized data sets their presence can lead to either under or overspecification of the orders of $\theta(B)$ and $\phi(B)$.

To develop the technique for detecting outliers, we define the residuals as

$$e_t = \frac{\phi(B)}{\theta(B)} x_t = \pi(B)\, x_t,$$

so that we have

$$\text{(AO)} \quad e_t = \omega \varepsilon_t^{(T)} + a_t \tag{12.10}$$

and

$$\text{(IO)} \quad e_t = \omega \pi(B)\, \varepsilon_t^{(T)} + a_t. \tag{12.11}$$

From least squares theory, the magnitude of an outlier can be estimated by

$$\text{(AO)} \quad \hat{\omega}_{A,T} = \eta^2 \pi(B^{-1})\, e_t$$

$$= \eta^2 (e_T - \pi_1 e_{T+1} - \ldots - \pi_{n-T} e_n)$$

and

$$\text{(IO)} \quad \hat{\omega}_{I,T} = e_T,$$

where $\eta^2 = (1 + \pi_1^2 + \pi_2^2 + \ldots + \pi_{n-T}^2)^{-1}$. The variances of these estimates are $V(\hat{\omega}_{A,T}) = \eta^2 \sigma_a^2$ and $V(\hat{\omega}_{I,T}) = \sigma_a^2$ respectively. Thus the best estimate of an IO at time T is the residual e_T, while the best estimate of an AO is a linear combination of e_T and the future residuals $e_{T+1}, e_{T+2}, \ldots, e_n$, with weights depending on the structure of the model for N_t.

Under the null hypothesis that there is no outlier at time T, likelihood ratio statistics for the alternatives of an AO or IO are given by

$$\text{(AO)} \quad \lambda_{A,T} = \hat{\omega}_{A,T} / \eta \sigma_a$$

and

$$\text{(IO)} \quad \lambda_{I,T} = \hat{\omega}_{I,T} / \sigma_a$$

respectively. Under this null hypothesis, both statistics have the standard normal distribution and therefore can readily be used in practice once the unknown parameters $\pi(B)$ and σ_a are replaced by consistent estimators.

These results enable us to develop an iterative method for detecting outliers which is a modification of that proposed by Tsay (1986a) (see also

Tiao (1985)). An iterative method is required because the existence of outliers can seriously bias the estimates of all the model parameters, and this contains the following steps:

(i) Model x_t by supposing that there are, initially, no outliers, i.e. $x_t = N_t$, and from the estimated model compute the residuals

$$\hat{e}_t = \hat{\pi}(B)\,x_t,$$

letting

$$\hat{\sigma}_a^2 = \frac{1}{n}\sum_{t=1}^{n} \hat{e}_t^2$$

be the initial estimate of σ_a^2.

(ii) Compute $\hat{\lambda}_{A,t}$ and $\hat{\lambda}_{I,t}$ for $t = 1,\ldots,n$, using $\hat{e}_t$, $\hat{\pi}(B)$ and $\hat{\sigma}_a$. If both $\hat{\lambda}_{A,t}$ and $\hat{\lambda}_{I,t}$ are less than some critical value C, usually set at 3.5, for all t, then there are no outliers, the detection procedure is terminated, and we go directly to step (iii). If, however, we find that

$$|\lambda_T| = \max_{(t)}(|\lambda_{A,t}|, |\lambda_{I,t}|)$$

exceeds C then there is an outlier at T; if $\lambda_{A,T} > \lambda_{I,T}$ an AO has been found, otherwise an IO is identified. In this latter case, the IO effect can be removed by modifying the observed data set according to equation (12.7); i.e. by adjusting x_t by

$$\begin{aligned} \tilde{x}_t &= x_t, & 1 \leqslant t \leqslant T-1 \\ &= x_t - \psi_{t-T}\,\hat{\omega}_{I,T}, & T \leqslant t \leqslant n, \end{aligned}$$

where ψ_i is the coefficient of B^i in $\psi(B) = \hat{\theta}(B)/\hat{\phi}(B)$. If, on the other hand, an AO is found, the data is modified according to

$$\begin{aligned} \hat{x}_t &= x_t, & t \neq T \\ &= x_T - \hat{\omega}_{A,T}, & t = T. \end{aligned}$$

After the effect of the outlier has been removed, the procedure may be iterated by going to step (i) with the modified data.

(iii) After J outliers have been detected, the $(J+1)$th iteration of steps (i) and (ii) will fail to identify an outlier. The order (p,q) in this last iteration of step (i) is the tentative order of the outlier-free series N_t, and the overall model can then be specified as

$$\begin{aligned} x_t &= \sum_{j=1}^{J} \omega_j\, v_j(B)\,\varepsilon_t^{(T_j)} + N_t \\ N_t &= \frac{\theta(B)}{\phi(B)}\,a_t, \end{aligned} \qquad (12.12)$$

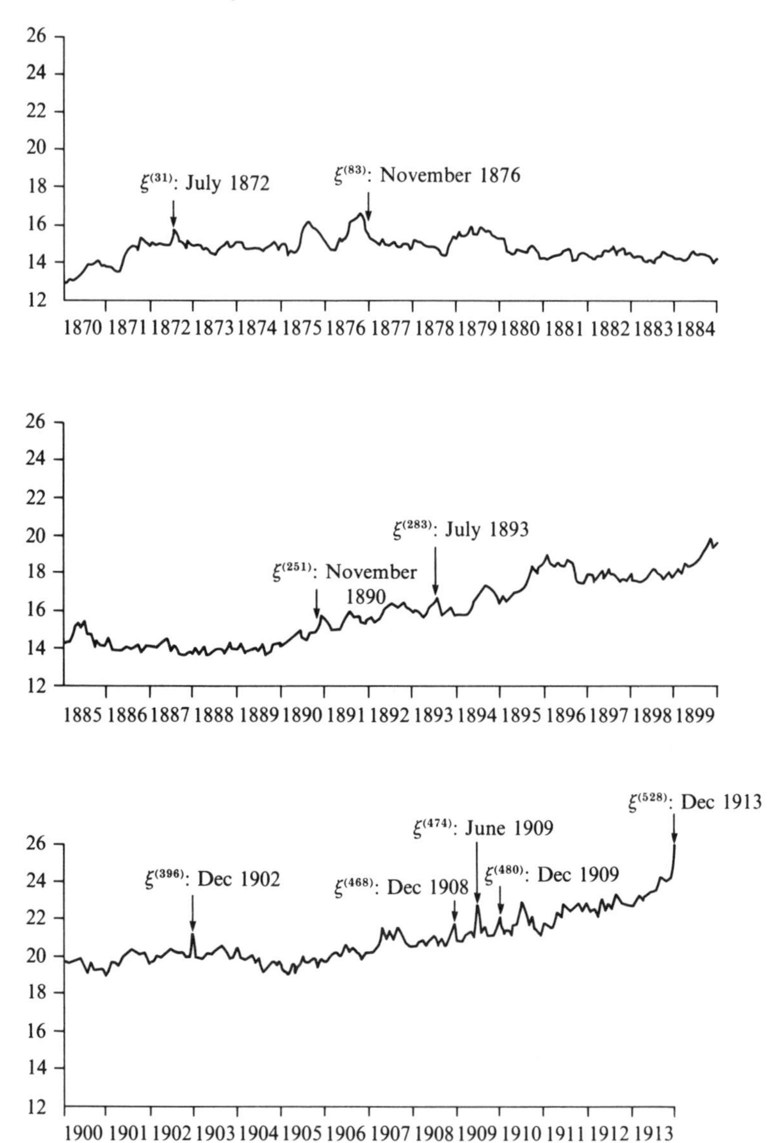

Exhibit 12.2 Money base (high powered money) January 1870 to December 1913 (£10 million) observed series

where $v_j(B) = 1$ if an AO is detected at T_j, and $v_j(B) = [\theta(B)/\phi(B)]$ if an IO is detected. After such a model has been tentatively specified, the original data should be examined and possible causes of the identified outliers proposed. Tsay (1986a) suggests that, after the model (12.12) has been specified, the original data should be used in any further analysis, such as estimation, checking and forecasting.

Example 12.4: Detecting outliers in the UK money base

This approach to outlier detection was employed on the UK money base series for the Gold Standard period from 1870 to 1913. Monthly observations are available in Capie and Webber (1985), the length of the series thus being $n = 528$. The levels are plotted in Exhibit 12.2, and are consistent with a logarithmic transformation being employed. Step (i) of the detection procedure suggested that these logarithms could be adequately modelled by the 'airline model'

$$x_t = \frac{(1-\theta B)(1-\Theta B^{12})}{\nabla\nabla_{12}} a_t.$$

Exhibit 12.3 presents details of the iteration procedure, initiated by using the above model for the first iteration. $J = 9$ outliers were detected, seven being of AO form and the remaining two therefore being IOs. The form of the ARIMA process for N_t remained the same throughout the iterations with the estimate of the seasonal parameter Θ being almost unchanged. The estimate of θ, however, is halved in value during the iterative procedure, thus providing an illustration of the bias that can be induced by outliers.

Given these outliers, the following model was initially fitted

$$\begin{aligned} x_t = {} & \frac{(1-\theta B)(1-\Theta B^{12})}{\nabla\nabla_{12}} a_t + \omega_1 \varepsilon_t^{(31)} + \omega_2 v_2(B)\, \varepsilon_t^{(83)} \\ & + \omega_3 v_3(B)\, \varepsilon_t^{(251)} \\ & + \omega_4 \varepsilon_t^{(283)} + \omega_5 \varepsilon_t^{(396)} + \omega_6 \varepsilon_t^{(468)} \\ & + \omega_7 \varepsilon_t^{(474)} + \omega_8 \varepsilon_t^{(480)} + \omega_9 \varepsilon_t^{(528)}. \end{aligned}$$

After some refitting, the following model was arrived at

$$\begin{aligned} x_t = {} & \frac{(1-\underset{(0.045)}{0.079}B)(1-\underset{(0.024)}{0.848}B^{12})}{\nabla\nabla_{12}} a_t \\ & + \underset{(0.011)}{0.042}\varepsilon_t^{(31)} - \underset{(0.015)}{0.054}\nabla^{-1}\varepsilon_t^{(83)} + \underset{(0.015)}{0.053}\nabla^{-1}\varepsilon_t^{(251)} \\ & + \underset{(0.011)}{0.036}\varepsilon_t^{(283)} + \underset{(0.011)}{0.061}\varepsilon_t^{(396)} + \underset{(0.011)}{0.039}\varepsilon_t^{(468)} \\ & + \underset{(0.011)}{0.056}\varepsilon_t^{(474)} + \underset{(0.011)}{0.036}\varepsilon_t^{(480)} + \underset{(0.016)}{0.072}\varepsilon_t^{(528)} \end{aligned}$$

$\hat{\sigma}_a = 0.0155; Q_{12}(10) = 11.7.$

Exhibit 12.3 *Iterative outlier detection results*

	Parameter estimates		Outlier		
Iteration	θ	Θ	Type	Time	Size
1	0.151	0.845	AO	474	0.055
2	0.129	0.848	AO	396	0.053
3	0.108	0.842	AO	528	0.066
4	0.110	0.847	AO	31	0.039
5	0.096	0.845	IO	251	0.054
6	0.099	0.839	IO	83	−0.053
7	0.104	0.842	AO	283	0.037
8	0.090	0.842	AO	468	0.033
9	0.087	0.845	AO	480	0.035
10	0.079	0.847			

All the ω_j's are estimated precisely and the two IOs enter in similar form. Recalling that $\nabla^{-1} = (1+B+B^2+\ldots)$, both outliers have effects that persist, rather than die out, through time. The detected outliers are indicated on Exhibit 12.2, and the two IOs are indeed seen to have these persistent effects, the former in November 1876 presaging a slow decline in the level of the series, the latter in November 1890 marking the onset of a steady increase. The remaining AOs all flag unusual, but transitory, increases in the money base, either in mid-summer or in December.

12.4 Detecting level shifts and variance changes

Tsay (1988) has recently extended the above approach to deal with level shifts, both permanent and transient, and variance changes in time series. A permanent level change (LC) is defined as

$$x_t = N_t + \frac{\omega}{(1-B)}\varepsilon_t^{(T)},$$

so that a level shift of magnitude ω occurs at time $t = T$, the change being permanent, while a transient level change (TC) is correspondingly defined as

$$x_t = N_t + \frac{\omega}{(1-\delta B)}\varepsilon_t^{(T)},$$

which models a disturbance that effects N_t for all $t \geqslant T$, but in which the effect decays exponentially at a rate δ after the initial impact ω. As Tsay shows, detection of LCs and TCs can easily be incorporated into the iterative detection procedure discussed above by a simple modification.

For the variance change model, Tsay defines

$$b_t = \frac{\phi(B)}{\theta(B)} x_t,$$

so that combining this with equation (12.9) yields

$$b_t = \begin{cases} a_t & \text{if} \quad t < T \\ a_t(1+\omega) & \text{if} \quad t \geqslant T. \end{cases}$$

Consequently, the variance ratio of b_t before and after the time point T is

$$r_t = \frac{(T-1) \sum_{t=T}^{n} b_t^2}{(n-T+1) \sum_{t=1}^{T-1} b_t^2},$$

where it is understood that $(T-1) > 0$ and $(n-T+1) > 0$. This ratio is an estimate of $(1+\omega^2)$ and is the likelihood ratio test statistic of a variance change at the given time point T under the assumption of normality. Tsay (1988) provides a second iterative procedure for detecting such variance changes, and the reader is referred to this paper for details and examples.

13 Transfer function-noise models

13.1 Introduction

Having extended the univariate ARIMA formulation of time series processes to include intervention variables modelling external events, this chapter develops the natural extension of introducing additional explanatory variables as inputs into the model. The next section discusses the identification, estimation and checking of single input transfer function-noise models and section 13.3 then extends this analysis to multiple inputs. Throughout this development it is assumed that all series are stationary (after appropriate transformation) and that there is no feedback from the output to the input variables: the latter are 'exogenous'. Section 13.4 investigates the important question of the consequences for model building of using nonstationary variables, and relates this to the notion of long-run equilibrium through the concept of cointegrability. The final section considers some popular special cases of transfer function-noise models that are regularly in use in econometrics, and also refers to some of the important techniques of dynamic model specification that have been developed in recent years.

13.2 Single input transfer function-noise models

We begin our discussion by considering the situation in which an output or, to use terminology more familiar to economists, endogenous variable Y_t is related to a single input, or exogenous, variable X_t through the dynamic model

$$Y_t = \upsilon(B) X_t + N_t, \qquad (13.1)$$

where the lag polynomial $\upsilon(B) = \upsilon_0 + \upsilon_1 B + \upsilon_2 B^2 + \ldots$, first introduced in Chapter 12, allows X to influence Y via a distributed lag, $\upsilon(B)$ often being referred to as the *transfer function* and the coefficients υ_i as the *impulse response weights*. The relationship between Y and X will not be deterministic; rather, it will be contaminated by noise, this being captured by the stochastic process N_t, which will generally be serially correlated. The crucial assumption made in the *transfer function-noise* model (13.1) is that X_t and N_t are independent, so that past X's influence future Y's, but not vice-versa. The full implications of this assumption, which rules out *feedback* from Y to X, are considered in Chapter 14, along with other related concepts of exogeneity.

In general $\upsilon(B)$ is of infinite order, and hence some restrictions must be placed on the transfer function before empirical implementation becomes feasible. The typical way in which restrictions are imposed is analogous to the approximation of the linear filter representation of a stochastic process by a ratio of low order polynomials in B, which leads to the familiar ARMA model (see Chapter 5). In the present circumstances, $\upsilon(B)$ is written as the *rational distribution lag*

$$\upsilon(B) = \frac{\omega(B)\,B^b}{\delta(B)}. \tag{13.2}$$

Here, the numerator and denominator polynomials are defined, as in Chapter 12, as

$$\omega(B) = \omega_0 - \omega_1 B - \ldots - \omega_s B^s$$

and

$$\delta(B) = 1 - \delta_1 B - \ldots - \delta_r B^r,$$

with the roots of $\delta(B)$ all assumed to be less than unity. The possibility that there may be a delay of b periods before X begins to influence Y is allowed for by the factorisation of the numerator in (13.2): if there is a contemporaneous relationship, $b = 0$. The parameter b is sometimes referred to as the *dead-time* of the model. The relation between the impulse response weights υ_i and the parameters $\omega_0, \ldots, \omega_s$, $\delta_1, \ldots, \delta_r$ and b can be obtained by equating the coefficients of B^j in

$$\delta(B)\,\upsilon(B) = \omega(B)\,B^b.$$

Exhibit 13.1 presents a variety of simple cases for illustration.

The construction of transfer function-noise models of the type (13.1) can be split into the familiar steps of identification, estimation and checking.

Exhibit 13.1 *Transfer functions and impulse response weights*

r	s	b	Transfer function	Impulse response, v_j
0	0	b	$v(B) = \omega_0 B^b$	$0 \quad j<b$ $\omega_0 \quad j=b$ $0 \quad j>b$
0	1	b	$v(B) = (\omega_0 - \omega_1 B) B^b$	$0 \quad j<b; \quad -\omega_1 \quad j=b+1$ $\omega_0 \quad j=b; \quad 0 \quad j>b+1$
0	2	b	$v(B) = (\omega_0 - \omega_1 B - \omega_2 B^2) B^b$	$0 \quad j<b \quad ; \quad -\omega_2 \quad j=b+2$ $\omega_0 \quad j=b \quad ; \quad 0 \quad j>b+2$ $-\omega_1 \quad j=b+1;$
1	0	b	$v(B) = \dfrac{\omega_0 B^b}{1-\delta_1 B}$	$0 \quad j<b$ $\omega_0 \quad j=b$ $\delta_1 v_{j-1} \quad j>b$
1	1	b	$v(B) = \dfrac{(\omega_0 - \omega_1 B) B^b}{1-\delta_1 B}$	$0 \quad j<b; \quad \delta_1 \omega_0 - \omega_1 \quad j=b+1$ $\omega_0 \quad j=b; \quad \delta_1 v_{j-1} \quad j>b+1$
1	2	b	$v(B) = \dfrac{(\omega_0 - \omega_1 B - \omega_2 B^2) B^b}{1-\delta_1 B}$	$0 \quad j<b \quad ; \quad \delta_1^2 \omega_0 - \delta_1 \omega_1 - \omega_2 \quad j=b+2$ $\omega_0 \quad j=b \quad ; \quad \delta_1 v_{j-1} \quad j>b+2$ $\delta_1 \omega_0 - \omega_1 \quad j=b+1;$
2	0	b	$v(B) = \dfrac{\omega_0 B^b}{1-\delta_1 B - \delta_2 B^2}$	$0 \quad j<b$ $\omega_0 \quad j=b$ $\delta_1 v_{j-1} + \delta_2 v_{j-2} \quad j>b$
2	1	b	$v(B) = \dfrac{(\omega_0 - \omega_1 B) B^b}{1-\delta_1 B - \delta_2 B^2}$	$0 \quad j<b; \quad \delta_1 \omega_0 - \omega_1 \quad j=b+1$ $\omega_0 \quad j=b; \quad \delta_1 v_{j-1} + \delta_2 v_{j-2} \quad j>b+1$
2	2	b	$v(B) = \dfrac{(\omega_0 - \omega_1 B - \omega_2 B^2) B^b}{1-\delta_1 B - \delta_2 B^2}$	$0 \quad j<b \quad ; \quad (\delta_1^2 + \delta_2) \omega_0 - \delta_1 \omega_1 - \omega_2 \quad j=b+2$ $\omega_0 \quad j=b \quad ;$ $\delta_1 \omega_0 - \omega_1 \quad j=b+1; \quad \delta_1 v_{j-1} + \delta_2 v_{j-2} \quad j>b+2$

13.2.1 *Identification*

The identification strategy proposed by Box and Jenkins (1976, chapter 11) revolves around the *cross correlation* function (CCF) between Y and X, defined from the *cross covariance* coefficients

$$\gamma_{XY}(k) = E[(X_t - \mu_X)(Y_{t+k} - \mu_Y)], \quad k = 0, 1, 2, \ldots$$

as

$$\rho_{XY}(k) = \frac{\gamma_{XY}(k)}{\sigma_X \sigma_Y},$$

where (μ_X, σ_X^2) and (μ_Y, σ_Y^2) are the mean and variance of X and Y respectively. Recalling the discussion of Chapter 5, where it was only possible to consider the mean, variance and autocorrelations to be constant through time if the univariate stochastic process was stationary, use of the cross covariance and correlation functions require, analogously, that the bivariate stochastic process (X_t, Y_t) be jointly stationary. Obviously, this assumption may not be satisfied for the levels of the two series and Box and Jenkins (1976, chapter 11) emphasise the need for prior differencing of X_t and Y_t to render the series jointly stationary. In general, different operators, $U_1(B)$ and $U_2(B)$ say, where $U_i(B) = \nabla^{d_i}\nabla_s^{D_i}$, may be employed to ensure that $(U_1(B)X_t, U_2(B)Y_t)$ is jointly stationary (see Abraham (1985)), but we will follow Box and Jenkins (1976, page 372) by assuming, for ease of exposition, that the process (x_t, y_t), where $x_t = \nabla^d X_t$ and $y_t = \nabla^d Y_t$, is stationary. Premultiplying equation (13.1) by ∇^d then yields

$$y_t = \upsilon(B)x_t + n_t, \tag{13.3}$$

where $n_t = \nabla^d N_t$. If we then assume that n_t follows an ARMA(p, q) process,

$$n_t = \frac{\theta(B)}{\phi(B)} a_t, \tag{13.4}$$

the combined transfer function-noise model can then equivalently be written as

$$y_t = \upsilon(B)x_t + \frac{\theta(B)}{\phi(B)} a_t \tag{13.5}$$

or

$$Y_t = \upsilon(B)X_t + \frac{\theta(B)}{\nabla^d \phi(B)} a_t, \tag{13.6}$$

showing that the observed relationship between the nonstationary Y_t and X_t is contaminated by a nonstationary noise process, although it is possible, under certain circumstances, for this noise to be stationary: see the discussion of cointegration in section 13.4 below.

The cross-correlation between x and y at lag k may then be defined as

$$\rho_{xy}(k) = \frac{\gamma_{xy}(k)}{\sigma_x \sigma_y}, \quad k = 0, 1, 2, \ldots$$

and estimates of the associated CCF from a paired sample of observations $(x_1, y_1), (x_2, y_2), \ldots, (x_n, y_n)$ may be calculated as

$$r_{xy}(k) = \frac{c_{xy}(k)}{s_x s_y}, \quad k = 0, 1, 2, \ldots$$

where

$$c_{xy}(k) = n^{-1} \sum_{t=1}^{n-k} (x_t - \bar{x})(y_{t+k} - \bar{y}), \quad k = 0, 1, 2, \ldots$$

$$s_x^2 = n^{-1} \sum_{t=1}^{n} (x_t - \bar{x})^2, s_y^2 = n^{-1} \sum_{t=1}^{n} (y_t - \bar{y})^2,$$

$$\bar{x} = n^{-1} \sum_{t=1}^{n} x_t, \bar{y} = n^{-1} \sum_{t=1}^{n} y_t.$$

In general, the variances and covariances of the sample cross-correlation function (the SCCF) are not easy to calculate. In two special cases, however, useful approximations are available:

(i) If x_t is a white noise sequence, y_t is an autocorrelated series with autocorrelations $\rho_y(k)$, and if y_t and x_t are uncorrelated at all lags, so that $\rho_{xy}(k) = 0$ for all k, then

$$V[r_{xy}(k)] \simeq (n-k)^{-1}.$$

Furthermore, the correlation between $r_{xy}(k)$ and $r_{xy}(k+m)$ is given by

$$\rho[r_{xy}(k), r_{xy}(k+m)] \simeq \rho_y(m).$$

(ii) If both x_t and y_t are white noise sequences and x_t and y_t are uncorrelated, then

$$\rho[r_{xy}(k), r_{xy}(k+m)] = 0.$$

Hence, even if x_t and y_t are *not* cross-correlated, the SCCF can be expected to vary about zero with standard deviation $(n-k)^{-\frac{1}{2}}$ in a systematic pattern typical of the behaviour of the ACF of y_t.

Given the combined transfer function-noise model of (13.3) and (13.4) and the calculated SCCF, values of the parameters r, s and b of the (rational approximation to the) transfer function $\upsilon(B)$ and the parameters p and q of the noise process can be obtained by the following procedure, which was suggested by Box and Jenkins (1976, chapter 11).

Initially, we might be tempted to specify the transfer function $\upsilon(B)$

directly from the cross-correlations $r_{xy}(k)$. As we have seen, however, the sampling properties of the SCCF from autocorrelated series can be difficult to assess and the computation of the υ_k weights from $r_{xy}(k)$ is not always easy. Fortunately, if the input series is white noise, simplifications are possible. This leads to the technique of *prewhitening* as a means of identifying the appropriate values of r, s and b in the transfer function. This procedure may be described as follows:

Step 1: Prewhiten the input series by first obtaining the ARMA model

$$\phi_x(B)\,x_t = \theta_x(B)\,\alpha_t$$

and then generating the white noise series α_t, with variance σ_α^2, as

$$\alpha_t = \theta_x^{-1}(B)\,\phi_x(B)\,x_t.$$

Step 2: Transform the output series using the *same filter*, i.e. generate the series

$$\beta_t = \theta_x^{-1}(B)\,\phi_x(B)\,y_t.$$

The cross-correlations between α_t and β_t play an important role in the specification of the transfer function. To see this, note that if the filter $\theta_x^{-1}(B)\,\phi_x(B)$ is applied to both sides of equation (13.3) we get

$$\beta_t = \upsilon(B)\,\alpha_t + n_t^*, \tag{13.7}$$

where $n_t^* = \theta_x^{-1}(B)\,\phi_x(B)\,n_t$. This implies that the transfer function between α and β is the same as that between X and Y. Multiplying both sides of (13.7) by α_{t-k} and taking expectations obtains

$$E(\alpha_{t-k}\beta_t) = \{\upsilon_0\,E(\alpha_{t-k}\,\alpha_t) + \ldots + \upsilon_k\,E(\alpha_{t-k}^2) + \ldots\} + E(\alpha_{t-k}\,n_t^*).$$

Since $E(\alpha_{t-k}\,\alpha_{t-j}) = 0$ for $j \neq k$ and $E(\alpha_{t-k}\,n_t^*) = 0$ (as x_t and n_t are uncorrelated), this yields

$$\gamma_{\alpha\beta}(k) = \upsilon_k\,\sigma_\alpha^2.$$

Thus

$$\upsilon_k = \frac{\gamma_{\alpha\beta}(k)}{\sigma_\alpha^2} = \frac{\sigma_\beta}{\sigma_\alpha}\frac{\gamma_{\alpha\beta}(k)}{\sigma_\alpha\,\sigma_\beta} = \frac{\sigma_\beta}{\sigma_\alpha}\rho_{\alpha\beta}(k), \quad k = 0, 1, 2, \ldots,$$

where σ_β^2 is the variance of the β_t process. The coefficient υ_k is therefore proportional to the cross-correlation $\rho_{\alpha\beta}(k)$ between α_t and β_t at lag k. Hence υ_k can be estimated by

$$\hat{\upsilon}_k = \frac{s_\beta}{s_\alpha} r_{\alpha\beta}(k), \tag{13.8}$$

where s_α and s_β are the estimated standard deviations of the α_t and β_t series and $r_{\alpha\beta}(k)$ is the SCCF between α_t and β_t.

Step 3: Calculate the SCCF and estimate the weights $\hat{\upsilon}_k$ from (13.8), assessing the significance of the cross-correlations, or equivalently of the $\hat{\upsilon}_k$, by comparing $r_{\alpha\beta}(k)$ with its standard error $(n-k)^{-\frac{1}{2}}$. The relationship between the $\hat{\upsilon}_k$ and the parameters b, $\omega_0, \ldots, \omega_s$ and $\delta_1, \ldots, \delta_r$ can be obtained by comparing coefficients of B^k in

$$(1-\delta_1 B - \ldots - \delta_r B^r)(\upsilon_0 + \upsilon_1 B + \ldots) = (\omega_0 - \omega_1 B - \ldots - \omega_s B^s) B^b,$$

leading to

$$\upsilon_k = \begin{cases} 0 & k = \ldots, -1, 0, 1, \ldots, b-1 \\ \delta_1 \upsilon_{k-1} + \delta_2 \upsilon_{k-2} + \ldots + \delta_r \upsilon_{k-r} + \omega_0, & k = b \\ \delta_1 \upsilon_{k-1} + \delta_2 \upsilon_{k-2} + \ldots + \delta_r \upsilon_{k-r} - \omega_{k-b}, & k = b+1, \ldots, b+s \\ \delta_1 \upsilon_{k-1} + \delta_2 \upsilon_{k-2} + \ldots + \delta_r \upsilon_{k-r}, & k = b+s+1, \ldots \end{cases} \quad (13.9)$$

Thus the weights are described by:

(i) b zero values $\upsilon_0, \upsilon_1, \ldots, \upsilon_{b-1}$,

(ii) an additional $s-r+1$ values $\upsilon_b, \upsilon_{b+1}, \ldots, \upsilon_{b+s-r}$ following no fixed pattern, and

(iii) for $k \geqslant b = s-r+1$, values υ_k following an rth order difference equation with starting values $\upsilon_{b+s-r+1}, \ldots \upsilon_{b+s}$.

The orders (r, s, b) can be determined from either the patterns in the estimated impulse response weights $\hat{\upsilon}_k$ or the patterns in the cross-correlations $r_{\alpha\beta}(k)$. Once these orders are chosen, preliminary estimates of the $\hat{\omega}_i$ and $\hat{\delta}_j$ can be obtained from (13.9) or by estimating

$$y_t = \frac{\omega(B) B^b}{\delta(B)} + n_t$$

with n_t assumed to be white noise.

We now need to determine the parameters p and q of the $\theta(B)$ and $\phi(B)$ polynomials in the noise model (13.4). This can be done as follows:

Step 4: Generate the noise series

$$\hat{n}_t = y_t - \hat{\upsilon}(B) x_t = y_t - \hat{\delta}^{-1}(B) \hat{\omega}(B) x_{t-b}.$$

Step 5: Calculate the SACF and SPACF of $\hat{n}_t$ and specify an appropriate ARMA model

$$\phi(B) \hat{n}_t = \theta(B) a_t.$$

Combining the transfer function and noise models then leads to

$$y_t = \frac{\omega(B)}{\delta(B)} x_{t-b} + \frac{\theta(B)}{\phi(B)} a_t. \quad (13.10)$$

13.2.2 *Estimation*

The parameters

$$\beta' = (\omega', \delta', \phi', \theta') = (\omega_0, \omega_1, \ldots, \omega_s; \delta_1, \ldots, \delta_r; \phi_1, \ldots, \phi_p; \theta_1, \ldots, \theta_q)$$

in (13.10) must now be estimated from the sample $(x_t, y_t), t = 1, 2, \ldots, n$. This model can also be written as

$$\delta_r(B)\,\phi_p(B)\,y_t = \phi_p(B)\,\omega_s(B)\,x_{t-b} + \delta_r(B)\,\theta_q(B)\,a_t,$$

where, as in Chapter 10, the orders of each lag polynomial have been appended as subscripts. This implies that

$$\begin{aligned} a_t = y_t + d_1 y_{t-1} + \ldots + d_{p+r} y_{t-p-r} + c_0 x_{t-b} + c_1 x_{t-b-1} \\ + \ldots + c_{p-s} x_{t-b-p-s} \\ + b_1 a_{t-1} + b_2 a_{t-2} + \ldots + b_{r+q} a_{t-r-q}, \end{aligned} \qquad (13.11)$$

where the coefficients in $\boldsymbol{d} = (d_1, \ldots, d_{p+r})$ are obtained from $\boldsymbol{\delta}$ and $\boldsymbol{\phi}$ by equating coefficients of B^j in $\delta(B)\,\phi(B) = d(B)$, and the coefficients in $\boldsymbol{c} = (c_0, \ldots, c_{p-s})$ and $\boldsymbol{b} = (b_1, \ldots, b_{r+q})$ are obtained from $\boldsymbol{\omega}$ and $\boldsymbol{\phi}$, and $\boldsymbol{\delta}$ and $\boldsymbol{\theta}$, respectively, in a similar fashion.

Assuming that $a_t \sim N(0, \sigma^2)$, the conditional likelihood function can be written as

$$L(\boldsymbol{\beta}, \sigma^2 | \boldsymbol{x}, \boldsymbol{y}, \boldsymbol{x}_0, \boldsymbol{y}_0, \boldsymbol{a}_0) \propto \sigma^{-n} \exp\left[\frac{1}{2\sigma^2} \sum_{t=1}^{n} a_t^2(\boldsymbol{\beta})\right],$$

where $\boldsymbol{x} = (x_1, \ldots, x_n)'$ and $\boldsymbol{y} = (y_1, \ldots, y_n)'$. Conditional on the starting values $\boldsymbol{x}_0 = (x_{1-b-(p+s)}, \ldots, x_0)'$, $\boldsymbol{y}_0 = (y_{1-p-r}, \ldots, y_0)'$, and $\boldsymbol{a}_0 = (a_{1-r-q}, \ldots, a_0)'$, $a_t(\boldsymbol{\beta})$ may be calculated from (13.11). Usually this calculation is begun from $t_* = \max\{p+r+1, b+p+s+1\}$ and the initial a_t's are set equal to zero. Conditional ML or LS estimates $\hat{\boldsymbol{\beta}}$ are then obtained by minimising

$$S = \sum_{t=t_*}^{n} a_t^2(\boldsymbol{\beta}).$$

An estimate of σ^2 is given by

$$\hat{\sigma}^2 = m^{-1} \sum_{t=t_*}^{n} \hat{a}_t^2,$$

where the $\hat{a}_t(t = t_*, \ldots, n)$ are the $m = n - t_* + 1$ residuals from the fitted model.

Pierce (1972) shows that the estimates $\hat{\boldsymbol{\beta}}$ will possess a limiting normal distribution with mean $\boldsymbol{\beta}$ and that the transfer function estimates $\hat{\boldsymbol{\beta}}_1' = (\hat{\omega}', \hat{\delta}')$ are asymptotically independent of the noise estimates $\hat{\boldsymbol{\beta}}_2' = (\hat{\boldsymbol{\phi}}', \hat{\boldsymbol{\theta}}')$.

It therefore follows that the information matrix

$$I(\hat{\boldsymbol{\beta}}) = -E\left[\frac{\partial^2 \ln L}{\partial \boldsymbol{\beta}\,\partial \boldsymbol{\beta}'}\bigg|_{\boldsymbol{\beta}=\hat{\boldsymbol{\beta}}}\right]$$

can be partitioned as

$$I(\hat{\boldsymbol{\beta}}) = -E\left[\begin{array}{c:c} \dfrac{\partial^2 \ln L}{\partial \boldsymbol{\beta}_1\,\partial \boldsymbol{\beta}_1'} & 0 \\ \hdashline 0 & \dfrac{\partial^2 \ln L}{\partial \boldsymbol{\beta}_2\,\partial \boldsymbol{\beta}_2'} \end{array}\Bigg|_{\boldsymbol{\beta}=\hat{\boldsymbol{\beta}}}\right],$$

the inverse of which can be used as an estimator of the asymptotic covariance matrix of $\hat{\boldsymbol{\beta}}' = (\hat{\boldsymbol{\beta}}_1', \hat{\boldsymbol{\beta}}_2')$. The structure of the information matrix, which allows the specification of the transfer function and noise terms to be investigated separately, has been exploited by Poskitt and Tremayne (1981) to develop a formal hypothesis testing procedure for model identification: details of this may be found in the above reference.

13.2.3 *Diagnostic checking*

After (13.10) has been estimated, various checks may be performed to determine its adequacy. The fitted model can be inadequate if (a) the noise model, (b) the transfer function, or (c) both noise and transfer function models, are incorrect. Box and Jenkins (1976, pages 392–3) show that

(i) if the noise model is inadequate, then $\rho_a(k) \neq 0$ for some k and $\rho_{\alpha a}(k) = 0$ for all k, where α_t is the prewhitened input series,

(ii) if the transfer function model is inadequate, then $\rho_a(k) \neq 0$ and $\rho_{\alpha a}(k) \neq 0$ for some k.

Thus, if the residuals are still autocorrelated, but the cross-correlations $r_{\alpha\hat{a}}(k)$ are negligible compared to their standard errors $(n-k)^{-\frac{1}{2}}$, the noise model needs to be modified. If, in addition, significant cross-correlations $r_{\alpha\hat{a}}(k)$ are observed, then the transfer function model, and possibly the noise model, have to be respecified.

In addition, various portmanteau tests can be carried out. To test whether the residuals are still autocorrelated, we may construct (Abraham and Ledolter (1983, page 344))

$$Q_1 = m(m+2)\sum_{k=1}^{K}\frac{r_{\hat{a}}^2(k)}{m-k} \sim \chi^2(K-p-q),$$

where the $r_{\hat{a}}(k)$ $(k = 1, 2, \ldots, K)$ denote the first K sample autocorrelations

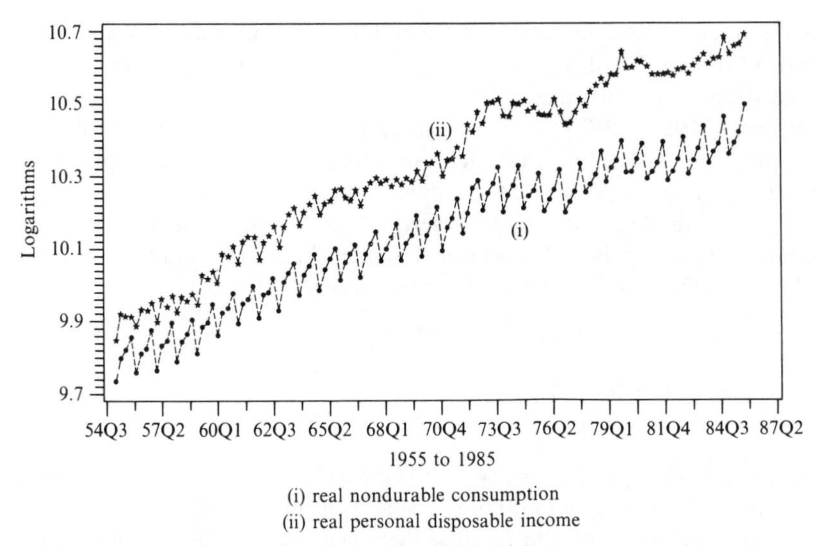

Exhibit 13.2 Logarithms of real consumption and income

from the m available residuals. To test whether the input and residuals are correlated, we calculate

$$Q_2 = m(m+2)\sum_{k=0}^{K}\frac{r^2_{\alpha\hat{a}}(k)}{m-k} \sim \chi^2(K-r-s).$$

A number of other identification methods have been proposed. Haugh and Box (1977) and Granger and Newbold (1977) propose prewhitening both input and output series using univariate models fitted to each of them, then combining these with the fitted transfer function (13.7). This is useful when the assumption of no feedback from Y to X is untenable (see Chapter 14), but appears unnecessarily to complicate matters in the simple case considered here. Liu and Hanssens (1982), Hanssens and Liu (1983), Lii (1985) and Tsay (1985) all propose the use of the 'corner' method (briefly mentioned in Chapter 8.7) for determining the form of the transfer function, but since this approach is especially useful when dealing with models containing more than one input, discussion of it will be postponed until Section 13.3.

Example 13.1: The relationship between UK consumption and income

As an illustration of the 'Box–Jenkins' approach to transfer function modelling, we consider the development of a transfer function–

noise model of the relationship between UK real nondurable consumption expenditure and real personal disposable income. Seasonally unadjusted logarithmically transformed quarterly observations on these two series for the period 1955 to 1985 are plotted in Exhibit 13.2. On the (as we will later see, invalid) assumption that there is no feedback from consumption (Y) to income (X), we wish to build a model of the form (13.1). Conventional univariate analysis of the two series found that both needed prior differencing using the $\nabla\nabla_4$ operator to render them individually stationary. The univariate models chosen for these series were

$$y_t = (1-0.193B)(1-0.598B^4)\, a_{1t}, \quad s_1 = 0.0116$$

$$x_t = \frac{(1-0.748B^4)}{(1+0.301B)}\alpha_t, \qquad s_\alpha = 0.0212,$$

where $y_t = \nabla\nabla_4 Y_t$ and $x_t = \nabla\nabla_4 X_t$. The SCCF between α_t and β_t, where $\beta_t = (1-0.748B^4)^{-1}(1+0.301B)\, y_t$, is shown in Exhibit 13.3(a), along with the estimates of the impulse response weights $\hat{\upsilon}_j$ obtained using equation (13.8). It is clear that only $\hat{\upsilon}_0$ and $\hat{\upsilon}_1$ are significantly different from zero, suggesting that $b = r = 0$ and $s = 1$; i.e. $\upsilon(B) = \omega_0 - \omega_1 B$. (An alternative specification with $b = s = 0$ and $r = 1$, i.e. $\upsilon(B) = \omega_0/(1-\delta B)$ was also investigated, but fitted slightly worse than the above model.)

CLS estimation assuming that n_t is white noise obtained

$$y_t = (0.207+0.164B)\, x_t + \hat{n}_t,$$

and the resulting SACF of $\hat{n}_t$ is shown in Exhibit 13.3(b). From this a multiplicative seasonal moving average noise model

$$\hat{n}_t = (1-\theta_1 B)(1-\theta_4 B^4)\, a_t$$

is identified, consistent with the univariate structure of y_t, and combining this with the above transfer function yielded, on CLS estimation

$$y_t = \left(\begin{matrix} 0.284+0.169B \\ (0.049)\;(0.042) \end{matrix}\right) x_t + \left(\begin{matrix} 1-0.578B \\ \quad(0.098) \end{matrix}\right)\left(\begin{matrix} 1-0.470B^4 \\ \quad(0.087) \end{matrix}\right) a_t$$

$$s_{\hat{a}} = 0.0098;\, R^2_* = 0.29;\, Q_1(4) = 8.8;\, Q_2(4) = 4.9,$$

where $Q_1(v)$ and $Q_2(v)$ are the portmanteau statistics defined in section 13.2.3, both with v degrees of freedom. All parameters are precisely estimated, the inclusion of x_t as an input reduces the residual variation in y_t by almost thirty per cent (R^2_* is Pierce's (1979b) measure defined here as $R^2_* = 1-(\hat{s}^2_a/\hat{s}^2_1)$), and neither portmanteau statistic indicates serious misspecification of either the transfer function or noise parts of the model.

More detailed investigation of the residuals revealed two large values at

Exhibit 13.3

(a) *SCCF between* α_t *and* β_t *and estimates of* $\hat{\upsilon}_k$

k	0	1	2	3	4	5	6	7	8
$r_{\alpha\beta}(k)$	0.379	0.246	0.006	0.057	−0.069	−0.026	−0.026	−0.019	−0.082
	(0.092)	(0.092)	(0.092)	(0.093)	(0.093)	(0.094)	(0.094)	(0.094)	(0.095)
$\hat{\upsilon}_k$	0.213	0.138	0.003	0.032	−0.039	−0.015	0.039	−0.011	0.046

Standard errors of $r_{\alpha\beta}(k)$ shown in parentheses; $(n-k)^{-\frac{1}{2}}, n = 119$. $s_\alpha = 0.0212$; $s_\beta = 0.0118$.

(b) *SACF of noise series* $\hat{n}_t$

k	1	2	3	4	5	6	7	8
$\hat{r}_k$	−0.491	0.119	0.125	−0.410	0.377	−0.250	0.068	0.103

1968 Q1 and 1973 Q1. These correspond to occasions when advance warning was given of purchase tax increases in the Budgets of these years, which caused a considerable amount of expenditure switching from the second to the first quarter. Pulse interventions were therefore defined as

$$\varepsilon_{1t} = \begin{cases} 1 \text{ for } 1968\,\text{Q}1 \\ 0 \text{ elsewhere} \end{cases} \quad \text{and} \quad \varepsilon_{2t} = \begin{cases} 1 \text{ for } 1973\,\text{Q}1 \\ 0 \text{ elsewhere} \end{cases}$$

and included as additional inputs in the transfer function-noise model. After some re-estimation, the following model was arrived at; now written in terms of the undifferenced variables Y_t and X_t

$$Y_t = \left(\underset{(0.047)}{0.269} + \underset{(0.041)}{0.162}B\right)X_t + \underset{(0.007)}{0.015}\varepsilon_{1t} + \underset{(0.007)}{0.037}\varepsilon_{2t} + \frac{\left(1-\underset{(0.099)}{0.553}B\right)\left(1-\underset{(0.090)}{0.420}B^4\right)}{\nabla\nabla_4}a_t$$

$s_a = 0.0095$; $Q_1(4) = 7.7$, $Q_2(4) = 4.4$.

The 1968 intervention increases consumption expenditure by 1.5% in the first quarter of that year, while the 1973 intervention increases consumption expenditure correspondingly by 3.7%.

Since the variables are logarithms, the transfer function implies a short-run income elasticity of 0.269 and a long-run elasticity, reached after just one quarter, of 0.431.

13.3 Multiple input transfer function-noise models

Having developed a methodology for identifying a transfer function-noise model having a single input, the natural extension is to consider the identification of a multiple input model, i.e.

$$\begin{aligned} Y_t &= \sum_{i=1}^{M} v_i(B)\,X_{it} + N_t \\ &= \sum_{i=1}^{M} \frac{\omega_i(B)\,B^{b_i}}{\delta_i(B)} X_{it} + N_t, \end{aligned} \tag{13.12}$$

where

$$\omega_i(B) = \omega_{i0} - \omega_{i1}B - \ldots - \omega_{ir_i}B^{r_i}$$

and

$$\delta_i(B) = 1 - \delta_{i1}B - \ldots - \delta_{is_i}B^{s_i}.$$

The simplest way to proceed is to use the Box–Jenkins approach in a 'piece-meal' fashion, identifying a set of single input transfer functions between Y and X_1, Y and X_2, etc., and then combining them to identify the noise model. Estimation and diagnostic checking can then be carried out by standard extensions of the procedures discussed in the previous section.

The problem here is that the SCCFs between each prewhitened input and the correspondingly filtered output may be misleading if the X_i's are intercorrelated. However, it should be noted that, typically, these input series will have been differenced to induce stationarity, and it is well known that such filtering reduces the correlation between series. Thus it may often be the case that the stationary x_{it} are only weakly related and, analogous to the situation of independent regressors in multiple regression analysis, this piece-meal approach may work quite well, particularly since it is being used only to identify possible values of the parameters (b_i, r_i, s_i). Joint estimation of the combined model and judicious use of diagnostic checking procedures may then allow an adequately specified model to be obtained.

Example 13.2: Including the price level as a determinant of consumption

Deaton (1977) has argued that the rate of inflation is an important determinant of consumption expenditure and Davidson et al. (1978) have provided empirical evidence supporting this for UK data. As a first step towards such a specification, the logarithm of the implicit consumption expenditure deflator was considered as an additional input to the consumption–income model developed in Example 13.1. Denoting this variable as X_2, and thus redefining income as X_1, the following univariate model for the price series was obtained, with $x_{2t} = \nabla\nabla_4 X_{2t}$:

$$x_{2t} = \frac{(1-0.637B^4)}{(1-0.383B-0.246B^2)}\alpha_{2t}, \quad s_{\alpha_2} = 0.0103.$$

On computing the SCCF between α_{2t} and $\beta_{2t} = (1-0.383B-0.246B^2)(1-0.637B^4)^{-1}y_t$, the only significant correlation was at lag 0, implying that $b_2 = r_2 = s_2 = 0$ and $\upsilon_2(B) = \omega_{20}$. Including this transfer function into the model of Example 13.1 produced, on estimation,

$$\begin{aligned} Y_t = & \left(\underset{(0.048)}{0.237} + \underset{(0.037)}{0.154B}\right)X_{1t} - \underset{(0.041)}{0.182}X_{2t} + \underset{(0.007)}{0.013}\varepsilon_{1t} \\ & + \underset{(0.007)}{0.016}\varepsilon_{2t} + N_t, \end{aligned}$$

where

$$\nabla\nabla_4 N_t = \left(1 - \underset{(0.090)}{0.746B}\right)\left(1 - \underset{(0.092)}{0.407^4}\right)a_t$$

$$s_a = 0.0086; \quad Q_1(4) = 6.9; \quad Q_{21}(4) = 3.1; \quad Q_{22}(5) = 2.6.$$

The price coefficient is correctly signed and precisely estimated, the residual standard error is substantially decreased, and all diagnostic checks are passed (the extension of the notation is obvious).

13.3.1 *Identification by the 'corner method'*

If it is felt that correlation between the inputs may seriously affect the set of bivariate SCCFs, then the following procedure may be used, which is based upon that proposed by Liu and Hanssens (1982) and Hanssens and Liu (1983) (see also Tsay (1985), Lii (1985) and Edlund (1984)).

Following a similar procedure to that of the single input case, each series can be differenced to stationarity, leading to

$$y_t = \sum_{i=1}^{M} v_i(B)\, x_{it} + n_t. \tag{13.13}$$

By defining K_i to be the maximum lag length of the $v_i(B)$ polynomial that it is thought sensible to consider, equation (13.13) can be estimated by least squares. Two well known problems arise in such estimation. The first is that of multicollinearity, brought about in this case if one, or more, of the input series contains an autoregressive factor with a root close to unity, so that a number of the sample autocorrelations will be close to one. This will result in the regressor cross product matrix being ill-conditioned for inversion. Liu and Hanssens (1982) suggest using a common filter on both the input and output series, although overdifferencing the offending input series would seem to be the easiest solution.

The second problem is that, typically, n_t will be serially correlated. We may therefore follow a procedure similar to that outlined for the single input case: estimate (13.13) by least squares under the assumption that n_t is white noise, identify an appropriate ARMA model for n_t from the resulting residuals, and re-estimate with this specification of the noise model.

We then wish parsimoniously to parameterise the transfer functions $v_i(B)$ in terms of the rational polynomials $\omega_i(B)\, B^{b_i}/\delta_i(B)$, i.e. to find values for b_i, r_i and s_i. Both Liu and Hanssens (1982) and Tsay (1985) propose using a modification of the corner method for ARMA identification (Beguin et al., 1980) which, they argue, is preferable to the method outlined in Step 3 of the single input procedure when the v_{ij}'s do not die out quickly (and, of course, which can be used in place of Step 3 even if they do).

Dropping the i subscript for convenience, let $v_{\max}$ be the maximum value of $|v_j|, (j = 0, 1, \ldots, K)$, and let $\eta_j = v_j/v_{\max}$. Then for each input

Exhibit 13.4 (a) *The C–array*

f \ g	0	1	2	3	...
0	$\Delta(0,0)$	$\Delta(0,1)$	$\Delta(0,2)$	$\Delta(0,3)$	
1	$\Delta(1,0)$	$\Delta(1,1)$	$\Delta(1,2)$	$\Delta(1,3)$	...
2	$\Delta(2,0)$	$\Delta(2,1)$	$\Delta(2,2)$	$\Delta(2,3)$	
3		.			
4		.			
⋮		⋮			

(b) *The pattern of the C–array*

f \ g	0	1 ...	$r-1$	r	$r+1$	$r+2$	...	
0	0	0 ...	0	0	0	0	...	b rows
⋮	⋮					⋮		
$b-1$	0	0 ...	0	0	0	0	...	
b	×	× ...	×	×	×	×	...	$s+1$ rows
⋮	⋮					⋮		
$b+s$	*	* ...	×	×	×	×	...	
$b+s+1$	*	* ...	×	0	0	0	...	
$b+s+2$	*	* ...	×	0	0	0		
⋮	⋮					⋮		

Note: 0, ×, *and* * denote zero, nonzero, and no-pattern terms, respectively.

variable we can define $\Delta(f, g)$ to be the determinant of a $(g+1) \times (g+1)$ matrix of η_j's:

$$\Delta(f,g) = \begin{vmatrix} \eta_f & \eta_{f-1} & \cdots & \eta_{f-g} \\ \eta_{f+1} & \eta_f & \cdots & \eta_{f-g+1} \\ \vdots & \vdots & \ddots & \vdots \\ \eta_{f+g} & \eta_{f+g-1} & \cdots & \eta_f \end{vmatrix}, \qquad (13.14)$$

where $f, g > 0$ and $\eta_j = 0$ if $j < 0$. These determinants can be arranged in a two-way table known as the C-array, shown in Exhibit 13.4(a). From the patterns of the impulse response weights v_j given in equation (13.9), it is easy to see that, for given values of b, s and r, the C-array will exhibit a special pattern that can be used to specify these parameters in practice. This special pattern is shown in Exhibit 13.4(b). In general, there are b

rows (labelled $0, 1, \ldots, b-1$) of zeros at the beginning of the array, followed by $s+1$ rows of nonzero entries (labelled $b, b+1, \ldots, s+b$), and finally a rectangle formed by zeros at the lower-right corner of the array. The coordinates of the upper-left vertex of the rectangle are $(b+s+1, r)$, from which r and s can be identified.

In practice the estimated υ_j's are subject to sampling error and, consequently, rather than zeros in the C-array we will find small, but nonzero, values. Tsay (1985) has provided a method of determining the approximate sample standard deviations of the estimates of $\Delta(f, g)$ but, in most cases, estimated arrays will show a sufficiently sudden drop in values in the upper and lower right-hand corners to enable identification of b, r and s.

Example 13.3: Identifying the consumption transfer function by the corner method

The corner method was employed on the model analysed in Example 13.2. The maximum order of the transfer function polynomials was set at $K_i = 4$, $i = 1$ and 2 and, after initial estimation has allowed identification of a noise model, estimates of the impulse response coefficients were obtained as

j	0	1	2	3	4
$\hat{\upsilon}_{1j}$	0.257	0.126	−0.028	0.003	−0.013
$\hat{\upsilon}_{2j}$	−0.115	−0.086	0.089	−0.182	0.094

from which the C-arrays shown in Exhibit 13.5 were calculated. From these arrays the parameter values $b = s = 0$ and $r = 2$ are indicated for X_1, while for X_2 the values $b = s = 0$ and $r = 4$ are indicated. Estimation of the implied model found that all the denominator polynomial coefficients for X_{2t} were insignificant and, after respecification, the following model was obtained:

$$Y_t = \frac{\underset{(0.046)}{0.242}}{\left(1 - \underset{(0.162)}{0.560}B + \underset{(0.122)}{0.226}B^2\right)} X_{1t} - \underset{(0.043)}{0.189}X_{2t} + \underset{(0.007)}{0.012}\varepsilon_{1t} + \underset{(0.007)}{0.017}\varepsilon_{2t} + N_t,$$

where

$$\nabla\nabla_4 N_t = \left(1 - \underset{(0.091)}{0.719}B\right)\left(1 - \underset{(0.096)}{0.394}B^4\right)a_t.$$

$$s_a = 0.0088; \quad Q_1(4) = 6.0; \quad Q_{21}(3) = 4.3; \quad Q_{22}(5) = 3.6.$$

The fit of this model is almost identical to that of the model identified in

Exhibit 13.5 *C–array for input X_1*

f \ g	0	1	2	3	4
0	1.00	−0.49	−0.11	0.01	−0.05
1	1.00	0.35	0.01	−0.01	0.00
2	1.00	0.24	−0.02	0.00	−0.00
3	1.00	0.21	−0.01	0.00	0.00
4	1.00	0.16	−0.00	−0.00	−0.00

C–array for input X_2

f \ g	0	1	2	3	4
0	−0.63	−0.47	0.49	−1.00	0.51
1	0.40	0.53	−0.23	0.75	0.26
2	−0.25	−0.79	−0.70	−0.62	0.13
3	−0.16	−0.85	−0.01	−0.64	0.07
4	−0.10	−0.91	−0.77	−0.66	0.03

Example 13.2, but the models differ in the form of the transfer function between Y (consumption) and X_1 (income). In the previous model the response of Y to a change in X_1 is completed after one quarter, with a long-run elasticity of 0.41. In the above model, the time path of the response is cyclical (since the roots of $\delta_1(B)$ are complex), with a period of 6.7 quarters and a long-run elasticity of 0.36.

13.3.2 *Imposing restrictions from economic theory*

In the models of consumption expenditure developed in the above examples, little use has been made of economic theory, apart from directing attention towards those series that may be considered as likely 'inputs'. Theoretical considerations may, however, provide useful information in two respects. Deaton's (1977) theory of consumption behaviour suggests that the current level of inflation should also influence current consumption expenditure. Hence ∇x_{2t}, or equivalently x_{2t-1} (within $\omega_2(B)$), should be included as an additional input.

Furthermore, the standard theory of consumption behaviour states that, in the long run, the marginal propensity and average propensity to consume are identical, which implies that the long-run income elasticity of consumption is unity. Since logarithmic transformations of the variables have been used, this implies that $\sum \upsilon_j = 1$. This restriction can easily be imposed in the model

$$Y_t = \upsilon(B) X_t + N_t \tag{13.15}$$

by rewriting the equation as

$$Y_t - X_t = (\upsilon(B) - 1) X_t + N_t.$$

To ensure that $\sum \upsilon_j = \upsilon(1) = 1$, we can write

$$\upsilon(B) - 1 = (1 - B)\, \upsilon^*(B),$$

leading to the model

$$Y_t - X_t = \upsilon^*(B) \nabla X_t + N_t. \tag{13.16}$$

The relationship between the coefficients in $v^*(B)$ and those in $v(B)$ can easily be shown to be

$$v_k^* = \sum_{j=0}^{k} v_j - 1.$$

Thus the theoretical restriction that the long-run income elasticity is unity implies that the appropriate output variable is $Y_t - X_t$, which is the average propensity to consume, while the input variable becomes ∇X_t, the growth of income.

Example 13.4: Including inflation in the consumption function

Including ∇X_{2t} as an additional input to the model developed in Example 13.2, for example, obtained an estimated coefficient attached to it of 0.084 with a standard error of 0.089. It would therefore seem that inflation has no additional explanatory power once the price level has been included in the model.

Estimating the appropriate counterpart of equation (13.16) yielded the following model:

$$\begin{aligned} Y_t - X_{1t} = & -\begin{pmatrix} 0.601 + 0.184B \\ (0.060)\ (0.055) \end{pmatrix} \nabla X_{1t} - \underset{(0.106)}{0.045} X_{2t} \\ & + \underset{(0.019)}{0.016}\varepsilon_{1t} + \underset{(0.009)}{0.017}\varepsilon_{2t} + N_t, \end{aligned}$$

where

$$\nabla\nabla_4 N_t = \left(1 - \underset{(0.116)}{0.097}B\right)\left(1 - \underset{(0.086)}{0.554}B^4\right) a_t$$

$$s_a = 0.0132; \quad Q(4) = 12.8; \quad Q_{21}(4) = 53.6; \quad Q_{22}(5) = 8.2.$$

Comparing this model with that of Example 13.2, we see that s_a^2 has been increased by 125% and that the portmanteau statistics flag serious model misspecification: imposing the restriction $\upsilon(1) = 1$, that the long-run income elasticity is unity, is therefore invalid.

13.3.3 *Estimating the long-run multiplier*

Attention often focuses on the value of the long-run multiplier $v(1) = \sum v_j$, irrespective of any theoretical constraint placed upon it. While a point estimate of $v(1)$ is easily computed from the estimated transfer function (13.15), it is rather more difficult to estimate its variance by this method. The point estimate and its associated variance are, however, obtainable directly if (13.15) is rewritten as

$$\begin{aligned} Y_t &= v(1)X_t - (v(B) - v(1))X_t + N_t \\ &= v(1)X_t - (v_1\nabla_1 + v_2\nabla_2 + \ldots)X_t + N_t, \end{aligned}$$

where, as usual, ∇_i is the ith difference, $\nabla_i = (1 - B^i)$. Since $\nabla_i = \nabla(1 + B + B^2 + \ldots + B^{i-1})$, an estimable form of the transfer function is thus

$$Y_t = v(1)X_t - v^{**}(B)\nabla X_t + N_t,$$

where the relationship between the coefficients in $v^{**}(B)$ and $v(B)$ is

$$v_k^{**} = \sum_{j=k+1}^{\infty} v_j,$$

from which it is easily seen that imposing the restriction $v(1) = \sum v_j = 1$ leads directly to (13.16). Wickens and Breusch (1988) present a number of alternative formulations of dynamic models similar to the above which are capable of providing directly point estimates of long-run multipliers.

13.4 Differencing, cointegration and error correction mechanisms

The model building process outlined in the previous sections has been based upon the assumption that all the time series being considered, both output and inputs, are stationary. This, by a simple extension of the analysis of univariate time series, enables sample autocorrelation and cross-correlation functions to be employed usefully in model identification. Typically, stationarity of these series has been achieved by differencing, thus leading to models of the form (13.13).

In fact, the assumption of stationarity is by no means an innocuous one in the important context of inference in multiple time series regressions, of which equation (13.13) is but an example. To see this, consider the well known problem in econometrics of 'nonsense' or 'spurious' regressions. Granger and Newbold (1974) have examined the likely empirical consequences of such regressions, centering their attention on the standard textbook warning about the presence of serially correlated errors invalidating conventional procedures of inference. In particular, they were concerned with the specification of regression equations in terms of the levels of economic time series, which, as we have seen, are typically nonstationary and usually well represented by integrated processes of the ARIMA type, often appearing to be near random walks. Such regression equations, argue Granger and Newbold, frequently have high R^2 statistics yet also typically display highly autocorrelated residuals, indicated by very low Durbin–Watson (DW) statistics. They contend that, in such situations, the usual significance tests performed on the regression coefficients can be very misleading, and they provide Monte Carlo simulation evidence to show that the conventional significance tests are seriously biased towards rejection of the null hypothesis of no relationship, and hence towards acceptance of a spurious relationship, even when the series are generated as statistically independent random walks.

These findings led Granger and Newbold (1974) to suggest that, in the joint circumstances of a high R^2 and low DW statistic (a useful rule of thumb being $R^2 > \mathrm{DW}$), regressions should be run on the first differences of the variables, thus implying that all series in the regression are integrated processes of order one, denoted in Chapter 6 as $I(1)$ processes. Further empirical evidence in favour of first differencing in regression models is provided by Granger and Newbold (1977, pages 202–14) and Plosser and Schwert (1978), and both Nelson and Plosser (1982) and Schwert (1987), as discussed in Chapter 11, find that a great many economic time series are indeed adequately characterised as $I(1)$ processes.

These essentially empirical conclusions have been placed on a firm analytical foundation by Phillips (1986) and Phillips and Durlauf (1986). These papers demonstrate that, for regressions between general integrated random processes, the distributions of the conventional test statistics (e.g. t, F) are not at all like those derived under the assumption of stationarity: the regression coefficients do not converge in probability as the sample size increases, the distribution of the intercept diverges, and both the regression coefficients and R^2 statistic have non-degenerate distributions. Moreover, the distributions of both the t and F tests diverge, so that there are *no* asymptotically correct critical values for these conventional significance tests, and the DW statistic actually converges on zero.

Nevertheless, even given the strong implication from these studies that regression analysis using the levels of economic time series can only be undertaken with great care, many economists feel unhappy about analysing regressions fitted to first differences, which is the obvious 'time series analyst' solution to the nonstationarity problem. Phrases like 'throwing the baby out with the bath water' and, less prosaically, 'valuable long-run information being lost' are often heard. These worries centre around the existence of long-run, steady state equilibria, a concept that (primarily static) economic theory devotes considerable attention to.

To develop this argument, consider a dynamic model of the form

$$Y_t = \alpha + \beta X_t + \gamma X_{t-1} + \delta Y_{t-1} + u_t, \tag{13.17}$$

where, as usual, all variables are measured in logarithms. In steady state equilibrium, where $Y_t = Y_{t-1} = Y_e$, $X_t = X_{t-1} = X_e$ and $u_t = 0$, we have the solution

$$Y_e = \alpha' + \beta' X_e,$$

where $\alpha' = \alpha/(1-\delta)$ and $\beta' = (\beta+\gamma)/(1-\delta)$. If, on the other hand, the differenced model

$$\nabla Y_t = \beta \nabla X_t + \gamma \nabla X_{t-1} + \delta \nabla Y_{t-1} + v_t \tag{13.18}$$

is considered, all differences are zero in steady state equilibrium and so no solution is obtainable: we can say nothing about the long-run relationship between Y_t and X_t. Moreover, in a constant growth equilibrium, where $\nabla Y = \nabla X = g$, the levels model (13.17) has the solution

$$Y_e = \alpha'' + \beta' X_e,$$

where $\alpha'' = (\alpha - g(\gamma+\delta))/(1-\delta)$; this, of course, reducing to the steady state equilibrium solution when $g = 0$. The differenced equation (13.18) again has no such solution in terms of Y_e and X_e.

Of course, models relating series generated by integrated processes may indeed not provide any information about long-run relationships: such relationships may simply not exist. But it is obviously important to allow for their possibility when building models between economic time series, which simply using differenced variables will fail to do.

There is, in fact, a linkage between these two, seemingly diametrically opposed, views of the model building process. In the regression model relating the integrated processes Y_t and $\boldsymbol{X}_t = (X_{1t} \dots X_{mt})'$, Phillips (1986, p. 321) shows that there is one case in which the usual least squares theory of stationary processes actually holds. This is when the limiting covariance matrix of the model is singular, in which case there exists a linear relationship between Y_t and $\boldsymbol{X}_t$ which enables the least squares coefficient estimator to be consistent. This singularity is, in fact, a necessary condition for $(Y_t, \boldsymbol{X}_t)$ to be *cointegrated.*

The concept of cointegration is the link between relationships between integrated processes and the concept of (steady state) equilibrium. It was originally introduced by Granger (1981), and extended in Granger and

Weiss (1983) and Engle and Granger (1987); useful elementary discussion of the concept may be found in Granger (1986a) and the implications it has for econometric modelling are discussed by Hendry (1986).

For our purpose, we can develop the concept of cointegration by considering just $I(0)$ and $I(1)$ processes, although Engle and Granger (1987) provide a completely general analysis in terms of vector $I(d)$ processes, i.e. vector processes that require each component to be differenced d times to achieve stationarity. If a time series x_t is $I(0)$, then it is stationary, whereas if it is $I(1)$ its change, ∇x_t, is stationary. The properties of $I(0)$ and $I(1)$ processes have been discussed in various sections of Chapters 5, 6 and 11, but it is useful to bring them together and list them here.

If $x_t \sim I(0)$, using the notation introduced in Chapter 11.6, and has zero mean, then
(i) the variance of x_t is finite,
(ii) an innovation has only a temporary effect on the value of x_t,
(iii) the expected length of times between crossings of $x = 0$ is finite,
(iv) the autocorrelations, ρ_k, decrease steadily in magnitude for large enough k, so that their sum is finite.

If $x_t \sim I(1)$ with $x_0 = 0$, then
(i) the variance of x_t goes to infinity as t goes to infinity,
(ii) an innovation has a permanent effect on the value of x_t because x_t is the sum of all previous changes,
(iii) the expected time between crossings of $x = 0$ is infinite,
(iv) the autocorrelations, $\rho_k \to 1$ for all k as t goes to infinity.

This last property implies that an $I(1)$ series will be rather smooth, with dominant long swings, when compared to an $I(0)$ series. Because of the relative sizes of the variances, it is always true that the sum of an $I(0)$ and an $I(1)$ series will be $I(1)$. Furthermore, if a and b are constants, $b \neq 0$, and if $x_t \sim I(1)$, then $a+bx_t$ is also $I(1)$.

If x_t and y_t are both $I(1)$, then it is generally true that the linear combination

$$z_t = y_t - \alpha x_t \tag{13.19}$$

will also be $I(1)$. However, it is possible that $z_t \sim I(0)$ and, when this occurs, a special constraint operates on the long-run components of x_t and y_t. Since x_t and y_t are both $I(1)$, they will be dominated by 'long wave' components, but z_t, being $I(0)$, will not be: y_t and αx_t must therefore have long run components that virtually cancel out to produce z_t. In such circumstances, x_t and y_t are said to be cointegrated, with α being called the cointegrating parameter or, more generally, $(1, -\alpha)$ being called the cointegrating vector.

To relate this idea to the concept of long-run equilibrium, suppose that such an equilibrium is defined by the relationship

$$y_t = \alpha x_t$$

or

$$y_t - \alpha x_t = 0.$$

Thus z_t given by (13.19) measures the extent to which the system (x_t, y_t) is out of equilibrium, and can therefore be termed the 'equilibrium error'. Hence, if x_t and y_t are both $I(1)$, then the equilibrium error will be $I(0)$ and z_t will rarely drift far from zero, if it has zero mean, and will often cross the zero line. In other words, equilibrium will occasionally occur, at least to a close approximation, whereas if x_t and y_t are not cointegrated, so that $z_t \sim I(1)$, the equilibrium error can wander widely and zero-crossings would be very rare, suggesting that under such circumstances the concept of equilibrium has no practical implications.

It is this feature of cointegration that links it with the spurious regression analysis discussed earlier. As we have seen, a regression on the differences of time series provides no information about the long-run equilibrium relationship, which can only be provided by a regression estimated on the levels of the data. But, if the series are integrated, standard statistical inference on levels regression breaks down completely. In the special case when the covariance matrix is singular, however, inference is possible using the asymptotic theory developed by Stock (1987) and Phillips and Durlauf (1986). The linear combination of variables provoking this singularity is exactly the cointegrating vector defined above. Thus, only if integrated series are cointegrated can inference be carried out on models estimated in levels, and only if they are cointegrated is there a meaningful equilibrium relationship between them. If the series are not cointegrated, then there is no equilibrium relationship existing between them and analysis *should* therefore be undertaken on their differences.

The importance of testing whether a pair, or in general a vector, of integrated time series are cointegrated is thus obvious. Furthermore, given that such series are cointegrated, estimation of the cointegrating vector is then an essential second step. The twin problems of testing for cointegration and estimating the cointegrating vector are intimately related, as the following discussion will make clear.

As a prerequisite, it must first be determined whether x_t and y_t are $I(0)$ or $I(1)$ processes. The mechanics of doing this, which is equivalent to determining whether x_t and y_t contain unit roots, has already been set out in Chapter 8.4. Given that both x_t and y_t are indeed $I(1)$ processes, a convenient method of testing whether they are cointegrated is to estimate the 'cointegrating regression'

$$y_t = \beta_0 + \beta_1 x_t + u_t \tag{13.20}$$

and then test if the residual $\hat{u}_t$ appears to be $I(0)$ or not. Two simple tests of the implied null hypothesis that x_t and y_t are *not* cointegrated are available, both of which are discussed in Engle and Granger (1987). These authors, and Banerjee et al. (1986), also provide Monte Carlo simulation evidence concerning the performance of these test procedures. The first test is based on the DW statistic for (13.20) and tests, on the null hypothesis that u_t is $I(1)$, whether DW is significantly *greater than zero* using the critical values provided by Sargan and Bhargava (1983). The second test examines the residuals $\hat{u}_t$ directly by performing a unit root test of the type discussed earlier. This latter test was found by Engle and Granger (1987) to have the more stable critical values from their, admittedly small, simulation study. With 100 observations, approximate significance levels for the τ statistic testing $\rho = 0$ in the regression

$$\nabla \hat{u}_t = \rho \hat{u}_{t-1} + \sum_{i=1}^{p} \gamma_i \nabla \hat{u}_{t-i} + e_t \tag{13.21}$$

were found to be $\tau_{0.10} \simeq 2.88, \tau_{0.05} \simeq 3.17, \tau_{0.01} \simeq 3.75$. From their simulations of the model with $p = 0$, however, Banerjee et al. (1986) prefer the DW statistic on the grounds that its distribution is invariant to nuisance parameters such as a constant. Similar approximate significance levels for this statistic are $\mathrm{DW}_{0.10} \simeq 0.21, \mathrm{DW}_{0.05} \simeq 0.28, \mathrm{DW}_{0.01} \simeq 0.46$. A number of other tests, based upon the estimation of vector autoregressions (see Chapter 14), are also investigated by Engle and Granger (1987).

When x_t and y_t are cointegrated, the OLS estimate of the slope parameter β_1 in (13.20) should provide, in large samples, an excellent estimate of the true cointegration parameter α. The reasoning for this is as follows. If x_t and y_t are cointegrated for $\beta_1 = -\alpha$, then u_t will have a finite (or small) variance, otherwise $u_t \sim I(1)$ and will thus have a very large variance. All omitted dynamics in (13.20) can be reparameterised purely in terms of ∇y_{t-i}, ∇x_{t-j} and $(y_{t-k} - \hat{\alpha} x_{t-k})$, all of which are $I(0)$ under cointegration, and can therefore be subsumed within u_t. Thus α can be *consistently* estimated in (13.20), using the result of Stock (1987), despite the complete omission of all dynamics. Furthermore, it is also the case that OLS estimates of α are highly efficient, with $\hat{\alpha}(= -\hat{\beta}_1)$ converging rapidly to α at a rate that is much faster than that of standard econometric estimators; see Stock (1987). Of course, the conventional least squares formula for estimating the variance of $\hat{\alpha}$ remains invalid due to the autocorrelation in u_t.

It should therefore be relatively easy to establish a value of α for cointegrated series and hence numerically to parameterise equilibrium economic theories: indeed, the derived long-run solutions from dynamic models for $I(1)$ variables should essentially coincide with the estimates obtained from the static, cointegrating regression, thus providing a check on the validity of the dynamic modelling. One statistical caveat is worth mentioning, however: Monte Carlo studies carried out by Banerjee et al. (1986) show that the bias in estimating α from the cointegrating regression (13.20) can be large and may, in fact, decline much more slowly than the theoretical rate. This is particularly so when the R^2 associated with the cointegrating regression is not very high. On a related theme, the 'reverse regression' of x_t on y_t will give a consistent estimate of $1/\alpha$. Since the product of the two estimates is R^2, then if this statistic is nearly unity, virtually identical estimates of α will result.

Engle and Granger (1987) point out that some, seemingly obvious, methods of estimating α are inconsistent; for example, regression of ∇y_t on ∇x_t (or, of course, vice versa), and the use of Cochrane–Orcutt or some other serial correlation correction in the cointegrating regression.

13.4.1 *Cointegration and error correction mechanisms*

How, though, does the existence of cointegration between y_t and x_t enable equilibrium relationships to be modelled while at the same time avoiding difficult problems of statistical inference? It does so through the Granger Representation Theorem (Engle and Granger (1987)). This shows that if y_t and x_t are both $I(1)$ and cointegrated, then there exists an *error correction representation*, with $z_t = y_t - \alpha x_t$, of the form

$$\delta_1(B)\nabla y_t = \omega_1(B)\nabla x_t - \gamma_1 z_{t-1} + \theta(B)\varepsilon_t, \tag{13.22}$$

where ε_t is white noise and $\delta_{10} \equiv 1$. In fact, the theorem is presented in the general terms of a vector process in which all components may be endogenous. The univariate case presented as (13.22), with no feedback assumed from y_t to x_t, is, though, the typical case used in many empirical examples of error correcting behaviour; see, for example, Davidson et al. (1978) and Sargan (1964).

A special case of (13.22) often considered for pedagogical purposes is (Gilbert (1986)):

$$\nabla y_t = \omega_0 + \omega_1 \nabla x_t - \gamma(y_{t-1} - x_{t-1}) + \varepsilon_t. \tag{13.23}$$

Here the error term has no moving average part, the systematic dynamics are kept as simple as possible, and the *error correction mechanism* (ECM) here presupposes that the cointegrating parameter is unity. Such a formulation allows us to see quickly the essential properties of the model.

All terms in (13.23) are $I(0)$, so that no inferential difficulties arise. When $\nabla y_t = \nabla x_t = 0$ the 'no change' steady state equilibrium of

$$y_t - x_t = \frac{\omega_0}{\gamma}$$

is reproduced (i.e. long-run proportionality between the variables if, as assumed, y_t and x_t are measured in logarithms), and the steady state growth path, obtained when $\nabla y_t = \nabla x_t = g$, takes the form

$$y_t - x_t = \frac{\omega_0 - g(1-\omega_1)}{\gamma}.$$

Models of this particularly simple form have been shown to be capable of being generated by a variety of economic mechanisms based upon minimising adjustment costs in a partial manner (see Hendry and von Ungern-Sternberg (1981) and Nickell (1985)).

As a corollary, the Representation Theorem also implies that the converse of the above analysis holds: if x_t and y_t are both $I(1)$ and have an error correction representation, then they are necessarily cointegrated.

While the simple form (13.23) has often been employed in empirical examples of error correcting behaviour, the imposition of $\alpha = 1$, although suggested by many economic theories, can, if incorrect, lead to a very serious model misspecification. The appropriate ECM should therefore use the estimated cointegrating vector obtained from the cointegrating regression.

More general discussion of cointegration and error correction models is presented in Granger and Weiss (1983), Engle and Granger (1987), Granger (1986a) and Wickens and Breusch (1988), and the reader is urged to consult such references to obtain the full implications of these concepts for the modelling of integrated economic time series.

Example 13.5: Are consumption and income cointegrated?

This example examines cointegration between consumption and income, thus continuing the theme of the previous examples in this chapter. It is also important because Davidson et al. (1978) have presented impressive evidence in favour of an error correction model for UK consumption behaviour from both empirical and theoretical points of view. Unlike previous examples, we use the logarithms of *seasonally adjusted* quarterly data on real nondurable consumption and real personal disposable income so as to avoid unnecessary complications in the analysis, but again using the sample period 1955 to 1985.

Prior to testing for cointegration, it was established that both series were $I(1)$ by using the unit root testing procedure of Chapter 8.4. The

Exhibit 13.6 *Regression of consumption on income*

	I	II	III	IV	V
Dep. Var	C	∇EC	∇C	∇C	∇C
Y	0.79 (143)				
C_{-1}			0.19 (2.4)		
Y_{-1}			−0.15 (2.4)		
EC_{-1}		−0.57 (6.7)		0.12 (2.0)	
∇C_{-1}			−0.30 (2.1)	−0.17 (1.7)	−0.12 (1.2)
∇C_{-2}			−0.10 (0.8)		
∇C_{-3}			−0.09 (0.6)		
∇C_{-4}			−0.15 (1.4)		
∇Y_{-1}			0.25 (3.5)	0.18 (3.9)	0.14 (3.5)
∇Y_{-2}			0.11 (1.5)		
∇Y_{-3}			0.11 (1.7)		
∇Y_{-4}			0.04 (0.8)		
$(C-Y)_{-1}$					0.02 (1.5)
Constant	1.99 (35)		−0.34 (2.1)	0.005 (5.9)	0.009 (3.1)
$\hat{\sigma}$	0.01335	0.01100	0.00758	0.00754	0.00759
DW	1.13	2.21	1.95	2.04	1.99

Absolute t ratios are in parentheses.

estimated models considered in the analysis are shown in Exhibit 13.6. The cointegrating regression of consumption (C) on income (Y) and a constant was first run (Regression I). The DW statistic was found to be 1.13, which clearly rejects the null hypothesis of 'non-cointegration'. Regression II shows the results of regressing the change in the residuals from Regression I on past levels; again, the τ statistic of -6.75 is clearly significant, thus also rejecting non-cointegrability. Regression III estimates the regression of the change in consumption on four lags of consumption and income changes plus the lagged levels of consumption and income. Noting that the cointegrating parameter was estimated in I to be 0.79, the lagged levels are of the appropriate signs and sizes for an error correction term and are individually significant. Thus Regression IV includes the error correction term (EC) estimated from the cointegrating regression, along with the lagged changes in consumption and income. The standard error of this regression is marginally lower than that of the unrestricted Regression III, thus suggesting the efficiency of the parameter restrictions. Regression V includes the lagged difference $(C_{t-1} - Y_{t-1})$ as an error correction term and this provides a worse fit than when the term estimated from the cointegrating regression is used, thus implying that the long run relationship is not that of proportionality between consumption and income.

13.4.2 *Cointegration and feedback*

Throughout this chapter we have assumed that there is no feedback from the output to the inputs and, in our set of examples, that there is no feedback from consumption to income. The analysis of cointegration and error correction models in fact allows us to check the reasonableness of this assumption. From the Granger Representation Theorem, it follows that if x_t and y_t are both $I(1)$ and are cointegrated then there always exists a *pair* of error correcting mechanisms, requiring that (13.22) be supplemented by

$$\delta_2(B)\nabla x_t = \omega_2(B)\nabla y_t - \gamma_2 z_{t-1} + \theta(B)\varepsilon_t, \tag{13.24}$$

$\theta(B)$ being the same in both equations. Since $|\gamma_1| + |\gamma_2| \neq 0$, the error correction term z_{t-1} must enter into one or both equations, and if it enters into both then neither variable can be regarded as weakly exogenous (a detailed discussion of the alternative concepts of exogeneity is to be found in Chapter 14).

Example 13.6: Checking for feedback between consumption and income

This suggests that the presence of feedback from consumption to income can be checked by running the sequence of 'reverse regressions' of Y on C. These are set out in Exhibit 13.7 and the following points are noteworthy. The reverse cointegrating regression provides an estimate of $1/\alpha$ as 1.26, whose inverse is indeed $\hat{\alpha} = 0.79$ (this should be expected since the R^2 of this regression exceeds 0.99); the error correction term EY is certainly significant and the selected error correction model (Regression IV) contains three lagged changes of both consumption and income, along with EY_{t-1}; imposing proportionality again worsens the fit of the model. We thus conclude that there is indeed feedback between consumption and income and that an efficient analysis would need to treat the variables within a multivariate framework, as is done subsequently in Chapter 14.

While these last two examples do indeed show that consumption and income are cointegrated, it should not be assumed that cointegrating vectors are common. Granger (1986a) provides a brief review of the available empirical evidence on which pairs (or sets) of series have been found, or not found, to be cointegrated.

Example 13.7: Cointegration and the Gibson Paradox

As an illustration of a pair of series that are not cointegrated, we consider the relationship between the annual UK Consol yield and the price level during the Gold Standard period 1870 to 1913. This is an important relationship because the observed movements of these series during these years gave rise to the phenomenon known as the 'Gibson Paradox': an observed relationship between the *levels* of interest rates and prices, rather than between the level of interest rates and inflation (i.e. the changes in the logarithms of prices), as is suggested by the traditional Fisherian analysis (for an extended discussion of the underlying theoretical arguments, see Friedman and Schwartz (1982, chapter 10)).

The actual data used are discussed in detail in Mills and Wood (1987), where a much more extensive multivariate analysis is reported, but regressing the level of the Consol yield (R_t) on the logarithm of the price level (P_t) obtained the cointegrating regression

$$R_t = -10.37 + 2.97P_t, \quad \text{DW} = 0.39, \quad s = 0.1310.$$

Since the sample size is 44, use of Table 1 of Sargan and Barghava (1983) shows that such a value of the DW statistic does not allow rejection of the null of non-cointegrability. Similarly, regressing R_t on the changes of P_t obtains

$$R_t = 2.93 + 0.37\nabla P_t, \quad \text{DW} = 0.12, \quad s = 0.2279,$$

Exhibit 13.7 *Regression of income on consumption*

	I	II	III	IV	V
Dep. Var	Y	∇EY	∇Y	∇Y	∇Y
C	1.26 (143)				
C_{-1}			0.46 (2.7)		
Y_{-1}			−0.37 (2.8)		
EY_{-1}		−0.57 (6.8)		−0.34 (2.7)	
∇C_{-1}			0.41 (1.4)	0.55 (2.3)	0.92 (4.3)
∇C_{-2}			0.78 (2.7)	0.92 (4.0)	1.19 (5.1)
∇C_{-3}			0.45 (1.6)	0.55 (2.7)	0.63 (3.0)
∇C_{-4}			−0.01 (0.0)		
∇Y_{-1}			−0.25 (1.6)	−0.27 (2.1)	−0.51 (5.2)
∇Y_{-2}			−0.24 (1.6)	−0.26 (2.1)	−0.44 (4.0)
∇Y_{-3}			−0.28 (2.1)	−0.28 (2.7)	−0.38 (3.8)
∇Y_{-4}			−0.02 (0.2)		
$(C-Y)_{-1}$					−0.002 (0.2)
Constant	−2.45 (27)		−0.85 (2.5)		
$\hat{\sigma}$	0.01686	0.01516	0.01580	0.01566	0.01618
DW	1.13	2.21	1.99	2.02	2.06

Absolute t ratios in parentheses.

which again does not allow non-cointegrability to be rejected. Thus, neither the Gibson Paradox relationship nor the Fisherian model hold as equilibrium relationships: the appropriate relationship is, in fact, between the change in interest rates and inflation, as judged by the residual standard errors of the appropriate differenced versions of the two cointegrating regressions presented above.

Although a number of methodological questions still need answering (see Hendry (1986)), cointegration and the related concept of error correction models are extremely important developments in the modelling of economic time series. As Granger (1986a, pages 226–7) remarks '(w)hether or not cointegration occurs is an empirical question but the beliefs of economists do appear to support its existence and the usefulness of the concept appears to be rapidly gaining acceptance'.

13.5 Single equation distributed lag models

Consider again the single input transfer function-noise model with the delay parameter b conveniently set to zero:

$$y_t = \frac{\omega(B)}{\delta(B)} x_t + \frac{\theta(B)}{\phi(B)} a_t. \tag{13.25}$$

This general specification has often been simplified in the econometric literature. For example, if $\delta(B) = \phi(B)$, then (13.25) becomes

$$\delta(B) y_t = \omega(B) x_t + \theta(B) a_t, \tag{13.26}$$

which is often referred to as an ARMAX model. If, on the other hand, $\theta(B) = \phi(B)$, then we have

$$y_t = \frac{\omega(B)}{\delta(B)} x_t + a_t, \tag{13.27}$$

known as the Rational Distributed Lag (RDL) model. Finally, if $\theta(B) = 1$ in the ARMAX model (13.26), the resultant equation

$$\delta(B) y_t = \omega(B) x_t + a_t \tag{13.28}$$

is referred to as autoregressive dynamic (AD). Thus, ARMAX is RDL with ARMA errors or AD with MA errors. These special cases of the transfer function-noise model can all be straightforwardly extended to more than one input and have been analysed in great detail in many econometric texts and surveys; for the former see, for example, Judge et al. (1985) and Spanos (1986), for the latter see Griliches (1967) and Nerlove (1972) for early surveys, and Hendry et al. (1984) for a recent and comprehensive treatment.

There have been a number of recent developments in the specification of dynamic econometric models and the reader is especially referred to the following topics: 'general-to-specific' specification searches (Mizon (1977), Mills (1981)), related use of the COMFAC procedure (Hendry and Mizon (1978), Sargan (1980a), Mizon and Hendry (1980), Mills and Wood (1982)), and the specification of polynomial distributed lags (Sargan (1980b), Hendry et al. (1984)). Since detailed treatment of these topics can be found in the above references, further discussion is not entered into here.

14 Multiple time series modelling

14.1 Introduction

In this chapter we extend our treatment of time series to allow the *joint* modelling of a vector stochastic process. The next section begins by extending the analysis of stochastic processes to the vector case and then introduces the vector ARMA model. The familiar dynamic simultaneous equation model is shown in section 14.3 to be related to this vector ARMA model when an endogenous/exogenous classification of variables, corresponding to a set of parameter exclusion restrictions, is made. In this section, related concepts such as reduced and final forms, multipliers and impulse response functions are also introduced.

Section 14.4 discusses the various definitions of exogeneity that now exist in the econometric literature and relates them to the concept of Wiener–Granger causality, while section 14.5 introduces related measures of feedback and associated tests of causal orderings. Section 14.6 discusses the 'innovation accounting' approach to multiple time series modelling in which endogenous/exogenous variable distinctions are eschewed.

The effects of model misspecification, in particular the omission of variables, measurement error and temporal aggregation, are the topic of section 14.7. As a prelude to multivariate model building, section 14.8 discusses the estimation of vector ARMA models before section 14.9 introduces techniques for the identification and diagnostic checking of such models. Aspects of forecasting are considered in section 14.10, while the final section notes some computer software that is available for multiple time series modelling.

14.2 Stationary vector stochastic processes

Instead of having data on just a single variable, as in the analysis developed in Part II, we now assume that we can observe realisations on a k-dimensional vector of variables $\boldsymbol{y}_t = (y_{1t}, \ldots, y_{kt})'$. Thus $\boldsymbol{y}_t$ may be thought of as a vector stochastic process, in the same way as x_t was considered a univariate stochastic process in Chapter 5. By direct analogy with x_t, $\boldsymbol{y}_t$ will be stationary if

(i) $E(\boldsymbol{y}_t) = \boldsymbol{\mu} < \infty$ for all t,
(ii) $E[(\boldsymbol{y}_t - \boldsymbol{\mu})(\boldsymbol{y}_t - \boldsymbol{\mu})'] = \sum_y < \infty$ for all t,
(iii) $E[(\boldsymbol{y}_t - \boldsymbol{\mu})(\boldsymbol{y}_{t-h} - \boldsymbol{\mu})'] = \Gamma_y(h)$ for all t and h,

where $\boldsymbol{\mu}$ is a k-dimensional mean vector, $\sum_y$ is a $(k \times k)$ covariance matrix and $\Gamma_y(h)$ is a $(k \times k)$ autocovariance matrix: $\Gamma_y(0) = \sum_y$.

A direct extension of the univariate ARMA process allows $\boldsymbol{y}_t$ to be modelled as a vector ARMA(p, q) process, defined as

$$\boldsymbol{y}_t = \Phi_1 \boldsymbol{y}_{t-1} + \ldots + \Phi_p \boldsymbol{y}_{t-p} + \boldsymbol{v}_t + \Theta_1 \boldsymbol{v}_{t-1} + \ldots + \Theta_q \boldsymbol{v}_{t-q}, \tag{14.1}$$

where

$$\Phi_j = \begin{bmatrix} \phi_{11j} & \cdots & \phi_{ik,j} \\ \vdots & \ddots & \vdots \\ \phi_{k1,j} & \cdots & \phi_{kk,j} \end{bmatrix}, \quad j = 1, 2, \ldots, p$$

$$\Theta_j = \begin{bmatrix} \theta_{11,j} & \cdots & \theta_{ik,j} \\ \vdots & \ddots & \vdots \\ \theta_{k1,j} & \cdots & \theta_{kk,j} \end{bmatrix}, \quad j = 1, 2, \ldots, q$$

and $\boldsymbol{v}_t = (v_{1t}, \ldots, v_{kt})'$ is k-dimensional *vector white noise* defined by

$$E(\boldsymbol{v}_t) = 0; E(\boldsymbol{v}_t \boldsymbol{v}_s') = \begin{cases} \sum_v \text{ (positive definite)}, & t = s \\ 0, & t \neq s. \end{cases}$$

In more concise notation, (14.1) can be written as

$$\Phi_p(B) \boldsymbol{y}_t = \Theta_q(B) \boldsymbol{v}_t, \tag{14.2}$$

where

$$\Phi_p(B) = I - \Phi_1 B - \ldots - \Phi_p B^p$$

and

$$\Theta_q(B) = I + \Theta_1 B + \ldots + \Theta_q B^q$$

are *matrix* polynomials in B. The vector ARMA process (14.2) will be stationary if the roots of the determinental polynomial $|\Phi_p(B)|$ are all outside the unit circle, in which case the process has an infinite order vector MA representation

$$\boldsymbol{y}_t = \boldsymbol{v}_t + \psi_1 \boldsymbol{v}_{t-1} + \ldots = \psi(B) \boldsymbol{v}_t, \tag{14.3}$$

where the matrix polynomial $\psi(B)$ is given by $\psi(B) = \Phi_p^{-1}(B) \Theta_q(B)$. The coefficient matrices ψ_i can be evaluated using the following recursions:

$$\psi_0 = I$$

$$\psi_1 = \Theta_1 + \Phi_1$$

$$\psi_2 = \Theta_2 + \Phi_2 + \Phi_1 \psi_1$$

$$\vdots$$

$$\psi_i = \Theta_i + \sum_{j=1}^{\min(i,p)} \Phi_j \psi_{i-j}, \quad \text{if} \quad i \leqslant q,$$

$$\psi_i = \sum_{j=1}^{\min(i,p)} \Phi_j \psi_{i-j}, \quad \text{if} \quad i > q.$$

The autocovariance matrices of the process $\boldsymbol{y}_t$ can be shown to be given by

$$\Gamma_y(h) = \sum_{j=0}^{h} \psi_j \textstyle\sum_v \psi'_{j-h}, \quad h \geqslant 0,$$

and $\Gamma_y(h) = \Gamma'_y(-h)$ for $h < 0$. In fact, by a multivariate version of Wold's decomposition theorem (Hannan (1970)), any nondeterministic stationary vector stochastic process has an infinite vector MA representation.

The process $\boldsymbol{y}_t$ is said to be invertible if the roots of the determinental polynomial $|\Theta_q(B)|$ all lie outside the unit circle, in which case the process has an infinite order vector AR representation

$$\boldsymbol{y}_t - \pi_1 \boldsymbol{y}_{t-1} - \ldots = \pi(B)\boldsymbol{y}_t = \boldsymbol{v}_t, \tag{14.4}$$

where $\pi(B) = \Theta_q^{-1}(B)\Phi_p(B)$. The coefficient matrices π_i can be determined by a recursion similar to that for ψ_i.

The vector ARMA representation of $\boldsymbol{y}_t$ is not unique. For example, since $\Phi_p^{-1}(B) = \Phi_p^*(B)/|\Phi_p(B)|$, where $\Phi_p^*(B)$ is the adjoint of $\Phi_p(B)$, premultiplying (14.2) by this adjoint results in

$$|\Phi_p(B)|\,\boldsymbol{y}_t = \Phi_p^*(B)\,\Theta_q(B)\,\boldsymbol{v}_t. \tag{14.5}$$

Since $|\Phi_p(B)|$ and $\Phi_p^*(B)$ are of finite order, (14.5) is another finite order ARMA representation of $\boldsymbol{y}_t$. In fact, $|\Phi_p(B)|$ is a scalar operator and so the same AR operator is applied to each component of $\boldsymbol{y}_t$ in (14.5). Moreover, as Zellner and Palm (1974) point out, the ith equation of (14.5) is given by

$$|\Phi_p(B)|\, y_{it} = \boldsymbol{\theta}'_i \boldsymbol{v}_t, \quad i = 1, 2, \ldots, k \tag{14.6}$$

where $\boldsymbol{\theta}'_i$ is the ith row of $\Phi_p^*(B)\,\Theta_q(B)$. The set of equations in (14.6) are termed the *final equations*, and each is in univariate ARMA form. Thus, as Zellner and Palm (1974) argue (page 19) '...ARMA processes for individual variables are compatible with some, perhaps unknown, joint process for a set of random variables and are thus not necessarily "naive", "ad hoc" alternative models'. Unless $|\Phi_p(B)|$ contains factors common to those appearing in all elements of the vector $\boldsymbol{\theta}'_i$, when some cancelling will take place (for example, when $\Phi_p(B)$ is triangular, diagonal or block diagonal), the AR orders *and* parameters of each final equation will be the same. This has important implications for the building of multiple time series models, although Wallis (1977, page 1484) discusses a number of statistical problems that may occur when these implications are matched with empirically determined univariate ARMA models.

The representation (14.5) is not the only one that has been found to be useful. In general, multiplying (14.1) by a nonsingular $(k \times k)$ matrix P will not alter the correlation structure of $\boldsymbol{y}_t$ and will therefore result in an equivalent representation. For example, suppose we define P to be triangular and with a unit diagonal such that $P\sum_v P'$ is a diagonal matrix. In this case, premultiplying (14.1) by P leads to

$$\begin{aligned} P\boldsymbol{y}_t = P\Phi_1\boldsymbol{y}_{t-1}+\ldots+P\Phi_p\boldsymbol{y}_{t-p}+\boldsymbol{w}_t \\ +P\Theta_1 P^{-1}\boldsymbol{w}_{t-1}+\ldots+P\Theta_q^{-1}P^{-1}\boldsymbol{w}_{t-q}, \end{aligned} \quad (14.7)$$

which is a vector ARMA representation in which the white noise process $\boldsymbol{w}_t = P\boldsymbol{v}_t$ has a *diagonal* covariance matrix $\sum_w$.

Alternatively, if we define S such that $SS' = \sum_v$ and $\boldsymbol{u}_t = S^{-1}\boldsymbol{v}_t$, then $E(\boldsymbol{u}_t\boldsymbol{u}'_t) = \sum_u = I_k$ and

$$\boldsymbol{y}_t = \Phi_1\boldsymbol{y}_{t-1}+\ldots+\Phi_p\boldsymbol{y}_{t-p}+S\boldsymbol{u}_t+\psi_1 S\boldsymbol{u}_{t-1}+\ldots+\psi_q S\boldsymbol{u}_{t-q}. \quad (14.8)$$

This lack of uniqueness of the ARMA representation of a vector stochastic process – that is, the problem that two different vector ARMA models may represent processes with identical covariance structures – corresponds to the identification problem in the context of simultaneous equation models in econometrics. An accessible discussion of the conditions for the uniqueness of ARMA representations may be found in, for example, Granger and Newbold (1977, pages 219–24), with a more rigorous treatment being given in Diestler (1985).

Although the vector ARMA process is the basic model used throughout this chapter, other multivariate specifications of time series are available. For example, Abraham (1980) considers the incorporation of interventions into the vector ARMA model, while Jones (1966), Enns et al. (1982) and Harvey (1986) develop multivariate extensions of exponential smoothing type models.

Example 14.1: A bivariate example

Consider the following bivariate model (see Granger and Newbold (1977, page 223))

$$\begin{bmatrix} y_{1t} \\ y_{2t} \end{bmatrix} = \begin{bmatrix} 0.7 & 0.3 \\ 0.4 & 0.4 \end{bmatrix} \begin{bmatrix} y_{1t-1} \\ y_{2t-1} \end{bmatrix} + \begin{bmatrix} 1 & 0 \\ 0 & 1 \end{bmatrix} \begin{bmatrix} v_{1t} \\ v_{2t} \end{bmatrix} + \begin{bmatrix} 0 & 0.5 \\ 0 & 0.6 \end{bmatrix} \begin{bmatrix} v_{1t-1} \\ v_{2t-1} \end{bmatrix}$$

where

$$\sum_v = \begin{bmatrix} 1.09 & 0.7 \\ 0.7 & 1.16 \end{bmatrix}.$$

By defining

$$P = \begin{bmatrix} 1 & 0 \\ -0.642 & 1 \end{bmatrix}$$

and $\boldsymbol{w}_t = P\boldsymbol{v}_t$, so that $P \sum_v P' = \sum_w$, where

$$\sum_w = \begin{bmatrix} 1.09 & 0 \\ 0 & 0.71 \end{bmatrix},$$

premultiplying the model by P yields

$$\begin{bmatrix} 1 & 0 \\ -0.642 & 1 \end{bmatrix} \begin{bmatrix} y_{1t} \\ y_{2t} \end{bmatrix} = \begin{bmatrix} 0.7 & 0.3 \\ -0.049 & 0.207 \end{bmatrix} \begin{bmatrix} y_{1t-1} \\ y_{2t-1} \end{bmatrix} + \begin{bmatrix} 1 & 0 \\ 0 & 1 \end{bmatrix} \begin{bmatrix} w_{1t} \\ w_{2t} \end{bmatrix}$$

$$+ \begin{bmatrix} 0.321 & 0.5 \\ 0.179 & 0.279 \end{bmatrix} \begin{bmatrix} w_{1t-1} \\ w_{2t-1} \end{bmatrix}.$$

The model is now in recursive form with a diagonal covariance matrix, so that the errors w_{1t} and w_{2t} are mutually stochastically uncorrelated.

Alternatively, suppose we define S to be

$$S = \begin{bmatrix} 1 & 0.3 \\ 0.4 & 1 \end{bmatrix},$$

then $SS' = \sum_v$ and by substituting $\boldsymbol{v}_t = S\boldsymbol{u}_t$ into the model we obtain

$$\begin{bmatrix} y_{1t} \\ y_{2t} \end{bmatrix} = \begin{bmatrix} 0.7 & 0.3 \\ 0.4 & 0.4 \end{bmatrix} \begin{bmatrix} y_{1t-1} \\ y_{2t-1} \end{bmatrix} + \begin{bmatrix} 1 & 0.3 \\ 0.4 & 1 \end{bmatrix} \begin{bmatrix} u_{1t} \\ u_{2t} \end{bmatrix}$$

$$+ \begin{bmatrix} 0.2 & 0.5 \\ 0.24 & 0.6 \end{bmatrix} \begin{bmatrix} u_{1t-1} \\ u_{2t-1} \end{bmatrix},$$

so that $\Theta_0 \neq I_2$, but $\sum_u = I_2$.

14.3 Dynamic simultaneous equation models

The general vector ARMA model given by equation (14.2) is related to the dynamic simultaneous equation model (DSEM) that underlies most econometric models. To see this, suppose that the identifying restriction $\Phi_0 = I$ is relaxed. Now consider partitioning $\boldsymbol{y}_t$ as $\boldsymbol{y}'_t = (\boldsymbol{z}'_t, \boldsymbol{x}'_t)$, where $\boldsymbol{z}_t$ is of dimension g and $\boldsymbol{x}_t$ is of dimension l. A conformable partition of (14.2) is then

$$\begin{bmatrix} \Phi_{11}(B) & \Phi_{12}(B) \\ \Phi_{21}(B) & \Phi_{22}(B) \end{bmatrix} \begin{bmatrix} \boldsymbol{z}_t \\ \boldsymbol{x}_t \end{bmatrix} = \begin{bmatrix} \Theta_{11}(B) & \Theta_{12}(B) \\ \Theta_{21}(B) & \Theta_{22}(B) \end{bmatrix} \begin{bmatrix} \boldsymbol{v}_{1t} \\ \boldsymbol{v}_{2t} \end{bmatrix}, \tag{14.9}$$

where $\boldsymbol{v}_{1t}$ and $\boldsymbol{v}_{2t}$ are assumed to be jointly independent white noise processes. Suppose that the following restrictions are imposed:

$$\Phi_{21}(B) \equiv 0, \Theta_{12}(B) \equiv 0, \Theta_{21}(B) \equiv 0. \tag{14.10}$$

This set of restrictions is commonly thought of as implying that $\boldsymbol{z}_t$ is endogenous and $\boldsymbol{x}_t$ is exogenous to the model, although various other definitions of the term 'exogenous' will be met later, and allows (14.9) to be written as a set of *structural* equations

$$\Phi_{11}(B)\,\boldsymbol{z}_t + \Phi_{12}(B)\,\boldsymbol{x}_t = \Theta_{11}(B)\,\boldsymbol{v}_{1t}, \tag{14.11}$$

the identification conditions of which are summarised in Hendry et al. (1984, section 6), and a vector ARMA process generating the exogenous variables

$$\Phi_{22}(B)\,\boldsymbol{x}_t = \Theta_{22}(B)\,\boldsymbol{v}_{2t}. \tag{14.12}$$

The structural equations can be expressed as

$$\Phi_{11,0}\,\boldsymbol{z}_t + \ldots + \Phi_{11,r}\,\boldsymbol{z}_{t-r} + \Phi_{12,0}\,\boldsymbol{x}_t + \ldots + \Phi_{12,s}\,\boldsymbol{x}_{t-s} = \boldsymbol{e}_t, \tag{14.13}$$

where $\boldsymbol{e}_t = \Theta_{11}(B)\,\boldsymbol{v}_{1t}$. The 'simultaneity' of the model is a consequence of $\Phi_{11,0} \neq I$, and this can be removed by considering the *reduced form*, which expresses each endogenous variable as a function of exogenous and lagged endogenous (together known as predetermined) variables,

$$\boldsymbol{z}_t = -\Phi_{11,0}^{-1}\{\Phi_{11,1}\,\boldsymbol{z}_{t-1} + \ldots + \Phi_{11,r}\,\boldsymbol{z}_{t-r}\} - \Phi_{11,0}^{-1}\,\Phi_{12}(B)\,\boldsymbol{x}_t + \Phi_{11,0}^{-1}\,\boldsymbol{e}_t. \tag{14.14}$$

The *final form* is obtained by solving (14.11) to yield

$$\boldsymbol{z}_t = -\Phi_{11}^{-1}(B)\,\Phi_{12}(B)\,\boldsymbol{x}_t + \Phi_{11}^{-1}(B)\,\boldsymbol{e}_t, \tag{14.15}$$

in which each endogenous variable is expressed as an infinite distributed lag function of the exogenous variables.

In this form the infinite matrix polynomial

$$\Xi(B) = \sum_{i=0}^{\infty} \Xi_i B^i = -\Phi_{11}^{-1}(B)\,\Phi_{12}(B)$$

contains the *dynamic multipliers*, describing the response of z_t to unit shocks in the exogenous variables $\boldsymbol{x}_t$.

$$\Xi(1) = \sum_{i=0}^{\infty} \Xi_i$$

represents the total response of z_t to these unit shocks in $\boldsymbol{x}_t$ and is therefore called the matrix of *total multipliers*, while the partial sums

$$\Xi_j(1) = \sum_{i=0}^{j} \Xi_i$$

are referred to as the matrices of the *jth interim multipliers*. The matrix Ξ_0 contains the immediate, or contemporaneous, effects of a unit shock in $\boldsymbol{x}_t$ and is called the matrix of *impact multipliers*.

Premultiplying the final form by $|\Phi_{11}(B)|$, and denoting the adjoint matrix associated with $\Phi_{11}(B)$ as $\Phi_{11}^{*}(B)$, yields the set of *fundamental dynamic equations* (for other terms, see Zellner and Palm (1974, page 21))

$$|\Phi_{11}^{*}(B)|\, z_t = -\Phi_{11}^{*}(B)\,\Phi_{12}(B)\,\boldsymbol{x}_t + \Phi_{11}^{*}(B)\,\boldsymbol{e}_t, \qquad (14.16)$$

in which each endogenous variable depends only on its own lagged values and on the exogenous variables, with or without lags. As in the set of final equations (14.6), each endogenous variable will have autoregressive factors having identical orders and parameters.

At this stage of the analysis it is worthwhile considering the possible uses of these different equation systems, and also discussing the requirements that must be met for these uses. The final equations of (14.6) can forecast future values of some or all of the variables in z_t, given that ARMA processes have been fitted to them. These final equations cannot, of course, be used for structural analysis or for control purposes. The reduced form equations (14.14), final form equations (14.15) and fundamental dynamic equations (14.16) can all be used for both prediction and control, since the exogenous variables $\boldsymbol{x}_t$ appear as inputs in them. However, because the lag polynomials appearing in the equations are functions of the structural parameters, none of these equation systems can be employed for structural analysis, except when the structural equations are themselves in either reduced form ($\Phi_{11,0} = I$) or final form ($\Phi_{11} = I$). Furthermore, use of these equation systems implies that we have enough

Exhibit 14.1 *Uses and requirements for various equation systems*

Equation system	Prediction	Control	Uses Structural analysis	Requirements for use
Final equations (14.6)	Yes	No	No	Forms of ARMA processes and parameter estimates
Reduced form equations (14.14)	Yes	Yes	No	Endogenous–exogenous classification of variables, forms of equations and parameter estimates
Fundamental dynamic equations (14.16)	Yes	Yes	No	Endogenous–exogenous classification of variables, forms of equations, and parameter estimates
Final form equations (14.15)	Yes	Yes	No	Endogenous–exogenous classification of variables, forms of equations, and parameter estimates
Structural equations (14.13)	Yes	Yes	Yes	Endogenous–exogenous classification of variables, forms of equations, identifying information, parameter estimates

prior information to distinguish endogenous and exogenous variables. If data on some of the endogenous variables are unavailable, it may still be possible to use the final form (or fundamental dynamic) equations relating to those endogenous variables for which data are available, whereas the reduced form equations, which need all the endogenous variables, will be unusable. When the structural equation system (14.13) is available, it can be used for structural analysis and the associated 'restricted' reduced or final form equations can be employed for prediction and control. Not only do the endogenous and exogenous variables have to be classified correctly for this structural system to be used, but further prior information must be available to identify the structural parameters. Exhibit 14.1 summarises these considerations for the different available equation systems.

It should also be appreciated that, before each of the equation systems can be used, the form of its equations must be ascertained. The most stringent set of informational requirements is that needed for the structural equation system (14.13): endogenous/exogenous variable classification, lag distribution specification, error serial correlation properties and identifying restrictions. Since these are often difficult to obtain, it may well be the case that some of the simpler equation systems will be employed even though their uses are more limited than those of structural equation systems. Moreover, as Zellner and Palm (1974) are able to take advantage of, the other equation systems are useful in providing checks on the lag structures implied by structural model assumptions (see also Anderson et al. (1983) for a complementary analysis).

14.4 Concepts of exogeneity and causality

In Chapter 13 we referred to a variable as being 'exogenous', equating this term with 'lack of feedback'. In the previous section of this chapter we related the concept of exogeneity to certain matrix polynomial exclusion restrictions in the general vector ARMA model. These exclusion restrictions are equivalent, when an additional restriction is imposed, to assuming that $\boldsymbol{x}_t$ is *strictly exogenous* in the DSEM (14.11) (see Sims (1977), Geweke (1978), Sargent (1979, chapter XI)). Geweke (1984a) (see also Engle et al. (1983)) provides a formal definition of a strictly exogenous variable as one whose value in each time period is statistically independent of the values of all the random disturbances in the DSEM in all periods.

If we rewrite the DSEM of equation (14.11), using less opaque notation, as

$$A(B)\boldsymbol{z}_t + \Gamma(B)\boldsymbol{x}_t = \boldsymbol{e}_t$$

$$\alpha(B)\boldsymbol{e}_t = \boldsymbol{u}_t, \qquad (14.17)$$

then strict exogeneity of $\boldsymbol{x}_t$ requires that

$$E(\boldsymbol{u}_t \boldsymbol{x}_{t-s}) = 0, \quad \text{all } s. \qquad (14.18)$$

Strict exogeneity of $\boldsymbol{x}_t$ is useful because no information is lost by limiting attention to distributions conditional on $\boldsymbol{x}_t$, which will usually result in considerable simplifications in statistical inference; for example, instrumental variable techniques may be used in the presence of serially correlated disturbances.

A related concept is that of a variable being *predetermined*: y_t is predetermined at time t if all its current and past values $Y_t = \{y_{t-s}, s \geqslant 0\}$ are independent of the current disturbances $\boldsymbol{e}_t$, *and* these disturbances are serially independent (Geweke (1984a)). Thus, if $\alpha(B) \equiv I$ in (14.17) above, then $\boldsymbol{x}_t$ is predetermined, while if as well

$$E(\varepsilon_t z_{t-s}) = 0, \quad s > 0, \tag{14.19}$$

then z_{t-s} will also be predetermined.

A third concept is that of weak exogeneity (Engle et al. (1983)). A variable is characterised as *weakly exogenous* if inference for a set of parameters can be made conditionally on that variable without loss of information, which, of course, is useful because again statistical inference is simplified. In many cases, strictly exogenous variables will also be weakly exogenous in DSEMs, although one important class of exceptions is provided by rational expectations models, in which behavioural parameters are generally linked to the distributions of exogenous variables (Sargent (1981)). Similarly, predetermined variables will usually be weakly exogenous, except again in the case where there are cross-restrictions between the behavioural parameters and the parameters of the distribution of the predetermined variables.

The assertion of weak exogeneity always relies, to some extent, on a priori assumptions; for example, weak exogeneity of $\boldsymbol{x}_t$ will follow from strict exogeneity if all the parameters in the distribution of $X_t = \{\boldsymbol{x}_{t-s}, s \geqslant 0\}$ are nuisance parameters. However, although weak exogeneity is an attractive specification, it *cannot* be tested for in isolation; it can only be tested together with the other restrictions of the model, because it is a condition on *parameters*, not a restriction on joint distributions. Any consistent test of weak exogeneity must be a joint test of weak exogeneity and other hypotheses.

Strict exogeneity can, however, be tested in DSEMs. Geweke (1978) discusses two implications of strict exogeneity in such models. The first implication concerns the final form of the DSEM. This can generally be written as

$$\begin{aligned} z_t &= J(B)\,\boldsymbol{x}_t + \varepsilon_t \\ &= \sum_{s=0}^{\infty} J_s\,\boldsymbol{x}_{t-s} + \varepsilon_t, \end{aligned} \tag{14.20}$$

where $E(\varepsilon_t \boldsymbol{x}_{t-s}) = 0$ for all s, and ε_t is a stochastic process possessing an autoregressive representation. Geweke proves that in the regression of z_t on all current, lagged, and future values of $\boldsymbol{x}_t$,

$$z_t = K(B)\boldsymbol{x}_t + \varepsilon_t$$

$$= \sum_{s=-\infty}^{\infty} K_s \boldsymbol{x}_{t-s} + \varepsilon_t, \tag{14.21}$$

then if, and only if, the coefficients on future values of $\boldsymbol{x}_t$, i.e. $\boldsymbol{x}_{t-s}$, $s < 0$, are equal to zero will there exist a DSEM relating $\boldsymbol{x}_t$ and z_t in which $\boldsymbol{x}_t$ is strictly exogenous. The second implication of exogeneity supposes the existence of equations (14.20) and (14.21) and also the existence of

$$\boldsymbol{x}_t = \sum_{s=1}^{\infty} E_{2s} \boldsymbol{x}_{t-s} + \sum_{s=1}^{\infty} F_{2s} z_{t-s} + \boldsymbol{w}_{2t}, \tag{14.22}$$

in which $E(z_{t-s} \boldsymbol{w}'_{2t}) = 0$ for all t and $s > 0$. Geweke then proves that $\boldsymbol{x}_t$ will be strictly exogenous in a DSEM relating $\boldsymbol{x}_t$ and z_t if, and only if, the coefficients F_{2s}, $(s = 1, 2, \ldots)$, are all zero, in which case $K(B)$ in (14.21) will again be one-sided (zero coefficients on future values of $\boldsymbol{x}_t$).

Geweke (1978) shows that these implications of exogeneity hold under weaker conditions than those implicitly being assumed here; specifically, $\boldsymbol{x}_t$ need not be stationary. However, following on from the discussion in Chapter 13 concerning the difficulties of conducting statistical inference in the presence of integrated processes, testing of strict exogeneity is still best carried out under the assumption that the exogenous variables are indeed jointly covariance stationary.

14.4.1 *Granger causality*

The testing of strict exogeneity is intimately related to the concept of *Granger causality* (Granger (1969a); see also Sims (1972), Chamberlain (1982) and Florens and Mouchart (1982) for further theoretical development). Granger causality, often referred to more properly as Wiener–Granger causality in recognition of the earlier work of Wiener (1956), has prompted a great deal of debate among economists and even philosophers, with Zellner (1979) and Holland (1986) being prominent critics from these respective disciplines, and with Granger himself (1980b, 1986b) and Sims (1977) providing spirited responses to such criticisms.

There are many discussions and applications of Granger causality readily available in the literature, but for our purposes the formulation provided by Geweke (1982, 1984a) seems ideal. Thus we begin by expressing the vector autoregressive representation (14.4) of $\boldsymbol{y}_t$ as

$$y_t = \sum_{j=0}^{\infty} \pi_j y_{t-j} + v_t, \; V(v_t) = E(v_t v_s') = \Sigma_v. \tag{14.23}$$

If y_t is then partitioned into subvectors z_t and x_t as in equation (14.9), reflecting an interest in causal relationships between $Z = \{z_{t-s}, \text{ all } s\}$ and $X = \{x_{t-s}, \text{ all } s\}$, then both Z and X will also possess autoregressive representations, which we denote by

$$z_t = \sum_{s=1}^{\infty} B_{1s} z_{t-s} + u_{1t}, \; V(u_{1t}) = T_1 \tag{14.24}$$

and

$$x_t = \sum_{s=1}^{\infty} E_{1s} x_{t-s} + w_{1t}, \; V(w_{1t}) = \Upsilon_1. \tag{14.25}$$

The disturbance u_{1t} is the one-step-ahead error when z_t is forecast from its own past alone, and similarly for w_{1t} and x_t. These disturbance vectors are each serially uncorrelated, but may be correlated with each other contemporaneously and at various leads and lags. Since u_{1t} is uncorrelated with $Z_{t-1} = \{z_{t-s}, s \geqslant 1\}$, (14.24) denotes the *linear projection* of z_t on its own past, and likewise (14.25) denotes the linear projection of x_t on $X_{t-1} = \{x_{t-s}, s \geqslant 1\}$ (for a discussion of linear projection theory see, for example, Sargent (1979, chapter X)).

The linear projection of z_t on Z_{t-1} and X_{t-1}, and of x_t on Z_{t-1} and X_{t-1}, is given by (14.23), which may be partitioned as

$$z_t = \sum_{s=1}^{\infty} B_{2s} z_{t-s} + \sum_{s=1}^{\infty} C_{2s} x_{t-s} + u_{2t}, \; V(u_{2t}) = T_2 \tag{14.26}$$

$$x_t = \sum_{s=1}^{\infty} E_{2s} x_{t-s} + \sum_{s=1}^{\infty} F_{2s} z_{t-s} + w_{2t}, \; V(w_{2t}) = \Upsilon_2, \tag{14.27}$$

where $v_t' = (u_{2t}', w_{2t}')$, this second projection being identical to equation (14.22). Σ_v can be correspondingly partitioned as

$$\Sigma_v = \begin{bmatrix} T_2 & C \\ C' & \Upsilon_2 \end{bmatrix},$$

where $C = E(u_{2t} w_{2t}')$, so that although the disturbance vectors u_{2t} and w_{2t} are each serially uncorrelated, they can be correlated with each other contemporaneously, although at no other lag. If the system (14.26)–(14.27) is premultiplied by the matrix

$$\begin{bmatrix} I_g & -C\Upsilon_2^{-1} \\ -C'T_2^{-1} & I_l \end{bmatrix},$$

then in the first g equations of the new system, z_t is a linear function of

Z_{t-1}, X_t, and a disturbance $\boldsymbol{u}_{2t} - C\Upsilon_2^{-1}\boldsymbol{w}_{2t}$. Since this disturbance is uncorrelated with $\boldsymbol{w}_{2t}$, it is uncorrelated with $\boldsymbol{x}_t$, as well as Z_{t-1} and X_{t-1}. Hence the linear projection of z_t on Z_{t-1} and X_t,

$$z_t = \sum_{s=1}^{\infty} B_{3s} z_{t-s} + \sum_{s=0}^{\infty} C_{3s} \boldsymbol{x}_{t-s} + \boldsymbol{u}_{3t},\ V(\boldsymbol{u}_{3t}) = T_3 \quad (14.28)$$

is provided by the first g equations of the new system. Similarly, the linear projection of $\boldsymbol{x}_t$ on Z_t and X_{t-1} is provided by the last l equations:

$$\boldsymbol{x}_t = \sum_{s=1}^{\infty} E_{3s} \boldsymbol{x}_{t-s} + \sum_{s=0}^{\infty} F_{3s} z_{t-s} + \boldsymbol{w}_{3t},\ V(\boldsymbol{w}_{3t}) = \Upsilon_3. \quad (14.29)$$

Finally, it can be shown that the linear projections of z_t on Z_{t-1} and X, and $\boldsymbol{x}_t$ on Z and X_{t-1}, are given by

$$z_t = \sum_{s=1}^{\infty} B_{4s} z_{t-s} + \sum_{s=-\infty}^{\infty} C_{4s} \boldsymbol{x}_{t-s} + \boldsymbol{u}_{4t},\ V(\boldsymbol{u}_{4t}) = T_4 \quad (14.30)$$

and

$$\boldsymbol{x}_t = \sum_{s=1}^{\infty} E_{4s} \boldsymbol{x}_{t-s} + \sum_{s=-\infty}^{\infty} F_{4s} z_{t-s} + \boldsymbol{w}_{4t},\ V(\boldsymbol{w}_{4t}) = \Upsilon_4. \quad (14.31)$$

This set of linear projections is termed by Geweke (1984a) as the *canonical form* of the stationary (vector) time series $\boldsymbol{y}_t' = (z_t', \boldsymbol{x}_t')$, and is displayed, for convenience, as Exhibit 14.2. Given this canonical form, the Wiener–Granger definition of causality may be stated in terms of the linear projection parameters. Thus

(i) *Z causes X* if, and only if,

$F_{2s} \not\equiv 0$ for all s or, equivalently, $|\Upsilon_1| > |\Upsilon_2|$

and

$C_{3s} \not\equiv C_{4s}$ for all s or, equivalently, $|T_3| > |T_4|$

(ii) *X causes Z* if, and only if,

$C_{2s} \not\equiv 0$ for all s or, equivalently, $|T_1| > |T_2|$

and

$F_{3s} \not\equiv F_{4s}$ for all s or, equivalently, $|\Upsilon_3| > |\Upsilon_4|$

Alternatively

(iii) *Z does not cause X* if, and only if,

$F_{2s} \equiv 0$ for all s or, equivalently, $|\Upsilon_1| = |\Upsilon_2|$

and

$C_{3s} \equiv C_{4s}$ for all s or, equivalently, $|T_3| = |T_4|$

Exhibit 14.2 *A canonical form for* $\boldsymbol{y}_t' = (\boldsymbol{z}_t', \boldsymbol{x}_t')$

$$\boldsymbol{z}_t = \sum_{s=1}^{\infty} B_{1s}\boldsymbol{z}_{t-s} + \boldsymbol{u}_{1t} \tag{14.24}$$

$$\boldsymbol{x}_t = \sum_{s=1}^{\infty} E_{1s}\boldsymbol{x}_{t-s} + \boldsymbol{w}_{1t} \tag{14.25}$$

$$\boldsymbol{z}_t = \sum_{s=1}^{\infty} B_{2s}\boldsymbol{z}_{t-s} + \sum_{s=1}^{\infty} C_{2s}\boldsymbol{x}_{t-s} + \boldsymbol{u}_{2t} \tag{14.26}$$

$$\boldsymbol{x}_t = \sum_{s=1}^{\infty} E_{2s}\boldsymbol{x}_{t-s} + \sum_{s=1}^{\infty} F_{2s}\boldsymbol{z}_{t-s} + \boldsymbol{w}_{2t} \tag{14.27}$$

$$\boldsymbol{z}_t = \sum_{s=1}^{\infty} B_{3s}\boldsymbol{z}_{t-s} + \sum_{s=0}^{\infty} C_{3s}\boldsymbol{x}_{t-s} + \boldsymbol{u}_{3t} \tag{14.28}$$

$$\boldsymbol{x}_t = \sum_{s=1}^{\infty} E_{3s}\boldsymbol{x}_{t-s} + \sum_{s=0}^{\infty} F_{3s}\boldsymbol{z}_{t-s} + \boldsymbol{w}_{3t} \tag{14.29}$$

$$\boldsymbol{z}_t = \sum_{s=1}^{\infty} B_{4s}\boldsymbol{z}_{t-s} + \sum_{s=-\infty}^{\infty} C_{4s}\boldsymbol{x}_{t-s} + \boldsymbol{u}_{4t} \tag{14.30}$$

$$\boldsymbol{x}_t = \sum_{s=1}^{\infty} E_{4s}\boldsymbol{x}_{t-s} + \sum_{s=-\infty}^{\infty} F_{4s}\boldsymbol{z}_{t-s} + \boldsymbol{w}_{4t} \tag{14.31}$$

$$V(\boldsymbol{u}_{jt}) = T_j \quad V(\boldsymbol{w}_{jt}) = \Upsilon_j \quad \operatorname{cov}(\boldsymbol{u}_{2t}\boldsymbol{w}_{2t}') = C$$

$$\Sigma_v = \begin{bmatrix} T_2 & C \\ C' & \Upsilon_2 \end{bmatrix}$$

(iv) *X does not cause Z* if, and only if,

$$C_{2s} \equiv 0 \text{ for all } s \text{ or, equivalently, } |T_1| = |T_2|$$

and

$$F_{3s} \equiv F_{4s} \text{ for all } s \text{ or, equivalently, } |\Upsilon_3| = |\Upsilon_4|$$

(v) There is *instantaneous causality* between Z and X if, and only if,

$$C_{2s} \not\equiv C_{3s} \text{ for all } s \text{ or, equivalently, } |T_2| > |T_3|$$

and

$$F_{2s} \not\equiv F_{3s} \text{ for all } s \text{ or, equivalently, } |\Upsilon_2| > |\Upsilon_3|.$$

These definitions of Wiener–Granger causality hold under rather less restrictive conditions than those assumed here, which are that Y is stationary and only linear relations are of interest. Hosoya (1977) has shown that the definitions remain pertinent even if Y is nonstationary, although any deterministic components have to be treated carefully. Chamberlain (1982) and Florens and Mouchart (1982) extend the analysis to include nonlinear relationships, and show that although much of the above carries over straightforwardly, some subtle issues of equivalence arise. Kang (1981) discusses the necessary and sufficient conditions for Wiener–Granger causality to hold in vector ARMA models, rather than just the vector AR case considered here, and the reader is referred to all these papers for further discussion.

14.4.2 *The relationship between causality and exogeneity*

Since the conditions for Z not causing X correspond to the zero restrictions implied by the strict exogeneity of X, it is tempting to regard the two concepts as being equivalent. This is not the case, however. As Geweke (1984a) points out, if X is strictly exogenous in the DSEM (14.17), then Z does not cause X, where Z is endogenous in that model. However, if Z does not cause X, then there exists *a* DSEM with Z endogenous and X strictly exogenous, in the sense that there will exist systems of equations formally similar to (14.17), *but* none of these systems need necessarily satisfy the overidentifying restrictions of that specific model. This implies that tests for the absence of a causal ordering can be used to refute the strict exogeneity specification in a given DSEM, as is proposed by Geweke (1978), but such tests cannot be used to establish it.

Furthermore, as we have already discussed, statistical inference may be carried out conditionally on a subset of variables that are not strictly exogenous; all that we require is that they be weakly exogenous. Thus, unidirectional causality is neither necessary nor sufficient for inference to proceed conditional on a subset of variables. However, if $\boldsymbol{x}_t$ is weakly exogenous for a set of parameters of interest and if $\boldsymbol{x}_t$ is not caused by z_t, $\boldsymbol{x}_t$ is termed by Engle et al. (1983) as being *strongly exogenous*.

More formally, $\boldsymbol{x}_t$ will be weakly exogenous if, when the joint density of $y_t = (z_t', \boldsymbol{x}_t')'$, conditional on the past, is factorised as the conditional density of z_t given $\boldsymbol{x}_t$ times the marginal density of $\boldsymbol{x}_t$; (a) the parameters λ_1 and λ_2 of these conditional and marginal densities are not subject to cross-restrictions, and (b) the parameters of interest (denoted by ψ) can be uniquely determined from the parameters of the conditional model alone (i.e. $\psi = f(\lambda_1)$). Under these conditions $\boldsymbol{x}_t$ may be treated 'as if' it were determined outside the conditional model for z_t.

Conditions (a) and (b) are not sufficient, however, to treat $\boldsymbol{x}_t$ as if it were fixed in repeated samples. If, though, z_t does not cause $\boldsymbol{x}_t$, then $\{\boldsymbol{x}_t\}$ may be treated as if it were fixed and hence $\boldsymbol{x}_t$ will be strongly exogenous, so that, for example, forecasts could be made conditional on fixed future $\boldsymbol{x}$'s.

Engle et al. (1983) define yet a further concept, that of *super exogeneity*: $\boldsymbol{x}_t$ is super exogenous for ψ if (a) and (b) hold and also that λ_1 is invariant to changes in λ_2. Super exogeneity requires, in general, that the conditional distribution is invariant to any change in the marginal distribution, in which case conditional policy experiments for fixed λ_1 can be performed. If policy parameters change, then false super exogeneity assumptions are liable to produce predictive failures in conditional models, and this is the basis of the 'Lucas critique' of macroeconometric models (Lucas (1976)).

14.5 Measures of feedback and tests of causal orderings

From the canonical representation of $\boldsymbol{y}_t$, Geweke (1982) defines *a measure of linear feedback from Z to X* as

$$F_{z\to x} = \ln(|\Upsilon_1|/|\Upsilon_2|) = \ln(|T_3|/|T_4|),$$

so that the statement 'Z does not cause X' is equivalent to $F_{z\to x} = 0$. Symmetrically, X does not cause Z if, and only if, the *measure of linear feedback from X to Z*,

$$F_{x\to z} = \ln(|T_1|/|T_2|) = \ln(|\Upsilon_3|/|\Upsilon_4|),$$

is zero. The existence of instantaneous causality between Z and X amounts to a nonzero measure of *instantaneous linear feedback*,

$$F_{z\cdot x} = \ln(|T_2|/|T_3|) = \ln(|\Upsilon_2|/|\Upsilon_3|).$$

A concept closely related to the idea of linear feedback is that of *linear dependence*, a measure of which is given by

$$F_{z,x} = \ln(|\Upsilon_1|/|\Upsilon_4|) = \ln(|T_1|/|T_4|).$$

From these measures it is easily seen that

$$F_{z,x} = F_{z\to x} + F_{x\to z} + F_{z\cdot x},$$

so that linear dependence can be decomposed additively into three kinds of feedback. Absence of a particular causal ordering is then equivalent to one of these feedback measures being zero.

One drawback with these measures is that they range from 0 to ∞. However, as Pierce (1982) comments, they can easily be transformed into an 'R^2' measure, ranging from 0 to 1, by defining

$$R^2 = 1 - e^{-F}$$

for each F measure, and such a measure is then identical to the one proposed by Pierce himself (1979b) and termed R_*^2 in Chapter 13.

To make these measures operational, and to enable tests of null hypotheses asserting the absence of one or more causal orderings to be computed, the infinite lag lengths in the canonical form must be truncated. We shall suppose that these lag lengths have been truncated at lag p and each of the equations in the canonical form have been estimated by least squares regression. We can then form estimates

$$\hat{\Upsilon}_j = \sum_{t=1}^{n} \hat{\boldsymbol{w}}_{jt} \hat{\boldsymbol{w}}'_{jt}/n$$

and

$$\hat{T}_j = \sum_{t=1}^{n} \hat{\boldsymbol{u}}_{jt} \hat{\boldsymbol{u}}'_{jt}/n, \quad j = 1, 2, 3, 4,$$

where $\hat{\boldsymbol{w}}_{jt}$ and $\hat{\boldsymbol{u}}_{jt}$ are the vectors of residuals corresponding to the disturbance vectors $\boldsymbol{w}_{jt}$ and $\boldsymbol{u}_{jt}$ respectively and n denotes the sample size. From these estimates, we can then compute the various F measures: for example

$$\hat{F}_{z\to x} = \ln(|\hat{\Upsilon}_1|/|\hat{\Upsilon}_2|). \tag{14.32}$$

It then follows that the likelihood ratio (LR) test statistic of the null hypothesis H_{01}: $F_{z\to x} = 0$ (Z does not cause X) is

$$n\hat{F}_{z\to x} \sim \chi^2(glp).$$

Similarly, the null H_{02}: $F_{x\to z} = 0$ is tested by

$$n\hat{F}_{x\to z} \sim \chi^2(glp)$$

and H_{03}: $F_{x\cdot z} = 0$ by

$$n\hat{F}_{x\cdot z} \sim \chi^2(gl).$$

Since these are tests of nested hypotheses, $\hat{F}_{z\to x}$, $\hat{F}_{x\to z}$ and $\hat{F}_{x\cdot z}$ are asymptotically independent. All three restrictions can be tested at once since

$$n\hat{F}_{z,x} \sim \chi^2(gl(2p+1))$$

on H_{04}: $F_{z,x} = 0$.

Geweke (1982) also considers the construction of confidence intervals. An approximate 95% confidence interval for $F_{z\to x}$, for example, is

$$\left\{\left[\left(\hat{F}_{z\to x} - \frac{glp-1}{3n}\right)^{\frac{1}{2}} - \frac{1.96}{\sqrt{n}}\right]^2 - \frac{2glp+1}{3n},\right.$$

$$\left.\left[\left(\hat{F}_{z\to x} - \frac{glp-1}{3n}\right)^{\frac{1}{2}} + \frac{1.96}{\sqrt{n}}\right] - \frac{2glp+1}{3n}\right\}.$$

Geweke (1984a) provides Wald (W) and Lagrange multiplier (LM) test statistics to accompany the above LR test, defined for $H_{01}: F_{x \to z} = 0$ as

$$\mathrm{W}: n[\mathrm{tr}(\hat{\Upsilon}_1 \hat{\Upsilon}_2^{-1}) - g] \sim \chi^2(glp)$$

and

$$\mathrm{LM}: n[g - \mathrm{tr}(\hat{\Upsilon}_2 \hat{\Upsilon}_1^{-1})] \sim \chi^2(glp)$$

respectively.

When testing the null that Z does not cause X, test statistics based upon $\hat{\Upsilon}_1$ and $\hat{\Upsilon}_2$, i.e. from a comparison of the linear projections (14.25) and (14.27), are often referred to as 'Granger tests', since they emerge directly from Granger's (1969a) definition of causality. Test statistics based upon $\hat{T}_3$ and $\hat{T}_4$, obtained from the comparison of the projections (14.28) and (14.30), are known as 'Sims tests', since the restriction under the null of noncausality was first noted, in a slightly different form, by Sims (1972). These test statistics take the form

$$\mathrm{LR}: n[\ln(|\hat{T}_3|/|\hat{T}_4|)],$$

$$\mathrm{W}: n[\mathrm{tr}(\hat{T}_3 \hat{T}_4^{-1}) - 1],$$

and

$$\mathrm{LM}: n[1 - \mathrm{tr}(\hat{T}_3^{-1} \hat{T}_4)]$$

respectively. It is important to note that, in a finite sample, there is no numerical relationship between, for example, the Granger LR test and the corresponding Sims test statistic: the null may well be rejected using one test but not the other.

It is fair to point out that the above Sims test is not the one that has usually been employed in the literature on causality testing. The most popular form of this test follows Sims (1972) by using the projections of z_t on X_t (the final form of z_t):

$$z_t = \sum_{s=0}^{\infty} J_{1s} \boldsymbol{x}_{t-s} + \varepsilon_{1t} \tag{14.33}$$

and z_t on X,

$$z_t = \sum_{s=-\infty}^{\infty} J_{2s} \boldsymbol{x}_{t-s} + \varepsilon_{2t}, \tag{14.34}$$

and testing the restriction $J_{1s} \equiv J_{2s}$ for all s. A problem here is that the disturbance terms in (14.33) and (14.34) will be serially correlated. Corrections for this serial correlation can be made, but consistent estimation of the pattern of serial correlation is computationally demanding. Further tests, also requiring considerable computation, have been proposed by Pierce (1977) and Pierce and Haugh (1977).

Exhibit 14.3 *Consumption and income feedback*

	LM	LR	W	df
Granger				
$C \to Y$	29.85	34.15	39.31	6
$Y \to C$	13.13	13.88	14.68	6
$C.Y$	31.17	35.90	41.64	1
C,Y	74.15	83.93	95.63	13
Sims				
$C \to Y$	34.37	40.25	47.55	6
$Y \to C$	20.00	21.81	23.85	6
$C.Y$	31.02	35.69	41.36	1
C,Y	85.39	97.75	112.76	13

All statistics $\sim \chi^2(df)$.

In any event, simulation studies carried out by Geweke et al. (1983), Guilkey and Salemi (1982) and Nelson and Schwert (1982) reach a consensus that inference should be carried out using either of the test procedures discussed in detail above, these being found to combine the greatest reliability with computational ease.

A number of practical problems regarding the implementation of these tests do still remain, however. Of particular importance is the appropriate determination of the lag length p and the problem of how to deal with non-autoregressive processes. For a discussion of these topics, and others, the reader is referred to Geweke (1984a).

Example 14.2: Feedback measures for consumption and income

In this example we continue the analysis of the relationship between consumption and income begun in the sequence of examples in Chapter 13. There we found evidence of feedback from consumption to income and formal tests of causality confirm this here. As we have a bivariate system, $g = l = 1$, the lag length p was set at six, in accordance with the suggestion of Geweke (1982, page 323). The error covariance matrices $\hat{T}_j$ and $\hat{\Upsilon}_j$ are therefore scalars, being the (maximum likelihood) estimators of the error variances. Both levels and first differences of the (logarithms of the) two variables were investigated, but since the test statistics follow essentially the same pattern, only those for the first differenced series are reported in Exhibit 14.3. Both the Granger and Sims forms of the test provide evidence of significant feedback from consumption to income and, indeed, the feedback measure $F_{c \to y}$ is substantially larger than $F_{y \to c}$. There is also a strong instantaneous feedback between the variables. These findings thus confirm the earlier analysis that neither variable can be regarded as exogenous in a bivariate characterisation of consumption and income.

Example 14.3: Feedback measures for a small model of the UK under the Gold Standard

In this example we consider a four variable system containing first differences of the logarithms of output, the money base and the price level, and the first difference of the interest rate, using annual UK data during the late Gold Standard era of 1870 to 1913. A detailed discussion of the data is to be found in Mills and Wood (1987). Dwyer (1985) describes two hypotheses concerning the behaviour of the United Kingdom economy during the period that yield testable endogenous/exogenous classifications of the four variables. The first is the 'small country' hypothesis, in which the price level and the interest rate are exogenous and the money base and output are endogenous; the second is the 'representative economy' hypothesis, which has the money base exogenously determined and the other three variables endogenous.

Denoting these variables as Q, M, P and R respectively, then $\boldsymbol{y}_t' = (Q_t, M_t, P_t, R_t)'$ with, for the first hypothesis, $\boldsymbol{z}_{1t}' = (Q_t, M_t)'$, $\boldsymbol{x}_{1t}' = (P_t, R_t)'$, and for the second hypothesis $\boldsymbol{z}_{2t}' = (Q_t, P_t, R_t)'$, $\boldsymbol{x}_{2t}' = (M_t)'$. More detailed analysis (see Example 14.6) suggests that the lag lengths could be set at $p = 1$ and Exhibit 14.4 shows the alternative test statistics for these hypothesised classifications (or causal orderings). In all cases the well known inequality $\mathrm{LM} \leqslant \mathrm{LR} \leqslant \mathrm{W}$ is confirmed and the assumption of independence between the hypothesised endogenous and exogenous subvectors is rejected. Both the Granger and Sims tests reject the 'small country' exogeneity hypothesis; there is significant feedback from money and output to the price level and interest rate. Indeed, the Granger tests do not indicate significant feedback in the reverse direction. The 'representative economy' hypothesis fares somewhat better. The Sims tests cannot reject the hypothesis that money is exogenous and neither can the Granger LM test at the 5% significance level, although the other Granger tests do reject this exogeneity assumption. Another anomaly is that the Sims tests show significant instantaneous causality, whereas the Granger tests fail to do so.

All-in-all, these tests show that neither of the maintained hypotheses is particularly suitable for the multivariate process being analysed here.

Exhibit 14.4 *Causal orderings for the gold standard system*

(a) '*Small country*' *hypothesis*

$z_1' = (Q, M)'$, $x_1' = (P, R)$

	LM	LR	W	df
Granger				
$Z \to X$	14.84	18.04	22.26	4
$X \to Z$	7.61	8.27	9.01	4
$Z.X$	5.91	6.23	6.57	4
Z,X	28.36	32.54	37.84	12
Sims				
$Z \to X$	13.02	15.46	18.57	4
$X \to Z$	9.55	10.56	11.72	4
$Z.X$	5.91	6.23	6.57	4
Z,X	28.48	32.25	36.86	12

(b) '*Representative economy*' *hypothesis*

$z_2' = (Q, P, R)'$, $x_2 = (M)$

	LM	LR	W	df
Granger				
$Z \to X$	7.71	8.49	9.39	3
$X \to Z$	10.68	12.27	14.20	3
$Z.X$	5.50	5.89	6.32	3
Z,X	23.89	26.65	29.91	9
Sims				
$Z \to X$	2.17	2.23	2.29	3
$X \to Z$	8.17	9.06	10.09	3
$Z.X$	8.66	9.67	10.85	3
Z,X	19.00	20.96	23.23	9

All statistics $\sim \chi^2(\text{df})$.

14.5.1 *Conditional feedback measures*

It is straightforward to extend the analysis so as to condition the feedback measures on other variables (see Geweke (1984b)). For example, the measures of feedback from Z to X conditional on $R_t = \{\boldsymbol{r}_{t-s}, s \geqslant 0\}$ can be calculated by replacing (14.25), (14.27), (14.29) and (14.31) by

$$\boldsymbol{x}_t = \sum_{s=1}^{\infty} E_{1s} \boldsymbol{x}_{t-s} + \sum_{s=0}^{\infty} H_{1s} \boldsymbol{r}_{t-s} + \boldsymbol{w}_{1t}$$

Exhibit 14.5 *Consumption and income feedback conditional on prices*

	LM	LR	W	df
Granger				
$C \rightarrow Y$	32.33	37.46	43.73	6
$Y \rightarrow C$	11.60	12.17	12.79	6
$C.Y$	26.99	30.44	34.50	1
C, Y	70.92	80.07	91.02	13
Sims				
$C \rightarrow Y$	31.36	36.15	41.97	6
$Y \rightarrow C$	16.78	18.03	19.41	6
$C.Y$	26.99	30.44	34.50	1
C, Y	75.13	84.62	95.88	13

All statistics $\sim \chi^2(\text{df})$.

$$\boldsymbol{x}_t = \sum_{s=1}^{\infty} E_{2s}\boldsymbol{x}_{t-s} + \sum_{s=1}^{\infty} F_{2s}\boldsymbol{z}_{t-s} + \sum_{s=0}^{\infty} H_{2s}\boldsymbol{r}_{t-s} + \boldsymbol{w}_{2t}$$

$$\boldsymbol{x}_t = \sum_{s=1}^{\infty} E_{3s}\boldsymbol{x}_{t-s} + \sum_{s=0}^{\infty} F_{3s}\boldsymbol{z}_{t-s} + \sum_{s=0}^{\infty} H_{3s}\boldsymbol{r}_{t-s} + \boldsymbol{w}_{3t}$$

$$\boldsymbol{x}_t = \sum_{s=1}^{\infty} E_{3s}\boldsymbol{x}_{t-s} + \sum_{s=-\infty}^{\infty} F_{4s}\boldsymbol{z}_{t-s} + \sum_{s=0}^{\infty} H_{4s}\boldsymbol{r}_{t-s} + \boldsymbol{w}_{4t}$$

respectively.

Example 14.4: Conditioning consumption and income feedback measures on prices

An obvious conditioning variable in the consumption–income model is the price level, given the findings in the examples of Chapter 13. Exhibit 14.5 shows the feedback measures obtained when the first differences of the price level and six lags were included in the sequence of regressions. One interesting feature is that the Granger LM and LR test statistics for income causing consumption are reduced below the critical 5% significance level. The finding that increasing the number of variables analysed can alter causal orderings is well known, and will be discussed further, along with other misspecifications, in section 14.7.

14.6 Innovation accounting

Consider again the moving average representation of $\boldsymbol{y}_t$:

$$\boldsymbol{y}_t = \psi(B)\,\boldsymbol{v}_t = \sum_{s=0}^{\infty} \psi_s \boldsymbol{v}_{t-s}. \tag{14.35}$$

When no distinction is made between endogenous and (strictly) exogenous variables, or if there are no exogenous variables, the ψ_i matrices can be interpreted as the dynamic multipliers of the system since they represent the model's response to a unit shock in each of the variables. The response of y_m to a unit shock in y_n is thus given by the sequence

$$\psi_{mn,0}, \psi_{mn,1}, \psi_{mn,2}, \ldots.$$

If a variable, or block of variables, are strictly exogenous, then the implied zero restrictions ensure that these variables do not react to a shock in any of the endogenous variables, as is easily confirmed if (14.9), subject to the restrictions (14.10), is written in MA form. As has been discussed, however, the above MA representation will not be unique, and in this particular form has the disadvantage that the components of $\boldsymbol{v}_t$ are contemporaneously correlated. If these correlations are high, simulation of a shock to y_n, while all other components of $\boldsymbol{y}$ are held constant, could be misleading.

Sims (1980, 1981) therefore renormalises the moving average representation (14.35) into the recursive form

$$\boldsymbol{y}_t = \sum_{s=0}^{\infty} \psi_s^* \boldsymbol{u}_{t-s} \tag{14.36}$$

by using the lower triangular matrix S (see equation (14.8)), so that $\psi_i^* = \psi_i S^{-1}$ and $\boldsymbol{u}_t = S\boldsymbol{v}_t$ with $\sum_u = I$. The impulse response function of y_m to a unit shock y_n is then given by the sequence

$$\psi^*_{mn,0}, \psi^*_{mn,1}, \psi^*_{mn,2}, \ldots.$$

The uncorrelatedness of the $\boldsymbol{u}$'s allows the error variance of the $H+1$ step-ahead forecast of y_m to be decomposed into components accounted for by these shocks, or 'innovations', hence the phrase coined by Sims (1981) for this technique, that of *innovation accounting*. In particular, the components of this error variance accounted for by innovations in y_n is given by

$$\sum_{i=0}^{H+1} \psi^{*2}_{mn,i}.$$

For large H, this decomposition allows the isolation of relative contributions to movements in variables which are, intuitively, 'persistent'. Further details of this technique are provided by Sims (1982) and Doan et al. (1984). As these references point out, however, there is an important disadvantage to the technique; the choice of the S matrix is not unique, so that a different ordering of the y variables will alter S, thus altering the $\psi^*_{mn,i}$ coefficients and hence the impulse response functions and innovation accounting decompositions. The extent of these changes will depend upon the size of the contemporaneous correlations between the components of the $\boldsymbol{v}_t$ vector. Sims (1981) recommends that various orderings of the variables be tried, with the subsequent sets of impulse response functions and innovation accounting decompositions then being checked for robustness and consistency, arguing that choosing plausible orderings of variables based upon prior theorising amounts to trying various Wold (1960) causal chain forms for the model.

The innovation accounting methodology has generated a great deal of interest, and is the subject of much detailed analysis and criticism. The references above to Sims and his co-workers contain spirited defences of the technique, and Sims (1987) is the most recent exposition of its advantages. Critics include, in particular, Leamer (1985) and Cooley and LeRoy (1985), who draw attention to, amongst other things, the fact that vector ARMA models cannot be regarded as 'structural' in the traditional econometric sense; arguing that unless prior predeterminedness or weak exogeneity assumptions concerning the presence or otherwise of contemporaneous variables in structural equations are made, innovations cannot be uniquely identified with a particular variable, thus invalidating the computed impulse response functions and innovation accounting decompositions. Such criticisms appear to imply that formal theoretical considerations should dictate the choice of variable ordering before triangularisation, although how different this is from the Sims' recommendations above is a debatable point.

Example 14.5: Innovation accounting decompositions for the Gold Standard model

Exhibit 14.6 presents innovation accounting decompositions for the Gold Standard model of Example 14.3 using triangularisations based on the two hypotheses considered there: a 'small country' triangularisation (ordering) of $[P, R, Q, M]$, which also embodies the Gibson Paradox relationship between the price level and the interest rate (see Example 13.7) plus the exogeneity assumptions used by Mills and Wood (1982) in analysing the demand for money during this period, and a 'representative economy' ordering of $[M, Q, P, R]$. The decompositions appear to be relatively robust to the different orderings and in both cases

Exhibit 14.6 *Proportions of forecast error k years ahead accounted by each innovation*

Forecast error in	k	'Small country' ('Representative economy') triangularised innovation in P	R	Q	M
P	1	1 (0.98)	0 (0)	0 (0)	0 (0)
	5	0.87 (0.82)	0 (0)	0.10 (16)	0.03 (0.02)
R	1	0.64 (0.57)	0.36 (0.33)	0 (0.03)	0 (0.07)
	5	0.61 (0.47)	0.12 (0.11)	0.26 (0.37)	0.01 (0.05)
Q	1	0.02 (0)	0 (0)	0.98 (0.98)	0 (0.02)
	5	0.09 (0.09)	0 (0)	0.86 (0.85)	0.05 (0.06)
M	1	0.02 (0)	0.01 (0)	0.03 (0)	0.94 (1)
	5	0.08 (0.05)	0.01 (0)	0.13 (0.15)	0.78 (0.80)

responses are virtually complete within five years. We note that there is no feedback from interest rates to the other variables, although more detailed analysis, discussed in a future example, isolates a feedback into output. Price innovations initially explain a large proportion of the variance of interest rates, but this proportion decreases over time as output innovations become more important. Other effects appear to be relatively small although, as we shall later see, such small proportions do not preclude variables from entering significantly into a multivariate model.

14.7 Misspecification in vector ARMA models

It is well known that incorrectly specifying the dimension of $\boldsymbol{y}_t$ may lead to incorrect inferences being drawn concerning the causal patterns existing between the variables of the system, and hence to incorrect categorisations of endogenous and exogenous variables and to misleading innovation accounting decompositions. Sims (1981, pages 288–90), for example, provides an extended discussion of the problems brought about by omitting variables from a model and Lutkepohl (1982a) presents an empirical example of this type of misspecification. To show this formally, consider again the AR representation of the process $\boldsymbol{y}_t$, where now $\boldsymbol{y}_t$ is partitioned as $\boldsymbol{y}_t' = (\boldsymbol{y}_{1t}, \boldsymbol{y}_{2t})$:

$$\begin{bmatrix} \pi_{11}(B) & \pi_{12}(B) \\ \pi_{21}(B) & \pi_{22}(B) \end{bmatrix} \begin{bmatrix} \boldsymbol{y}_{1t} \\ \boldsymbol{y}_{2t} \end{bmatrix} = \begin{bmatrix} \boldsymbol{v}_{1t} \\ \boldsymbol{v}_{2t} \end{bmatrix}. \tag{14.37}$$

The corresponding partitioned MA representation is

$$\begin{bmatrix} y_{1t} \\ y_{2t} \end{bmatrix} = \begin{bmatrix} D^{-1}(B), & -D^{-1}(B)\,\pi_{12}(B)\,\pi_{22}^{-1}(B) \\ -\pi_{22}^{-1}(B)\,\pi_{21}(B)\,D^{-1}(B), & \pi_{22}^{-1}(B) \\ & \times[I+\pi_{21}(B)\,D^{-1}(B)\,\pi_{12}(B)\,\pi_{22}^{-1}(B)] \end{bmatrix} \begin{bmatrix} \boldsymbol{v}_{1t} \\ \boldsymbol{v}_{2t} \end{bmatrix} \quad (14.38)$$

where

$$D(B) = \pi_{11}(B) - \pi_{12}(B)\,\pi_{22}^{-1}(B)\,\pi_{21}(B),$$

from the rules for a partitioned inverse. Using the terminology of Lutkepohl (1982a, b), building a model for $\boldsymbol{y}_{1t}$ alone means modelling the *subprocess* of $\boldsymbol{y}_t$,

$$D(B)\,\boldsymbol{y}_{1t} = \boldsymbol{v}_{1t} - \pi_{12}(B)\,\pi_{22}^{-1}(B)\,\boldsymbol{v}_{2t}. \quad (14.39)$$

Since the moving average part of (14.39) can be written as $\Lambda(B)\,V_t$, say, from Wold's decomposition theorem, the subprocess is useful for forecasting. As it is a reduced form, however, it is not useful for structural analysis. For this we need the *partial process*

$$D(B)\,\boldsymbol{y}_{1t}^* = \boldsymbol{v}_{1t}, \quad (14.40)$$

which differs from the subprocess (14.39) unless $\pi_{12} = 0$, and which cannot be determined unless we have the model for the complete process (14.37).

Suppose now that $\boldsymbol{y}_{1t}' = (y_{1t}, y_{2t})'$ and $\boldsymbol{y}_{2t}' = (y_{3t})'$, so that we have a trivariate system partitioned as

$$\left[\begin{array}{cc|c} \alpha_{11}(B) & \alpha_{12}(B) & \alpha_{13}(B) \\ \alpha_{21}(B) & \alpha_{22}(B) & \alpha_{23}(B) \\ \hline \alpha_{31}(B) & \alpha_{32}(B) & \alpha_{33}(B) \end{array}\right] \begin{bmatrix} y_{1t} \\ y_{2t} \\ y_{3t} \end{bmatrix} = \begin{bmatrix} v_{1t} \\ v_{2t} \\ v_{3t} \end{bmatrix} \quad (14.41)$$

where

$$\pi_{11}(B) = \begin{bmatrix} \alpha_{11}(B) & \alpha_{12}(B) \\ \alpha_{21}(B) & \alpha_{22}(B) \end{bmatrix}, \quad \pi_{12}(B) = \begin{bmatrix} \alpha_{13}(B) \\ \alpha_{23}(B) \end{bmatrix}$$

$$\pi_{21}(B) = [\alpha_{31}(B) \quad \alpha_{32}(B)], \quad \pi_{22}(B) = \alpha_{33}(B)$$

$$\boldsymbol{v}_{1t}' = (v_{1t}, v_{2t})', \quad \boldsymbol{v}_{2t}' = (v_{3t})'.$$

In this case the subprocess for $\boldsymbol{y}_{1t}$ and corresponding partial process are

$$D(B)\begin{bmatrix} y_{1t} \\ y_{2t} \end{bmatrix} = \begin{bmatrix} v_{1t} - \alpha_{13}(B)\,\alpha_{33}^{-1}(B)\,v_{3t} \\ v_{2t} - \alpha_{23}(B)\,\alpha_{23}^{-1}(B)\,v_{3t} \end{bmatrix}$$

and

$$\begin{bmatrix} y^*_{1t} \\ y^*_{2t} \end{bmatrix} = \begin{bmatrix} v_{1t} \\ v_{2t} \end{bmatrix}$$

respectively, where

$$D(B) = \begin{bmatrix} \alpha_{11}(B) - \alpha_{13}(B)\,\alpha_{33}^{-1}(B)\,\alpha_{31}(B), & \alpha_{12}(B) - \alpha_{13}(B)\,\alpha_{33}^{-1}(B)\,\alpha_{32}(B) \\ \alpha_{21}(B) - \alpha_{23}(B)\,\alpha_{33}^{-1}(B)\,\alpha_{31}(B), & \alpha_{22}(B) - \alpha_{23}(B)\,\alpha_{33}^{-1}(B)\,\alpha_{32}(B) \end{bmatrix}.$$

If, in the complete process, y_{2t} does not cause y_{1t}, so that $\alpha_{12}(B) = 0$, it is possible in the subprocess for such causality to be observed; $\alpha_{13}(B) \neq 0$ and $\alpha_{32}(B) \neq 0$ will ensure this. Conversely, if $\alpha_{13}(B) = 0$ and $D_{21}(B) = 0$, then y_{2t} will not cause y_{1t} in the subprocess, but causation occurs in the complete process if $\alpha_{12}(B) \neq 0$.

A number of other misspecifications may result in incorrect inferences concerning causal orderings being drawn. Newbold (1978) shows, for example, that a one-way causal system may turn into a system containing feedback if measurement error is present. A similar effect is also found with seasonal adjustment; see Wallis (1974) and Sims (1974). Finally, Tiao and Wei (1976) show that temporal aggregation (see Chapter 11 for analysis in the univariate case) may alter causal structures. For example, if a variable is caused by another in the basic underlying system, temporal aggregation may produce a system in which no causal relationship between the variables is found, or it may result in a system exhibiting feedback.

We thus conclude that all the usual misspecifications commonly discussed in econometrics have consequences for causal orderings, endogenous/exogenous classifications and hence the interpretation of vector ARMA models.

14.8 Estimation of vector ARMA processes

It is recommended that the parameters of a vector ARMA process are estimated by ML procedures as they result in consistent, asymptotically efficient and normally distributed estimators under very general conditions. Generally, optimisation of the likelihood function results in a complicated nonlinear optimisation problem, although some special cases allow fairly straightforward estimation. The small sample properties of these ML procedures are, in general, unknown.

Anderson (1980), for example, contains a detailed discussion of ML estimation procedures for the vector ARMA model and, as in the univariate case, the associated inverse information matrix can be used as an estimate of the asymptotic covariance matrix of the vector of parameter estimates. These procedures can easily incorporate parameter constraints, perhaps given by economic theory, and nonstochastic exogenous variables, so that vector ARMAX processes are included.

If the process has no MA part, so that it is a pure vector AR model, then major simplifications result. If there are no parameter restrictions then the model has the form of a seemingly unrelated regression model with an equal number of regressors in each equation: OLS estimation of each equation then provides consistent and asymptotically efficient estimates. If there are zero constraints on the parameters, then there may be different regressors in the different equations. In this case either Zellner's (1962) SUR method can be used or a simple transformation can be applied to the model so that OLS estimation of each separate equation remains efficient.

Lutkepohl (1982c) discusses another subclass of ARMA processes that can be estimated efficiently by OLS; the class of autoregressive-discounted-polynomial processes. These models attempt to impose smoothness constraints on the AR polynomial, in the hope of alleviating distortions to the parameter estimates that are merely due to sample randomness rather than the actual underlying structure.

14.9 Vector ARMA model building

Many of the features of vector ARMA model building are extensions of those met in building univariate ARMA models; unfortunately, however, the increased dimensionality of the model can cause major difficulties. It is fair to say that such model building is very much in its infancy, with no consensus having yet been reached as to the appropriate method of identification. Consequently, we will provide only a fairly cursory discussion of this, admittedly crucial, aspect of multivariate time series analysis, although two detailed examples are provided.

As in univariate analysis, it is necessary to work with stationary series and differencing of the component series contained within $\boldsymbol{y}_t$ would seem natural. This is indeed the usual way to proceed, but as Lutkepohl (1982d) points out, this should be done with care, for differencing univariate series may lead to overdifferenced, and hence noninvertible, multiple time series. This, of course, is a phenomenon that is related to the discussion of cointegrability in Chapter 13. As we shall see, there are some descriptive devices, which extend the analysis of SACFs and PACFs to the multivariate case, that can be employed to alleviate such problems.

If we restrict attention to unconstrained vector AR processes then sequential testing procedures are the traditional tool to use. Sequences of likelihood ratio tests may be performed (Hannan (1970)) or, alternatively, Lagrange Multiplier tests might be applied (Hosking (1981b)). Multivariate extensions of the model selection criteria introduced in Chapter 8 may also be employed; see Lutkepohl (1985) for a survey and comparison of such criteria.

The determination of the order of the AR process by such criteria and resultant estimation of the unconstrained model will often lead to an overly generous parameterisation. Various methods have been proposed to reduce the number of parameters in an AR model chosen in such a way. Hsiao (1979) and Penm and Terrell (1982) consider methods that allow zero restrictions to be imposed within the AR polynomials, while Sargent and Sims (1977), Sims (1981) and Reinsel (1983) discuss the use of 'index models', which basically extend the idea of principle components to multiple time series.

Example 14.6: A vector AR Gold Standard model

We consider here the modelling of the UK Gold Standard data, previously analysed in Examples 14.3 and 14.5, as a vector AR process. If we let $\boldsymbol{y}_t' = (Q_t, M_t, P_t, R_t)'$, where the variables are as defined in Example 14.3, then the appropriate order p of the four-dimensional AR(p) process may be determined by, for example, any of the usual model selection criteria. Exhibit 14.7 shows Akaike's AIC criteria for $p = 0, 1, \ldots, 9$, which selects a value of $p = 1$. Although it is well known that AIC has a tendency to overestimate the correct order of p, finding $p = 1$ suggests that use of other, consistent, criteria would probably reach the same conclusion.

The OLS estimation of this AR(1) model produced

$$\begin{bmatrix} Q_t \\ \\ M_t \\ \\ P_t \\ \\ R_t \\ \end{bmatrix} = \begin{bmatrix} 0.060 & 0.273 & 0.224 & 0.130 \\ (0.138) & (0.130) & (0.155) & (0.054) \\ 0.303 & 0.100 & 0.113 & 0.085 \\ (0.156) & (0.147) & (0.176) & (0.061) \\ 0.385 & 0.274 & 0.308 & -0.000 \\ (0.122) & (0.115) & (0.137) & (0.047) \\ 0.383 & -0.104 & 0.541 & 0.432 \\ (0.391) & (0.368) & (0.441) & (0.152) \end{bmatrix} \begin{bmatrix} Q_{t-1} \\ \\ M_{t-1} \\ \\ P_{t-1} \\ \\ R_{t-1} \\ \end{bmatrix} + \begin{bmatrix} v_{1t} \\ \\ v_{2t} \\ \\ v_{3t} \\ \\ v_{4t} \\ \end{bmatrix}$$

$$\Sigma_v = 10^{-3} \begin{bmatrix} 0.565 & & & \\ 0.097 & 0.727 & & \\ -0.087 & -0.003 & 0.441 & \\ -0.233 & 0.275 & 0.591 & 4.555 \end{bmatrix},$$

Exhibit 14.7 *AICs for AR models fitted to* $\mathbf{y}_t = (Q_t, M_t, P_t, R_t)'$

Lag, p	AIC(p)
0	−1161.87
1	<u>−1177.18</u>
2	−1162.94
3	−1153.81
4	−1141.92
5	−1119.23
6	−1111.98
7	−1102.00
8	−1136.86
9	−1162.76

where all variables are measured as deviations from means, and standard errors are shown in parentheses.

We see that, even with $p = 1$, a number of parameters in the π_1 matrix are insignificant, and economic interpretation of the model is further complicated by $\sum_v$ being nondiagonal. Deleting insignificant coefficients and transforming to a diagonal error covariance matrix leads to the model (see Mills and Wood (1987) for details):

$$\begin{bmatrix} 1 & 0 & 0 & 0 \\ 0 & 1 & 0 & 0 \\ 0 & 0 & 1 & 0 \\ 0 & 0 & \underset{(0.386)}{-1.188} & 1 \end{bmatrix} \begin{bmatrix} Q_t \\ M_t \\ P_t \\ R_t \end{bmatrix} = \begin{bmatrix} 0 & \underset{(0.137)}{-0.272} & 0 & \underset{(0.052)}{-0.113} \\ \underset{(0.130)}{0.371} & 0 & 0 & 0 \\ \underset{(0.126)}{0.373} & \underset{(0.116)}{0.280} & \underset{(0.137)}{0.327} & 0 \\ 0 & 0 & 0 & \underset{(0.129)}{0.442} \end{bmatrix} \begin{bmatrix} Q_{t-1} \\ M_{t-1} \\ P_{t-1} \\ R_{t-1} \end{bmatrix} + \begin{bmatrix} w_{1t} \\ w_{2t} \\ w_{3t} \\ w_{4t} \end{bmatrix}$$

$$\Sigma_w = 10^{-5}\begin{bmatrix} 0.610 & & & \\ 0 & 0.778 & & \\ 0 & 0 & 0.462 & \\ 0 & 0 & 0 & 4.020 \end{bmatrix}.$$

No variable is strictly or strongly exogenous in the model but, given the analysis of Cooley and LeRoy (1985), Mills and Wood (1987) argue that the w_{it} innovations can be regarded as structural disturbances, in which case the impulse response functions calculated from the associated MA

Exhibit 14.8 *Proportions of forecast error k years ahead accounted by each innovation*

Forecast error in	k	'Small country' ('Representative economy') triangularised innovation in P	R	Q	M
P	1	1 (0.98)	0 (0)	0 (0)	0 (0)
	5	0.87 (0.82)	0 (0)	0.10 (16)	0.03 (0.02)
R	1	0.64 (0.57)	0.36 (0.33)	0 (0.03)	0 (0.07)
	5	0.61 (0.47)	0.12 (0.11)	0.26 (0.37)	0.01 (0.05)
Q	1	0.02 (0)	0 (0)	0.98 (0.98)	0 (0.02)
	5	0.09 (0.09)	0 (0)	0.86 (0.85)	0.05 (0.06)
M	1	0.02 (0)	0.01 (0)	0.03 (0)	0.94 (1)
	5	0.08 (0.05)	0.01 (0)	0.13 (0.15)	0.78 (0.80)

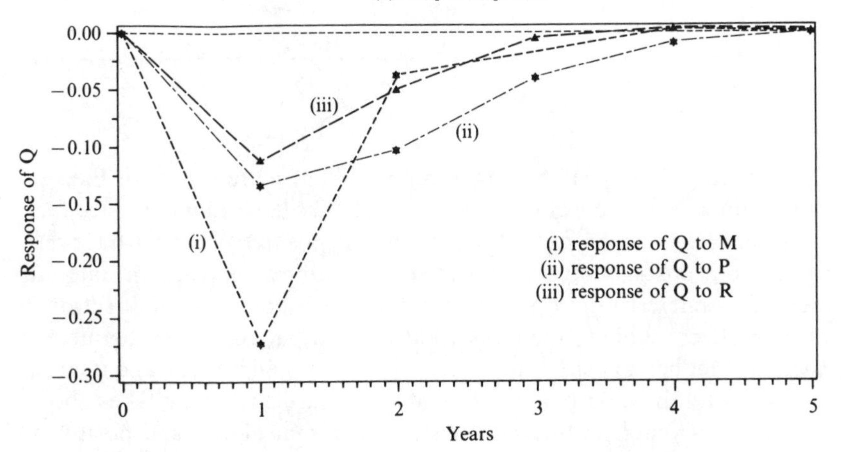

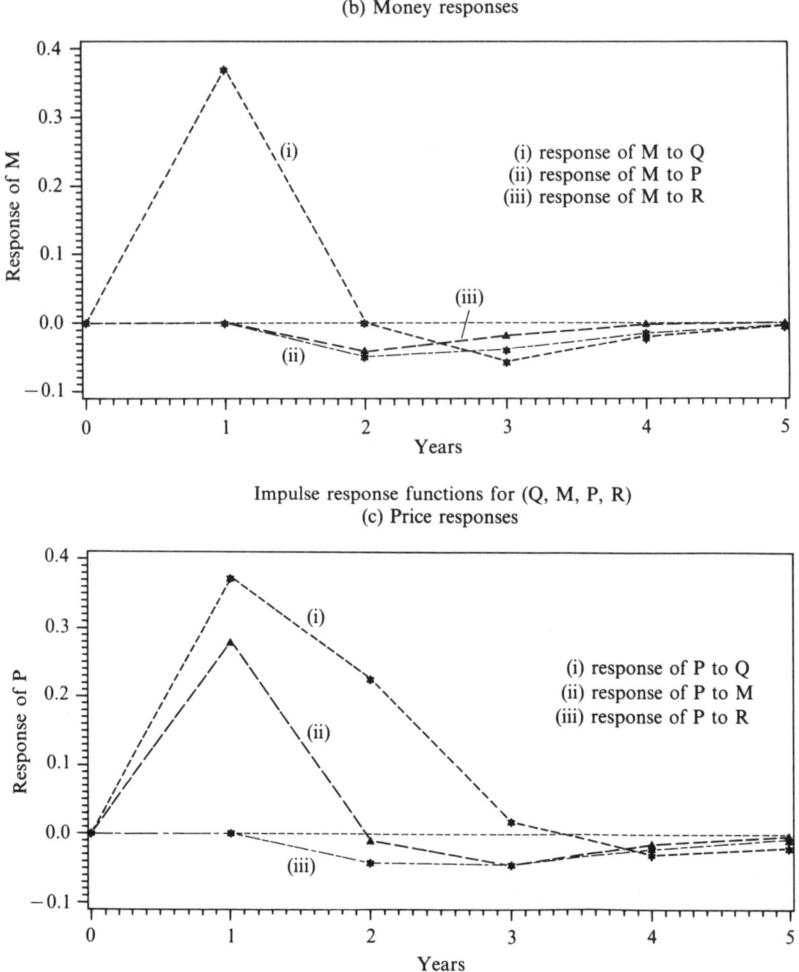

representation have a unique interpretation. These are shown in Exhibit 14.8, from which we conclude that money shocks have an initial once-for-all negative impact on output, a similar but positive effect on the price level, and a positive impact, spread over a number of years, on interest rates. This interest rate response to a monetary shock is an indirect one, with the effect running from money, through output and prices, to interest rates. Output has a positive impact on both money and prices, and interest rates respond to an output shock positively, but the response takes about five years to complete. Interest rates respond immediately and positively

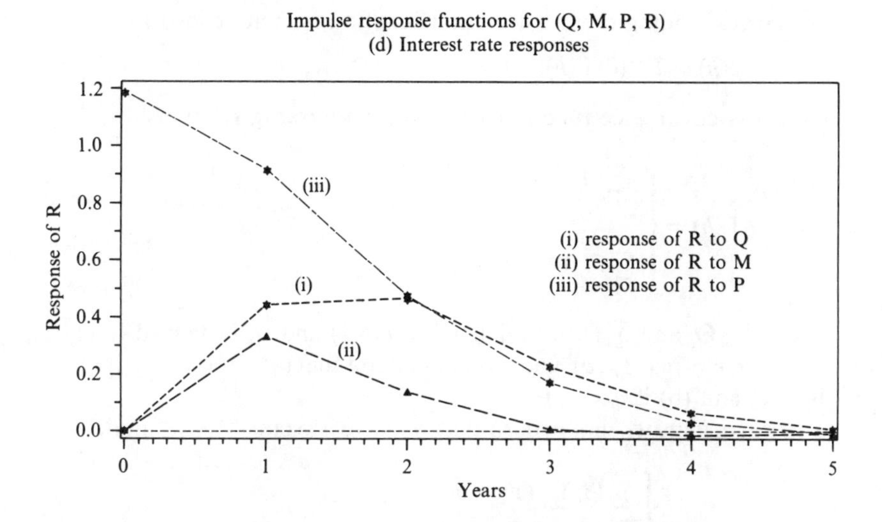

to a price shock, the response being powerful, roughly exponential in decline, and with much of the response being completed within four years. Price shocks also have a negative effect on output and money. For a detailed theoretical interpretation of these responses, see Mills and Wood (1987).

As a final comparison, univariate models fitted to the four series had error variances ($\times 10^{-5}$) of 0.807, 0.818, 0.615 and 4.733 respectively, so that analysing the series in a multivariate framework leads to decreases in error variances of the order of 5% for money, 15% for the interest rate, and 25% for output and the price level.

14.9.1 *Identification of vector ARMA models*

When the restriction to pure AR models is relaxed, the identification of vector ARMA processes becomes, once again, a very tentative process. A number of model building techniques have been suggested, for example those proposed by Granger and Newbold (1977), Wallis (1977) and Chan and Wallis (1978), but these are difficult to apply when vectors containing more than two variables are under analysis. The identification procedures that appear to be most promising are those of Tiao and Box (1981), Jenkins and Alavi (1981), and Tiao and Tsay (1983).

These all begin by considering, analogous to univariate identification, the properties of the cross-covariance function associated with the stationary vector time series $\boldsymbol{y}_t$, defined earlier as

$$\Gamma(h) = E[(\boldsymbol{y}_t - \boldsymbol{\mu})(\boldsymbol{y}_{t-h} - \boldsymbol{\mu}')], \quad h = 0, \pm 1, \pm 2, \ldots$$

The corresponding cross-correlation function is then defined as

$$\rho(h) = \Gamma^{-1}(0)\,\Gamma(h), \quad h = 0, \pm 1, \pm 2, \ldots.$$

The cross-covariance function satisfies the following relationship

$$\Gamma(h) = \begin{cases} \sum_{j=h-r}^{h-1} \Gamma(j)\,\Phi'_{h-j} - \sum_{j=0}^{r-h} \psi_j \Sigma_v \Theta'_{j+h}, & h = 0, 1, \ldots, r \\ \sum_{j=1}^{r} \Gamma(j)\,\Phi'_j, & h > r \end{cases}$$

where Φ_i, Θ_i and Σ_v are as defined in (14.1) and ψ_j is defined in (14.3), $\Theta_0 = -I$, $r = \max(p, q)$, and it is understood that (a) if $p < q$, $\Phi_{p+1} = \ldots = \Phi_r = 0$, and (b) if $q < p$, $\Theta_{q+1} = \ldots = \Theta_r = 0$.

In particular, when $p = 0$, so that we have a vector MA (q) model, then

$$\Gamma(h) = \begin{cases} \sum_{j=0}^{q-h} \Theta_j \Sigma_v \Theta'_{j+h} & h = 0, 1, \ldots, q \\ 0 & h > q. \end{cases}$$

Thus, all auto- and cross-correlations are zero when $h > q$. On the other hand, for a vector AR (p) model the $\Gamma(h)$ will decay gradually as h increases.

Analogous to the partial autocorrelation function for the univariate case, the partial autoregressive matrix function can be defined as having the property that if the model is AR (p) then

$$P(h) = \begin{cases} \Phi_h, & h = p \\ 0, & h > p, \end{cases}$$

an explicit formula for $P(h)$ being given in Tiao and Box (1981, page 805).

Sample estimates of $\Gamma(h)$ and $P(h)$ may then be used to identify possible models. However, the presentation of these matrices presents some difficulties when the dimension of k is greater than two. Tiao and Box (1981) recommend that, instead of displaying these matrices either in full or graphically, which would be prohibitively cumbersome, indicator symbols highlighting significant elements should be used. These are illustrated in the following example.

Example 14.7: Multivariate modelling of the consumption and price relationship

Here we consider the multivariate relationship between consumption, income and the price level; the variables being those analysed in previous examples. As has been mentioned above, there can be difficulties in determining the appropriate degrees of differencing of the

Exhibit 14.9 *Summary of cross-correlation matrices for* $\boldsymbol{y}_t' = (\nabla c_t, \nabla y_t, \nabla p_t)$

(a) *Matrices of indicators at lags* $h = 1, 2, \ldots, 10$

$$
\begin{array}{lccccc}
\text{Lag, } h & 1 & 2 & 3 & 4 & 5 \\
\begin{array}{l} \nabla c_t \\ \nabla y_t \\ \nabla p_t \end{array} &
\begin{bmatrix} \cdot & + & - \\ \cdot & - & \cdot \\ - & \cdot & + \end{bmatrix} &
\begin{bmatrix} \cdot & \cdot & - \\ \cdot & \cdot & \cdot \\ \cdot & \cdot & + \end{bmatrix} &
\begin{bmatrix} \cdot & - & \cdot \\ \cdot & \cdot & \cdot \\ \cdot & \cdot & + \end{bmatrix} &
\begin{bmatrix} \cdot & \cdot & - \\ \cdot & \cdot & \cdot \\ \cdot & \cdot & + \end{bmatrix} &
\begin{bmatrix} \cdot & \cdot & - \\ \cdot & \cdot & \cdot \\ \cdot & \cdot & + \end{bmatrix}
\end{array}
$$

$$
\begin{array}{lccccc}
\text{Lag, } h & 6 & 7 & 8 & 9 & 10 \\
\begin{array}{l} \nabla c_t \\ \nabla y_t \\ \nabla p_t \end{array} &
\begin{bmatrix} \cdot & \cdot & - \\ \cdot & \cdot & \cdot \\ \cdot & \cdot & + \end{bmatrix} &
\begin{bmatrix} \cdot & \cdot & \cdot \\ \cdot & \cdot & \cdot \\ \cdot & \cdot & + \end{bmatrix} &
\begin{bmatrix} \cdot & \cdot & \cdot \\ \cdot & \cdot & \cdot \\ \cdot & \cdot & + \end{bmatrix} &
\begin{bmatrix} \cdot & \cdot & \cdot \\ - & \cdot & \cdot \\ \cdot & \cdot & + \end{bmatrix} &
\begin{bmatrix} \cdot & \cdot & \cdot \\ \cdot & \cdot & \cdot \\ \cdot & \cdot & + \end{bmatrix}
\end{array}
$$

(b) *Indicators for the* (i, j) *position*

	∇c_t	∇y_t	∇p_t
∇c_t	....	+.−.......	−−−−−−....
∇y_t	........−.	−.........	....
∇p_t	−.........	....	++++++++++

component series which ensure that the multivariate process is both stationary *and* invertible.

The individual series analysed here all require differencing, the ARIMA models fitted to them being:

$$\nabla c_t = 0.0052 + a_{1t}, \qquad \hat{\sigma}_1 = 0.00787$$

$$\nabla y_t = 0.0064 + (1 - 0.209B)\, a_{2t}, \qquad \hat{\sigma}_2 = 0.01803$$

$$\nabla^2 p_t = (1 - 0.359B)\, a_{3t}, \qquad \hat{\sigma}_3 = 0.00768.$$

Note that the price series requires second differencing, but in view of the above discussion concerning the problems of nonstationarity in vector ARMA models, it was decided to begin multivariate analysis by using the first differences of all variables. If p_t does require second differencing, then this should be revealed in the model building (see Hillmer, Larcker and Schroeder (1983) for a similar approach and also Heyse and Wei (1985)).

Exhibit 14.9 shows the sample cross-correlation matrices for $h = 1, 2, \ldots, 10$ using the indicator symbols suggested by Tiao and Box (1981). These symbols are defined in the following way: in the (i, j)th position of the cross-correlation matrix at lag h a symbol $(+, -, \cdot)$ is placed where

$+$: denotes a value $> 2n^{-\frac{1}{2}}$

$-$: denotes a value $< -2n^{-\frac{1}{2}}$

$\cdot$: denotes a value $< |2n^{-\frac{1}{2}}|$.

Exhibit 14.10 *Indicator matrices for partial autoregressions and related statistics*

Lag, h	Indicator matrices	AIC (h)	Diagonal elements of $\sum_v (\times 10^{-4})$
1	$\begin{bmatrix} - & + & . \\ . & - & . \\ + & + & + \end{bmatrix}$	-3464.42	0.5 3.0 0.5
2	$\begin{bmatrix} . & . & . \\ + & - & . \\ . & . & + \end{bmatrix}$	-3465.29	0.5 2.8 0.5
3	$\begin{bmatrix} . & . & . \\ + & - & . \\ . & + & . \end{bmatrix}$	-3481.25	0.5 2.4 0.5
4	$\begin{bmatrix} . & . & . \\ . & . & . \\ . & . & . \end{bmatrix}$	-3468.78	0.5 2.4 0.5
5	$\begin{bmatrix} . & . & . \\ . & . & . \\ . & . & . \end{bmatrix}$	-3455.13	0.5 2.4 0.5
6	$\begin{bmatrix} . & . & . \\ . & . & . \\ . & . & . \end{bmatrix}$	-3442.00	0.5 2.4 0.5
7	$\begin{bmatrix} . & . & . \\ . & - & . \\ . & . & . \end{bmatrix}$	-3439.00	0.5 2.2 0.4
8	$\begin{bmatrix} . & . & . \\ . & . & . \\ . & . & . \end{bmatrix}$	-3447.18	0.4 2.1 0.4
9	$\begin{bmatrix} . & . & . \\ . & . & . \\ + & . & . \end{bmatrix}$	-3448.49	0.4 2.1 0.4
10	$\begin{bmatrix} . & . & . \\ . & . & . \\ . & . & . \end{bmatrix}$	-3438.12	0.4 2.0 0.4

If $\boldsymbol{y}_t$ is a vector white noise process, then the approximate standard error estimate of an element in the cross-correlation matrix is $n^{-\frac{1}{2}}$ (where n is the effective number of observations after differencing). Thus the above symbols indicate whether auto- and cross-correlations are non-zero at roughly a 5% level of significance. If the components of $\boldsymbol{y}_t$ (where $\boldsymbol{y}_t' = (\nabla c_t, \nabla y_t, \nabla p_t)$) are cross-correlated, then decisions based on these indicators will overestimate the actual level of significance. In Exhibit

14.9, the approximate standard error on which these indicators are based is $123^{-\frac{1}{2}} = 0.09$.

From this information, it is seen that there is a clear indication that ∇p_t may well be nonstationary and that the cross-correlations of this variable with ∇c_t are persistently negative. Further information is available from the partial autoregression matrices $P(h)$, estimates of which can be obtained by fitting autoregressive models of successively higher order $h = 1, 2, \ldots$. The estimated $P(h)$ matrices for $\mathbf{y}_t$ are shown in Exhibit 14.10, where again indicator symbols are used to flag significant elements. Also shown are AIC(h) values for $h = 1, 2, \ldots, 10$ and the diagonal elements of the sequence of error covariance matrices: these are useful to help determine the order of a pure AR model. However, given that the majority of significant elements tend to occur in the partial autoregression matrices for the first three lags, this suggests, in combination with the behaviour of the sample cross-correlation matrices, that a mixed vector ARMA model should be considered.

In fact a vector ARMA(2, 1) model was initially estimated and, after the deletion of numerous insignificant parameters, the following specification was finally arrived at:

$$\mathbf{y}_t = \hat{\Phi}_1 \mathbf{y}_{t-1} + \hat{\Phi}_2 \mathbf{y}_{t-2} + \mathbf{v}_t + \hat{\Theta}_1 \mathbf{v}_{t-1}$$

where

$$\mathbf{y}_t' = (\nabla c_t, \nabla y_t, \nabla p_t),$$

$$\hat{\Phi}_1 = \begin{bmatrix} \underset{(0.092)}{-0.213} & \underset{(0.038)}{0.103} & \underset{(0.049)}{-0.185} \\ 0 & \underset{(0.117)}{0.240} & 0 \\ \underset{(0.091)}{0.215} & \underset{(0.038)}{0.112} & \underset{(0.051)}{0.916} \end{bmatrix}$$

$$\hat{\Phi}_2 = \begin{bmatrix} 0 & 0 & 0 \\ 0 & 0 & \underset{(0.056)}{-0.157} \\ 0 & 0 & 0 \end{bmatrix}$$

$$\hat{\Theta}_1 = \begin{bmatrix} 0 & 0 & 0 \\ \underset{(0.197)}{1.119} & \underset{(0.202)}{-0.832} & 0 \\ 0 & 0 & 0 \end{bmatrix}$$

$$\hat{\Sigma}_v = 10^{-4}\begin{bmatrix} 0.522 & & \\ 0.577 & 2.600 & \\ -0.154 & -0.491 & 0.517 \end{bmatrix},$$

$$\hat{\rho}_v = \begin{bmatrix} 1 & & \\ 0.50 & 1 & \\ -0.30 & -0.42 & 1 \end{bmatrix}.$$

This model confirms that c_t and y_t exhibit feedback, and both 'cause' p_t; p_t, however, feedbacks into c_t, but not into y_t. There are also large contemporaneous correlations between all three variables as shown by $\hat{\rho}_v$. The error variances are decreased by 16%, 20% and 12% respectively when compared to the univariate models. We note, however, that $\hat{\Phi}_{33,1} = 0.916$, suggesting that second differencing of p_t may be required.

On re-identification and re-estimation, the following ARMA(1,1) specification for $\boldsymbol{y}_t^{*\prime} = (\nabla c_t, \nabla y_t, \nabla^2 p_t)$ was obtained:

$$\boldsymbol{y}_t^* = \hat{\Phi}_1^* \boldsymbol{y}_{t-1}^* + \boldsymbol{v}_t^* + \hat{\Theta}_1^* \boldsymbol{v}_{t-1}^*,$$

where

$$\hat{\Phi}_1^* = \begin{bmatrix} 0 & \underset{(0.037)}{0.107} & 0 \\ 0 & \underset{(0.124)}{0.258} & 0 \\ \underset{(0.083)}{0.235} & \underset{(0.039)}{0.094} & \underset{(0.080)}{-0.176} \end{bmatrix}$$

$$\hat{\Theta}_1^* = \begin{bmatrix} 0 & 0 & 0 \\ \underset{(0.183)}{1.114} & \underset{(0.202)}{-0.771} & 0 \\ 0 & 0 & 0 \end{bmatrix}$$

$$\hat{\Sigma}_v^* = 10^{-4}\begin{bmatrix} 0.586 & & \\ 0.579 & 2.682 & \\ -0.142 & -0.468 & 0.508 \end{bmatrix},$$

$$\hat{\rho}_v^* = \begin{bmatrix} 1 & & \\ 0.46 & 1 & \\ -0.26 & -0.40 & 1 \end{bmatrix}.$$

While the fit of the ∇c_t and ∇y_t equations are slightly inferior to those of the preceding model, that for $\nabla^2 p_t$ is substantially better. $\nabla^2 p_t$ does not feedback into ∇c_t, providing further support for the previous findings that inflation does not affect consumption. From a comparison of the two specifications, it would seem that the model containing ∇p_t is marginally preferable, thus showing that differencing vector time series on the basis of univariate analysis can indeed be a hazardous procedure.

14.9.2 *Model checking*

To guard against model misspecification and to search for possible directions of improvements, a detailed diagnostic analysis of the residual series

$$\hat{\boldsymbol{v}}_t = \boldsymbol{y}_t - \hat{\Phi}_1 \boldsymbol{y}_{t-1} - \dots - \hat{\Phi}_p \boldsymbol{y}_{t-p} - \hat{\Theta}_1 \hat{\boldsymbol{v}}_{t-1} - \dots - \hat{\Theta}_q \hat{\boldsymbol{v}}_{t-q}$$

should be performed. This may include plotting standardised residuals against time and analysing the estimated cross-correlation matrices of the residual series in a similar fashion to that discussed above. If the fitted model provides an adequate representation of the data, then the residuals will be neither auto- nor cross-correlated, and such analysis of the models fitted in the previous examples revealed no inadequacies.

Multivariate portmanteau and Lagrange multiplier statistics are also available (Hosking (1980, 1981b), Poskitt and Tremayne (1982)) but Tiao and Box (1981) emphasise that, particularly with vector time series, there is no substitute for detailed inspection of the residual correlation structure for revealing subtle relationships which may indicate important directions of improvement.

14.10 Forecasting from vector ARMA models

For the general vector ARMA process (14.1), we assume that observations $\boldsymbol{y}_n, \boldsymbol{y}_{n-1}, \dots$ are available and that we wish to forecast $\boldsymbol{y}_{n+l}$, $l \geqslant 1$. As in the univariate case the MMSE forecast of $\boldsymbol{y}_{n+l}$ from origin n, denoted here as $\hat{\boldsymbol{y}}_n(l)$, is the conditional expectation of $\boldsymbol{y}_{n+l}$:

$$\begin{aligned}\hat{\boldsymbol{y}}_n(l) = {} & \Phi_1 \hat{\boldsymbol{y}}_n(l-1) + \dots + \Phi_p \hat{\boldsymbol{y}}_n(l-p) \\ & + E_n(\boldsymbol{v}_{n+l}) + \dots + \Theta_q E_n(\boldsymbol{v}_{n+l-q})\end{aligned}$$

where

$$\hat{\boldsymbol{y}}_n(l) = \boldsymbol{y}_{n+l} \quad \text{for} \quad l \leqslant 0$$

and

$$\begin{aligned}E_n(\boldsymbol{v}_{n+l}) &= E(\boldsymbol{v}_{n+l} \mid \boldsymbol{y}_n, \boldsymbol{y}_{n-1}, \dots) \\ &= \begin{cases} \boldsymbol{v}_{n+l} & \text{for} \quad l \leqslant 0 \\ 0 & \text{for} \quad l > 0. \end{cases}\end{aligned}$$

The forecast error vector $\boldsymbol{e}_n(l) = \boldsymbol{y}_{n+l} - \hat{\boldsymbol{y}}(l)$ has covariance matrix

$$V(l) = E[\boldsymbol{e}_n(l)\,\boldsymbol{e}'_n(l)]$$

and, from Jenkins and Alavi (1982), it follows that

$$V(l) = \sum\nolimits_v + \psi_1 \sum\nolimits_v \psi'_1 + \ldots + \psi_{l-1} \sum\nolimits_v \psi'_{l-1},$$

where $\psi_1, \ldots, \psi_{l-1}$ are the moving average matrices from the vector MA representation (14.3).

Note that when the Φ_j and Θ_j matrices are all triangular, the relationships between the components of $\boldsymbol{y}_t$ can be represented by a set of transfer functions of the type discussed in Chapter 13. The above analysis of forecasting using vector ARMA models thus subsumes the topic of forecasting using transfer functions, although this is developed in detail in Pierce (1975b), with Abraham and Ledolter (1983, chapter 8) providing textbook discussion and examples.

In practice, the parameter matrices Φ_j and Θ_j will have to be replaced by their estimates. The influence of estimated parameters on forecasts from vector AR models has been considered by Baillie (1979) and Reinsel (1980), and it has been found that estimation error can have *less* serious effects in multivariate models than in the univariate case.

Example 14.8: Forecasting from the multivariate consumption, income and price model

The vector ARMA(2, 1) model for $\boldsymbol{y}_t = (\nabla c_t, \nabla y_t, \nabla p_t)$ fitted in Example 14.7 was used to forecast the seven observations available after the end of the sample period used for model fitting, 1986Q1 to 1987Q3. For comparison, the forecasts from the univariate ARMA models were also obtained. Hence, from the origin n of 1985Q4, l-step ahead forecasts ($l = 1, 2, \ldots, 7$) were obtained. Summary statistics of these forecasts are shown in Exhibit 14.11.

As follows from the smaller error variances obtained from the vector ARMA model, expected forecast standard errors are smaller for the multivariate model than for the corresponding univariate model at all values of l. Actual forecast performance, however, does not show a clear gain in accuracy for the vector ARMA model. In terms of root MSE, forecasts of y_t from the vector ARMA model substantially outperform those from the univariate model, the accuracy of the competing c_t forecasts are comparable, while the univariate forecasts of p_t are certainly superior to those from the multivariate model.

These findings compare closely with those of Riise and Tjostheim (1985), who conclude that forecasts from multivariate models can be inferior to those from univariate models because the former are more

Exhibit 14.11 *Forecast statistics for Example* 14.8

(a) *Expected forecast standard errors*

	c_t		y_t		p_t	
Lag l	Vector	Univariate	Vector	Univariate	Vector	Univariate
1	0.0072	0.0079	0.0161	0.0180	0.0072	0.0077
2	0.0102	0.0111	0.0205	0.0230	0.0147	0.0148
3	0.0127	0.0136	0.0244	0.0271	0.0233	0.0229
4	0.0149	0.0157	0.0278	0.0306	0.0325	0.0321
5	0.0169	0.0176	0.0310	0.0337	0.0419	0.0422
6	0.0188	0.0193	0.0339	0.0366	0.0515	0.0531
7	0.0206	0.0208	0.0367	0.0393	0.0611	0.0649

(b) *Actual forecasting performance*

	Vector RMSE	Univariate RMSE
c_t	0.0326	0.0323
y_t	0.0074	0.0226
p_t	0.0542	0.0237

sensitive to changes in structure. Such a structural change may have occurred here during the forecast period, with price increases being far less than would have been predicted from comparable movements in consumption and income.

14.11 Computing software for multiple time series models

Computer software for vector ARMA model building is somewhat limited. Of the major statistical packages, SAS/ETS's (SAS (1985c)) PROC STATESPACE contains most of the identification procedures discussed above and allows automatic model building using Akaike's (1976) canonical correlation technique. PROC SYSNLIN can also be adapted to estimate vector ARMA models, but this requires rather sophisticated knowledge of the SAS language. Multiple input, including intervention, transfer functions, on the other hand, are straightforwardly identified and estimated in PROC ARIMA.

Possibly the most complete packages for vector ARMA modelling are the SCA system (Liu and Hudak (1983)) and the WMTS-1 program (Tiao et al. (1979)).

Part IV

Nonlinear time series models

Throughout the book, we have concentrated on the analysis of linear models for time series processes, leading to the development, for example, of vector ARMA and multivariate exponential smoothing models. The final part of the book, comprising just two chapters, is concerned with nonlinear models.

Chapter 15 considers models in which the conditional variance of the process under analysis is allowed to vary through time, rather than being restricted to be constant as in all the models so far introduced. The basic model allowing non-constant conditional variances is the autoregressive conditional heteroskedastic process, commonly referred to as the ARCH process. This has created a great deal of interest amongst economists, and has resulted in a number of applications in the macroeconomic and financial areas. The basic ARCH model, and a variety of extensions, are discussed in detail in this chapter. Rather than deal explicitly with changing variances, however, many analysts prefer to attempt to induce constancy of variance through the well known Box–Cox power transformation, which has been introduced in earlier chapters. The implications of using power transformed series for forecasting, and the biases inherent in such a procedure, are also discussed in Chapter 15. Transforming a series is also a popular method of inducing normality, but the final section of the chapter considers an alternative explanation for a skewed marginal distribution: the series could have been generated by an asymmetric process. Such processes are briefly discussed, along with their potential application to financial time series.

The final chapter of the book provides an introductory discussion of the many explicitly nonlinear models that have been proposed in the literature for modelling time series: these include the bilinear, threshold and exponential autoregressive models. All of these can be shown to be special cases, along with the linear ARMA model, of a very general formulation, that of the state dependent model. Although such models have found a number of applications in engineering and hydrology, they have yet to make much of an impact on the analysis of economic time series. Because of this, and also because of the complexities of model identification and estimation, methods of testing whether a time series is linear or nonlinear are of obvious importance. The final section of the chapter therefore discusses a simple-to-compute test of nonlinearity based upon Tukey's general one degree of freedom test for non-additivity.

15 Conditional variance models and related topics

15.1 Introduction

Up to this point, the models and methods developed so far have concentrated on the conditional means of time series, implicitly assuming that the conditional variance remains constant. Much recent work in economic theory has been concerned with behaviour under uncertainty, so that economic agents have to make decisions based upon the distribution of a random variable at some future point in time. For risk averse agents having general utility functions, a measure of variability for this conditional distribution will be just as important as the conditional mean. Time series models are now available that allow quantitative measures of risk and uncertainty to be calculated by using time varying conditional variances, and it is to a discussion of such models that this chapter is primarily devoted.

Section 15.2 begins by introducing the Autoregressive Conditional Heteroskedastic (ARCH) process as a method of modelling conditional variances and then reviews the various extensions of the basic ARCH model. Sections 15.3 and 15.4 discuss the estimation and testing of ARCH models and provide examples of their application. Section 15.5 considers some alternative models for conditional variances that have been used primarily in analysing financial time series.

Section 15.6 considers further the use of power transformations to induce constancy in conditional variances and shows that care needs to be taken when forecasting such transformed series. The final section discusses asymmetric time series, which may easily be mistaken for series requiring power transformation, and shows how asymmetric models are potentially useful in modelling financial series.

15.2 ARCH models

In our previous discussions of forecasting, we have concentrated on the construction of point forecasts which, as we have set out formally in Chapter 7, are to be interpreted as conditional means. For example, the forecast of the value of a time series at time t, y_t, given information on the series up to time $t-1$, is

$$E(y_t \mid y_{t-1}, y_{t-2}, \ldots).$$

As a simple illustration, consider the (zero mean) AR(1) process

$$y_t = \phi y_{t-1} + \varepsilon_t, \tag{15.1}$$

where ε_t is white noise with $V(\varepsilon) = \sigma^2$. The conditional mean of y_t is ϕy_{t-1}, whereas the unconditional mean is zero. As Engle (1982, page 987) remarks, the vast improvements in forecasts due to the use of time series models clearly stems from their use of the conditional mean.

Associated with a conditional mean forecast will be a conditional forecast variance. In our AR(1) example, the one-step ahead conditional variance

$$V(y_t \mid y_{t-1}, y_{t-2}, \ldots) = E\{[y_t - E(y_t \mid y_{t-1}, y_{t-2}, \ldots)]^2 \mid y_{t-1}, y_{t-2}, \ldots\}$$

is simply σ^2, while the unconditional variance is $\sigma^2/(1-\phi^2)$. Neither the one-step ahead nor k-step ahead forecast variance depends upon the information set $(y_{t-1}, y_{t-2}, \ldots)$ and thus all forecast variances will be constant over the sample period. This will also be true for more general ARMA models for y_t and for regression type models relating y_t to a set of weakly exogenous variables $\boldsymbol{x}_t$. Hence it will not be possible for such models to measure changes in forecast variances, even though in practice one might prefer forecast variances that are affected by past information.

The standard econometric approach to dealing with a time changing variance, the 'problem' of heteroskedasticity, is to introduce an exogenous variable x_t which predicts the variance; for example

$$y_t = \varepsilon_t x_{t-1}, \tag{15.2}$$

in which the conditional variance of y_t is $\sigma^2 x_{t-1}^2$. The forecast intervals thus depend upon the evolution of an exogenous variable, but this is often regarded as unsatisfactory since it requires specifying a cause of the changing variance, rather than recognising that both conditional means and variances may evolve jointly over time.

The more general approach proposed by Engle (1982) allows the variance to depend upon the available information set. Assuming conditional normality, a general specification of the evolution of y_t would be

$$y_t \mid Y_{t-1}, X_t \sim N(g_t, h_t) \tag{15.3}$$

where $Y_{t-1} = \{y_{t-s}, s \geqslant 1\}$ and $X_t = \{\boldsymbol{x}_{t-s}, s \geqslant 0\}$, and where both g_t and h_t are functions of the variables in Y_{t-1} and X_t. Although this set-up contains the standard heteroskedasticity model as a special case, Engle prefers to consider the model with $g_t = \boldsymbol{z}_t' \boldsymbol{\beta}$, for some set of variables $\boldsymbol{z}_t$ in (Y_{t-1}, X_t), and where

$$h_t = \alpha_0 + \sum_{i=1}^{q} \alpha_i \varepsilon_{t-i}^2, \quad \varepsilon_t = y_t - g_t. \tag{15.4}$$

Equations (15.3) and (15.4) together are known as an autoregressive conditional heteroskedasticity (ARCH) regression model. By defining

$$\boldsymbol{w}_t' = (1, \varepsilon_{t-1}^2, \ldots, \varepsilon_{t-q}^2)$$

and

$$\boldsymbol{\alpha}' = (\alpha_0, \alpha_1, \ldots, \alpha_q),$$

the ARCH regression model can be conveniently written as

$$y_t \mid Y_{t-1}, X_t \sim N(\boldsymbol{z}_t' \boldsymbol{\beta}, \boldsymbol{w}_t' \boldsymbol{\alpha}). \tag{15.5}$$

This notation makes the connections with more conventional heteroskedastic models very clear, and (15.5) can easily be extended to the case where $\boldsymbol{w}_t$ includes directly observable variables that are contained in the information set.

The AR(1) model for y_t combined with ARCH(1) errors, obtained by setting $\boldsymbol{z}_t' \boldsymbol{\beta} = \phi y_{t-1}$ and $q = 1$ in (15.5):

$$y_t = \phi y_{t-1} + \varepsilon_t$$

$$E(\varepsilon_t \mid E_{t-1}) = 0$$

$$V(\varepsilon_t \mid E_{t-1}) = h_t = \alpha_0 + \alpha_1 \varepsilon_{t-1}^2, \tag{15.6}$$

where $E_t = \{\varepsilon_{t-s}, s \geqslant 0\}$, has several important properties which serve to illustrate the usefulness of ARCH models. We assume, as usual, that $|\phi| < 1$, so that y_t is stationary. To ensure that h_t is positive we must have $\alpha_0 > 0$ and $\alpha_1 \geqslant 0$. Engle (1982) shows that the unconditional variance of ε_t will be finite if $\alpha_1 < 1$, in which case it will be given by

$$V(\varepsilon_t) = \sigma^2 = \frac{\alpha_0}{1 - \alpha_1}.$$

Using this notation, the conditional variance of ε_t can be written as

$$h_t - \sigma^2 = \alpha_1(\varepsilon_{t-1}^2 - \sigma^2),$$

so that the conditional variance will be above the unconditional variance whenever the squared 'surprise', ε_{t-1}^2, exceeds its unconditional expectation, σ^2.

The errors ε_t, although serially uncorrelated through the white noise assumption, are *not* independent since they are related through their second moments. Although y_t is conditionally normal, it is not jointly normal, and neither is its marginal distribution. The marginal distribution of y_t will be symmetric, however, if the conditional distribution of ε_t is symmetric. If ε_t is conditionally normal, the fourth unconditional moment of ε_t will exceed $3\sigma^4$ so that the marginal distribution of ε_t exhibits fatter tails than the normal, while a finite fourth moment is ensured if $3\alpha_1^2 < 1$.

From (15.6) it follows that

$$E(y_t \mid Y_{t-1}) = \phi y_{t-1}$$

and

$$V(y_t \mid Y_{t-1}) = h_t = \alpha_0 + \alpha_1 (y_{t-1} - \phi y_{t-2})^2,$$

so that both the conditional mean and conditional variance of the one-step ahead forecast depend on the available information set. In particular, the conditional variance is increased by large 'surprises' in y_t.

From

$$y_{t+k} = \phi^k y_t + \sum_{i=1}^{k} \phi^{k-i} \varepsilon_{t+1},$$

k-step ahead conditional forecast variances from the AR(1)–ARCH(1) model can be computed from

$$V(y_{t+k} \mid Y_t) = \sum_{i=1}^{k} \phi^{2(k-i)} E(h_{t+i} \mid Y_t).$$

From Engle and Kraft (1983), if $\alpha_1 < 1$ and $j > 2$ then

$$E(h_{t+i} \mid Y_t) = \sigma^2 + \alpha_1 E[(h_{t+i-1} - \sigma^2) \mid Y_t],$$

so that the k-step ahead conditional forecast variance equals

$$V(y_{t+k} \mid Y_t) = \phi^2 \sum_{i=0}^{k-1} \sigma^{2i} + \alpha_1^{k-1} (h_{t+1} - \sigma^2) \sum_{i=0}^{k-1} \phi^{2i} \alpha_1^{-i}. \qquad (15.7)$$

The conditional variance is clearly not independent of the current information set, but for large k the dependence on $(h_{t+1} - \sigma^2)$ becomes negligible and (15.7) is then well approximated by

$$V(y_{t+k} \mid Y_t) = \sigma^2 \sum_{i=0}^{k-1} \phi^{2i},$$

the conventional conditional variance in the absence of ARCH errors.

Engle (1982) shows that the ARCH(q) model of (15.4) will have a finite but positive variance

$$V(\varepsilon_t) = \frac{\alpha_0}{1 - \sum_{i=1}^{q} \alpha_i}$$

if $\alpha_0 > 0, \alpha_1, \ldots, \alpha_q \geqslant 0$ and if all of the roots of the associated characteristic equation lie outside the unit circle, so that $\sum \alpha < 1$.

15.2.1 *Extensions of the standard ARCH model*

The standard ARCH model can be extended in various ways. Weiss (1984) assumes that y_t, where y_t is now possibly a differenced series, is generated by an ARMA model,

$$\phi(B)(y_t - \bar{y}) = \theta(B)\,\varepsilon_t,$$

and that the conditional variance equation is

$$h_t = \alpha_0 + \sum_{i=1}^{q} \alpha_i \varepsilon_{t-i}^2 + \sum_{i=1}^{p} \delta_i (y_{t-i} - \bar{y})^2 + \delta_0 (y_t - \varepsilon_t - \bar{y})^2, \qquad (15.8)$$

with $\delta_i \geqslant 0$, thus leading to the ARMA–ARCH model for y_t.

Bollerslev (1986, 1988), on the other hand, extends the ARCH regression model (15.5) by generalising h_t to

$$h_t = \alpha_0 + \sum_{i=1}^{q} \alpha_i \varepsilon_{t-i}^2 + \sum_{i=1}^{p} \beta_i h_{t-i}, \qquad (15.9)$$

where $\beta_i \geqslant 0$. This function for h_t, which allows lagged conditional variances to enter, is termed a Generalised ARCH(p, q) process, or GARCH(p, q), and when combined with $g_t = z_t' \boldsymbol{\beta}$ is known as the GARCH regression model. Bollerslev argues that in many applications ARCH(q) models with relatively large q are required, and to avoid problems with negative variance estimates, a fixed lag structure has typically to be imposed (Engle, 1982). The GARCH(p, q) model can thus be regarded as an extension of the ARCH class of models, allowing both a longer memory and a more flexible lag structure without having to impose, a priori, any fixed lag pattern.

The ARCH models so far considered concentrate on allowing the variance of y_t to depend on the information set, but they leave the usual specification of the mean unaltered. Engle et al. (1987) extend the ARCH model to allow the conditional variance to affect the mean, resulting in what is termed the ARCH-M (ARCH in mean) model:

$$y_t \mid Y_{t-1}, X_t \sim N(z'_{1t}\boldsymbol{\beta} + \delta h_t, h_t^2)$$

$$h_t^2 = \boldsymbol{w}'_t \boldsymbol{\alpha} + z'_{2t}\gamma, \tag{15.10}$$

where z_{1t} and z_{2t} are (possibly different) sets of variables in (Y_{t-1}, X_t).

References to recent research on multivariate extensions to these models may be found in Engle and Bollerslev (1986), who also introduce the concept of integrated GARCH (IGARCH) models, which arise when there is a unit root in the GARCH (p, q) process. Tsay (1987) also considers extensions to ARCH models.

15.2.2 *ARCH effects in economic time series*

Why should ARCH type effects be found in economic time series? To answer this, consider again the conditionally normal, white noise error process ε_t, such that $E(\varepsilon_t) = 0$, $E(\varepsilon_t^2) = \sigma^2$ and $E(\varepsilon_t \varepsilon_j) = 0$ for $t \neq j$. This series could be generated by the ARCH (q) process

$$\varepsilon_t \mid E_t \sim N(0, h_t)$$

$$h_t = \alpha_0 + \sum_{i=1}^{q} \alpha_i \varepsilon_{t-i}^2.$$

As we have seen, although ε_t is white noise, successive values are not independent because they are related through higher moments. Hence, large values of ε_t are likely to be followed by large values of ε_{t+1} of either sign, unlike an autoregression in the conditional mean, where (for positive coefficients) a positive value is likely to be followed by another positive one. As a consequence, we would expect a realisation of ε_t to exhibit behaviour in which clusters of large observations are followed by clusters of small ones: the plot of the daily differences of the interbank rate shown as Exhibit 3.8 illustrates this admirably. Examples of this type of behaviour are also found in, for example, Engle (1983), Engle et al. (1985), and Pagan et al. (1983). The empirical application to which ARCH models were first addressed was the measurement of the variability of inflation, and a second fruitful area of application has been that of testing for time varying risk premia in the term structure of interest rates and other financial assets (Bollerslev, 1987). More recently, ARCH models have been applied to the foreign exchange market, used to test whether long bonds satisfy Shiller's (1979) variance bounds, and employed to derive pricing formulae for financial assets. References to this burgeoning literature are again provided by Engle and Bollerslev (1986), who remark that these applications have a common focus, that of modelling the behaviour of economic agents in relation to risk within a time series context.

15.3 Estimation of ARCH models

If we consider the ARCH regression model of (15.5):

$$y_t \mid Y_{t-1}, X_t \sim N(z_t' \boldsymbol{\beta}, h_t)$$

$$h_t = \alpha_0 + \sum_{i=1}^{q} \alpha_i \varepsilon_{t-i}^2$$

$$\varepsilon_t = y_t - z_t' \boldsymbol{\beta},$$

then we can define the loglikelihood of the tth observation as

$$l_t = -\tfrac{1}{2}\log h_t - \tfrac{1}{2}\varepsilon_t^2 / h_t$$

and the average loglikelihood over T observations as

$$l = \frac{1}{T}\sum_{t=1}^{T} l_t.$$

This likelihood function can then be maximised with respect to the unknown parameters $\boldsymbol{\alpha}$ and $\boldsymbol{\beta}$. Engle (1982) shows that the associated information matrix is block diagonal, so that the estimation of $\boldsymbol{\alpha}$ and $\boldsymbol{\beta}$ can be considered separately without loss of asymptotic efficiency and, moreover, that either can be estimated based only on a consistent estimate of the other. He recommends that $\boldsymbol{\beta}$ be initially estimated by OLS, from the residuals of which a consistent estimate of $\boldsymbol{\alpha}$ can be constructed. Based upon these $\hat{\boldsymbol{\alpha}}$ estimates, efficient estimates of $\boldsymbol{\beta}$ are then found. Scoring algorithms are used to calculate the iterations in each step, these being comprised of a sequence of least squares regressions on transformed variables: see Engle (1982, pages 996–9) for details and for an examination of the gains in efficiency obtained from using ML estimation rather than OLS applied directly to

$$y_t = z_t' \boldsymbol{\beta} + \varepsilon_t \tag{15.11}$$

and

$$\hat{\varepsilon}_t^2 = \alpha_0 + \sum_{i=1}^{q} \alpha_1 \hat{\varepsilon}_{t-i}^2, \tag{15.12}$$

where the $\hat{\varepsilon}_t$ are the OLS residuals from estimation of (15.11).

ML estimation of ARMA–ARCH models is discussed in Weiss (1984), whose empirical work suggests that again ML is preferable to OLS. Bollerslev (1986) considers ML estimation of GARCH regression models.

Here the autoregressive terms in h_t complicate the iterative estimation procedure, so that the scoring algorithms cannot be expressed in terms of simple auxiliary regressions. He recommends the use of the Berndt et al. (1974) algorithm to maximise the likelihood function, and this is also used in the estimation of the ARCH–M model (see Engle et al. (1987) for details).

All these estimators are asymptotic, and their small-sample properties have not yet been analytically established. Engle et al. (1985) present an extensive Monte Carlo simulation study of the ARCH(1) regression model with a single strongly exogenous regressor. They found that a two-step ML approximation produced a substantial negative bias in the estimation of α_1, but that the natural procedure of first using OLS, then testing for ARCH errors, and, upon finding it, re-estimating the model with ARCH, produced results almost as good as using ARCH directly when there is ARCH present. This procedure is much better than just using OLS in these circumstances, yet little worse when ARCH is absent. Such a procedure requires the use of tests for the presence of ARCH, of course, and these are now discussed.

15.4 Testing for ARCH

As we have seen, the presence of ARCH enables conditional variances to alter through time, thus explicitly allowing changing uncertainty to be modelled and hence leading to more realistic forecast intervals. The presence of ARCH can also lead to serious model misspecification if it is ignored: as with all forms of heteroskedasticity, analysis assuming its absence will result in appropriate parameter standard errors, and these will typically be too small. For example, Weiss (1984) shows that ignoring ARCH will lead to the identification of ARMA models that are over parameterised.

It is therefore important that methods should be available for testing whether ARCH is present, particularly as estimation incorporating it requires expensive iterative techniques. Since ARCH processes are modelled as functions of lagged squared errors, it is natural to look at the autocorrelation behaviour of the squared residuals from, for example, a regression or ARMA model fitted to y_t. That this is theoretically justified is shown in Bollerslev (1986), who notes that the GARCH(p, q) process (15.9) can equivalently be written as

$$\varepsilon_t^2 = \alpha_0 + \sum_{i=1}^{q} \alpha_i \varepsilon_{t-i}^2 + \sum_{i=1}^{p} \beta_i \varepsilon_{t-i}^2 + \sum_{i=1}^{p} \beta_j v_{t-j} + v_t \qquad (15.13)$$

where

$$v_t = \varepsilon_t^2 - h_t.$$

Since, by definition, v_t is serially uncorrelated, the GARCH (p,q) process can be interpreted as an ARMA (m,p) process in ε_t^2, where the order $m = \max(p,q)$. Bollerslev goes on to show that the standard ARMA theory developed in Chapter 5 follows through in this case.

The squared residuals $\hat{\varepsilon}_t^2$ can then be used to identify m and p, and therefore q, in a fashion similar to the way the usual residuals are used in conventional ARMA modelling. McLeod and Li (1983) show that if the SACF of $\hat{\varepsilon}_t^2$ is estimated as

$$\hat{r}_{\varepsilon\varepsilon}(k) = \frac{\sum_{t=k+1}^{T} (\hat{\varepsilon}_t^2 - \hat{\sigma}^2)(\hat{\varepsilon}_{t-k}^2 - \hat{\sigma}^2)}{\sum_{t=1}^{T} (\hat{\varepsilon}_t^2 - \hat{\sigma}^2)^2}$$

where

$$\hat{\sigma}^2 = \sum_{t=1}^{T} \hat{\varepsilon}_t^2 / T,$$

then the asymptotic variance of $\hat{r}_{\varepsilon\varepsilon}(k)$ is T^{-1} and the portmanteau statistic (cf. Ljung and Box (1978) and equation (8.20)),

$$Q_{\varepsilon\varepsilon} = T(T+2) \sum_{k=1}^{M} (T-k)^{-1} \hat{r}_{\varepsilon\varepsilon}(k),$$

is asymptotically $\chi^2(M)$ if the ε_t^2 are independent. Such tests can thus be used to ascertain whether the null hypothesis of no ARCH should be replaced by one of ARCH being present.

Of course, as with the use of portmanteau statistics in conventional ARMA modelling, no specific alternative hypothesis is specified. Engle (1982) suggests that, under these circumstances, an LM test of the alternative hypothesis of ARCH (q) errors would seem desirable, since such a test can be computed from an auxiliary regression after OLS estimation with $p = 0$. Indeed, this auxiliary regression is just the regression of $\hat{\varepsilon}_t^2$ on an intercept and q lagged values of $\hat{\varepsilon}_t^2$, i.e. equation (15.12). The appropriate test statistic is $T.R^2$, which should be tested as $\chi^2(q)$ (cf. the LM tests developed in Chapter 8). Familiar problems arise if the alternative hypothesis is that of GARCH (p,q) errors. As Bollerslev (1986) notes, under a null of white noise, a general test of $p > 0, q > 0$ is not feasible, nor is a test of GARCH $(p+r_1, q+r_2)$ errors, where $r_1 > 0$ and $r_2 > 0$, when the null is GARCH (p,q). Furthermore, under this null, the LM test for GARCH (r,q) and ARCH $(q+r)$ alternatives coincide (again cf. the discussion of LM tests in Chapter 8.7). A further complication in the GARCH case is that the test statistic $T.R^2$ is calculated from the first iteration of the Berndt et al. (1974) algorithm for the general model, starting at the ML estimates under H_0.

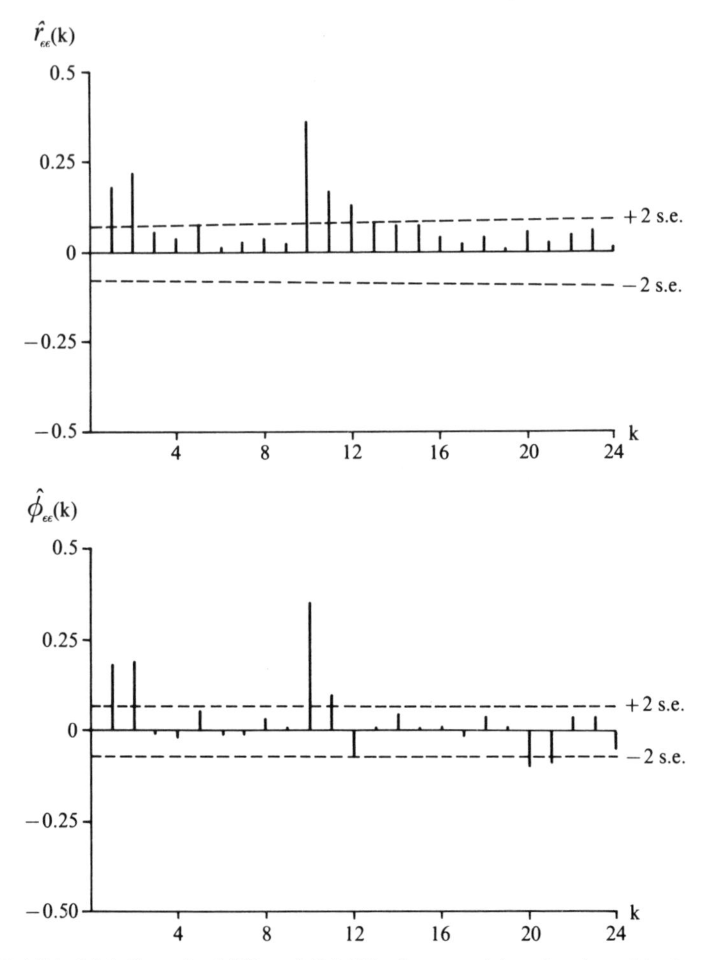

Exhibit 15.1 Sample ACF and PACF of squared interbank residuals

The small sample performance of the LM test for $q = 1$, along with the corresponding Wald statistic (the t-ratio associated with $\hat{\alpha}_1$), has been investigated by Engle et al. (1985), with fairly satisfactory results being obtained as long as the sample size is not too small.

Example 15.1: ARCH effects in daily interest rates

In this example we reconsider the 3-month interbank rate series, which was found in Chapter 8 to be adequately modelled by, for example, an ARIMA (2, 1, 0) process. It was pointed out in Example 8.8, however, that a significant sample autocorrelation appears at lag 10. Since the data

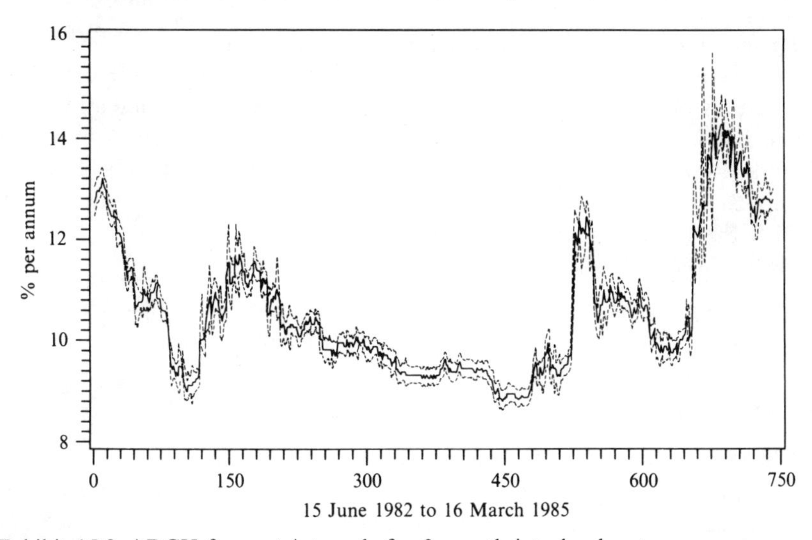

Exhibit 15.2 ARCH forecast intervals for 3-month interbank rate

is daily, such a lag corresponds to a fortnightly effect (there being a 5-day week). To account for this correlation, the following model was estimated:

$$\nabla y_t = \frac{\underset{(0.037)}{(1+0.149B^{10})}}{\underset{(0.036)\quad(0.037)}{(1-0.219B+0.137B^2)}}\varepsilon_t, \quad h_t = 0.0238.$$

From the plot of the first differences shown earlier as Exhibit 3.8, ARCH effects are clearly apparent. The SACF and PACF of the squared residuals from the above model are shown in Exhibit 15.1, with significant correlations being found at lags 1, 2, 10, 11 and 12. The LM test for ARCH(12) yields a test statistic of $T.R^2 = 155.8$, which is clearly significant. Least squares estimation of a suitably restricted form of the implied ARCH(12) process produced

$$h_t = \underset{(0.0121)}{0.0222} + \underset{(0.0360)}{0.0935}\varepsilon^2_{t-1} + \underset{(0.0335)}{0.2004}\varepsilon^2_{t-2} + \underset{(0.0335)}{0.3419}\varepsilon^2_{t-10} + \underset{(0.0360)}{0.0793}\varepsilon^2_{t-11}.$$

All coefficients are positive, thus ensuring $h_t > 0$. Although more efficient estimates of both parts of the model could be obtained by using iterative ML estimation (see Weiss (1984) for details), 95% one-step ahead forecast intervals computed using these LS estimates are shown in Exhibit 15.2; the changing uncertainty in the series is clearly revealed by the fluctuations in the width of the forecast intervals in response to movements in the series itself.

Example 15.2: Measuring inflation uncertainty

ARCH models have frequently been used to provide empirical measures of the uncertainty of inflation: see Engle (1982, 1983) and Engle and Kraft (1983). Bollerslev (1986) provides a recent example of such an application by analysing the quarterly rate of inflation of the US GNP implicit price deflator between 1948 and 1983. He begins by estimating by OLS the following model for this rate of inflation, denoted π_t, on the assumption of the absence of ARCH:

$$\begin{aligned} \pi_t &= \underset{(0.080)}{0.240} + \underset{(0.083)}{0.552}\pi_{t-1} + \underset{(0.089)}{0.177}\pi_{t-2} + \underset{(0.090)}{0.232}\pi_{t-3} - \underset{(0.080)}{0.209}\pi_{t-4} + \varepsilon_t \\ h_t &= \underset{(0.034)}{0.282}. \end{aligned}$$

None of the first ten autocorrelations or partial autocorrelations of $\hat{\varepsilon}_t$ are significant, but for $\hat{\varepsilon}_t^2$ the 1st, 3rd, 7th, 9th and 10th auto- and partial correlations exceed two standard errors. LM tests of ARCH(1), ARCH(4) and ARCH(8) are also highly significant. Following the earlier work of Engle and Kraft (1983), Bollerslev then fits the following ARCH(8) model by ML:

$$\begin{aligned} \pi_t &= \underset{(0.059)}{0.138} + \underset{(0.081)}{0.423}\pi_{t-1} + \underset{(0.108)}{0.222}\pi_{t-2} + \underset{(0.078)}{0.377}\pi_{t-3} - \underset{(0.104)}{0.175}\pi_{t-4} + \varepsilon_t \\ h_t &= \underset{(0.033)}{0.058} + \underset{(0.265)}{0.802} \sum_{i=1}^{8} \frac{(9-i)}{36} \varepsilon_{t-i}^2. \end{aligned}$$

An eighth-order linearly declining lag structure is chosen for the ARCH process which, although ad hoc, is motivated by the long memory in the conditional variance equation. Bollerslev presents evidence that h_{t-1} should be included in this equation, and then considers an alternative specification in which h_t is GARCH(1, 1):

$$\begin{aligned} \pi_t &= \underset{(0.060)}{0.141} + \underset{(0.081)}{0.433}\pi_{t-1} + \underset{(0.110)}{0.229}\pi_{t-2} + \underset{(0.077)}{0.349}\pi_{t-3} - \underset{(0.104)}{0.162}\pi_{t-4} + \varepsilon_t \\ h_t &= \underset{(0.006)}{0.007} + \underset{(0.070)}{0.135}\varepsilon_{t-1}^2 + \underset{(0.068)}{0.829}h_{t-1}. \end{aligned}$$

A variety of misspecification tests do not reveal any model inadequacy, and Bollerslev argues that this GARCH(1, 1) model exhibits both a slightly better fit and a more reasonable lag structure than the ARCH(8) model. He then presents graphs of the inflation rate along with 95% confidence intervals calculated from the one-step ahead forecast errors from the OLS model and the GARCH model: up to the mid-1950s the inflation rate was very volatile and hard to predict, being reflected by wide confidence intervals in the GARCH model. The 1960s and early 1970s had a stable and predictable inflation rate, with the OLS confidence interval appearing much too wide. From 1974 there is an increase in the uncertainty of the inflation rate, but it does not compare in magnitude to that found at the beginning of the sample period.

15.5 Alternative conditional variance models

There have been various other conditional variance models proposed over recent years. Hsu (1977, 1979) considers a simple model in which the variance alters at a single, but unknown, point in time, estimating this change-point by ML techniques; Bayesian analyses of the model are provided by Hsu (1982) and Menzefricke (1981), and robust estimation is discussed by Davis (1979). Ali and Giaccotto (1982) extend the model to contain more than one jump in variance while Tyssedal and Tjostheim (1982) allow the variance to depend on time in a continuous fashion.

Clark (1973) develops a model in which the conditional variance is a deterministic function of an exogenous variable, while Tauchen and Pitts (1983) extend this to a stochastic setting, with Taylor (1986) providing further extensions.

The typical area of application of these conditional variance models is that of financial markets, where daily time series of considerable length are available. Such models are felt to be needed because the early work of Mandelbrot (1963) and Fama (1965) found that conventional statistical analysis of daily price changes in financial markets was inadequate in that normal distributions do not satisfactorily model the observed data; as we have seen, conditional variance models do not force the distribution of y_t to be normal.

Financial time series have characteristics peculiar to themselves, and published research into their properties is diffuse and often relatively inaccessible. The recent book by Taylor (1986) is therefore a welcome addition to the literature and the reader is referred there for a state-of-the-art presentation of the modelling of financial time series.

15.6 Forecasting power-transformed series

In Chapter 6 we considered the class of Box–Cox (1964) power transformations for inducing homogeneity of variance and, possibly, normally distributed errors. Such a transformation may well be applied to the time series under analysis before models for the conditional variance are considered.

In general, the observed time series x_t may be transformed into another series $y_t = x_t^{(\lambda)}$ through the monotonic, instantaneous power transformation

$$
\begin{aligned}
x_t^{(\lambda)} &= (x_t^{\lambda} - 1)/\lambda, \quad \lambda \neq 0 \\
&= \ln x_t, \qquad \lambda = 0.
\end{aligned}
\tag{15.14}
$$

A model for y_t is then developed from which forecasts of y_t can be generated. It is typically the case, however, that forecasts of the originally observed series x_t are required. An obvious way of calculating forecasts of x_t is to apply the inverse of the power transformation (15.13) to the forecasts of y_t, i.e.

$$
\begin{aligned}
x_t &= (\lambda y_t + 1)^{1/\lambda}, \quad \lambda \neq 0 \\
&= \exp(y_t), \qquad \lambda = 0.
\end{aligned}
\tag{15.15}
$$

The 'naive' retransformation to obtain the forecast of x_{n+h}, say, will provide a minimum mean *absolute* error (MMAE) forecast, since it will equal the *median* of the conditional density function of x_{n+h}, and will thus be optimal for an *absolute* forecast error loss function (see Granger (1969b) for a discussion of the importance of knowing the forecast error loss function in any particular application). Typically, we assume this loss function to be *quadratic*; i.e., the cost of error is proportional to the squared error. In this case the optimal forecast of x_{n+h}, for any conditional density function, is the MMSE forecast, equal to the *mean* of the conditional density function. Thus, if a quadratic loss function is assumed, so that MMSE forecasts are optimal, two questions naturally arise: by how much do MMAE forecasts of x_t deviate from MMSE forecasts, and how can MMSE forecasts be found?

Granger and Newbold (1976) present general theoretical results for various transformations, but give results for only two cases, $\lambda = 0$ and $\lambda = 0.5$. They show that when $\lambda = 0$, the MMSE h-step ahead forecast made at origin n is given by

$$\hat{x}_{n+h} = \exp[\hat{y}_{n+h} + \tfrac{1}{2}\sigma_h^2],$$

where σ_h^2 is the h-step ahead forecast error variance associated with $\hat{y}_{n+h}$. The relative bias from using the MMAE forecast,

$$\tilde{x}_{n+h} = \exp(\hat{y}_{n+h}),$$

is therefore

$$B = (\tilde{x}_{n+h} - \hat{x}_{n+h})/\tilde{x}_{n+h} = \exp(-\tfrac{1}{2}\sigma_h^2) - 1.$$

When $\lambda = 0.5$, Granger and Newbold show that

$$\hat{x}_{n+h} = \hat{y}_{n+h}^2 + \sigma_h^2,$$

$$\tilde{x}_{n+h} = \hat{y}_{n+h}^2,$$

and so

$$B = -\frac{(\sigma_h^2/\hat{y}_{n+h}^2)}{[1+(\sigma_h^2/\hat{y}_{n+h}^2)]}.$$

These theoretical results suggest that using the naive retransformation may give forecasts of x_t that deviate substantially from MMSE forecasts. In an empirical study, however, Nelson and Granger (1979) found using $\hat{x}_{n+h}$ rather than $\tilde{x}_{n+h}$ to be only moderately worthwhile, giving better forecasts in about 60% of the time when 21 macroeconomic series were forecast over various horizons, with a similar performance being found in simulation experiments.

This issue has been recently reinvestigated by Pankratz and Dudley (1987). In general, the relationship between $\hat{x}_{n+h}$ and $\tilde{x}_{n+h}$ is given by

$$\hat{x}_{n+h} = G\tilde{x}_{n+h},$$

where the factor G is an integral with no closed form solution for a general value of λ. Pankratz and Dudley evaluate the integral G numerically, providing values of the percent bias

$$100B = \frac{100(1-G)}{G}$$

for various combinations of λ and $r = \sigma_h/(\hat{y}_{n+h}+\lambda^{-1})$, which is the ratio of the forecast standard deviation of y_{n+h} to the sum of the forecast and the inverse of the power transformation. For many combinations ($|\lambda| > 0.10, |r| < 0.10$), only a small bias is found, often in the region of ± 1% (Pankratz and Dudley show that we must have $|r| < 0.25$ for the transformation to produce a normal density for y_t having negligible truncation). For $|\lambda| < 0.10$ and $|r| > 0.10$, however, the bias can be enormous; for example, the bias associated with $\lambda = 0.10$ and $r = 0.25$ is -86%!

These theoretical findings also go a long way to explain the modest improvement in forecast accuracy reported by Nelson and Granger (1979) when a non-biasing procedure was used with the Box–Cox transformation. In 12 of their 21 series, they found $|\lambda| > 0.6$ and Pankratz and Dudley provide evidence to suggest that the associated values of $|r|$ were rather small: exactly the conditions under which little bias is found.

These results are important in that they allow us easily to obtain an idea of the bias inherent in using the naive transformation, and thus obtaining MMAE forecasts, when MMSE forecasts are desired.

15.7 Asymmetric time series

The above class of power transformations, as well as inducing homogeneity of variance, can also be used to transform a skewed marginal distribution of x_t into a more normal distribution. Many economic time series are such that $x_t > 0$, prices for example, and may thus be characterised by a lognormal marginal distribution; use of the logarithmic transformation ($\lambda = 0$) is then implied. Wecker (1981) argues that observed skewness in a time series may result from the series being generated by an *asymmetric* model, one in which x_t responds in a different fashion to the innovation ε_t depending on whether the innovation is positive or negative. This may be illustrated by the asymmetric MA(1) process, defined as

$$x_t = \varepsilon_t - \theta^+ \varepsilon_{t-1}^+ - \theta^- \varepsilon_{t-1}^-, \tag{15.16}$$

where

$$\varepsilon_t^+ = \max(\varepsilon_t, 0)$$

and

$$\varepsilon_t^- = \min(\varepsilon_t, 0)$$

are the positive and negative innovations respectively.

Wecker (1981) shows that, unlike the symmetric MA(1) process where the mean of x_t is zero, the mean of the asymmetric MA(1) process is

$$\mu = \frac{\theta^- - \theta^+}{\sqrt{2\pi}} \tag{15.17}$$

if we assume, for simplicity, that $\varepsilon_t \sim N(0, 1)$. The asymmetric series x_t has variance

$$\gamma_0 = 1 + \frac{(\theta^+)^2 + (\theta^-)^2}{2} - \mu^2, \tag{15.18}$$

and lag one autocovariance

$$\gamma_1 = -\frac{\theta^+ + \theta^-}{2}, \tag{15.19}$$

with $\gamma_k = 0$ for $k \geqslant 2$. The symmetric MA(1) process is obtained when $\theta^+ = \theta^- = \theta$, in which case the familiar results $\mu = 0$, $\gamma_0 = 1 + \theta^2$ and $\gamma_1 = -\theta$ are obtained. Moreover, if $\theta^- = -\theta^+, \gamma_1 = 0$ and the asymmetric process becomes indistinguishable from white noise. Using the mean $\mu = (-2\theta^+)/\sqrt{2\pi}$ to make forecasts will then result in a forecast error variance of

$$1 + \left(\frac{\pi - 2}{\pi}\right)(\theta^+)^2,$$

rather than the true innovation variance of one. For $-\theta^- = \theta^+ = -0.9$, this represents an increase in forecast error variance of about 30% over what would be possible if the forecasts were made using the true asymmetric model.

Note that the SACF will give no clue as to whether the series is being generated by an asymmetric or a symmetric process, and any observed skewness in the marginal distribution of x_t may well prompt the application of a power transformation. Wecker thus suggests that, unknown to analysts, asymmetric time series may abound, but are analysed as either white noise processes, if $\theta^- \simeq -\theta^+$, or as symmetric processes of power transformed series.

Wecker provides details of ML estimation and of a test for symmetry, and investigates the asymmetric MA(1) model on a variety of industrial price change series compiled by Stigler and Kindahl (1970), who argue intuitively that certain of the series, being quoted prices, should be asymmetric, while others, since they are transactions prices, should be symmetric. Wecker indeed confirms those predictions, and furthermore finds that for all asymmetric series the estimate of θ^- is large and positive, while that for θ^+ is small and nonpositive, thus providing support for Stigler and Kindahl's view that when market conditions change, quoted prices are not revised immediately, and this delay operates more strongly against reductions in price quotations than against increases.

These findings suggest that asymmetric time series models could be a useful addition to the set of techniques used by economists to model financial series.

16 State dependent models

16.1 Nonlinear time series models

In Chapter 5 we introduced Wold's decomposition, which states that any weakly stationary, non-deterministic time series can be written as a linear combination of a sequence of uncorrelated random variables. In the development of ARMA processes in that chapter, the further assumption was made that this 'white noise' sequence was independent, rather than just uncorrelated. Such a sequence has the property that the past contains no information on the future and hence the best forecast of a future value of the sequence is simply its (unconditional) mean of zero, and this property was used extensively in the theory of forecasting developed in Chapter 7.

Are there any consequences if we relax the assumption of independence and just allow the sequence to be uncorrelated, as in Wold's decomposition? It is still true that, if we restrict attention to forecasts which are *linear* functions of past observations, then the past contains no information on the future. However, the past may well contain useful information on the future if we allow *nonlinear* functions of past observations. For example, consider the process η_t defined as

$$\eta_t = a_t + \beta a_{t-1} a_{t-2}, \tag{16.1}$$

where a_t is a strictly independent process with zero mean and constant variance. It follows immediately that η_t has zero mean, constant variance and an autocovariance function given by

$$E(\eta_t \eta_{t-k}) = E(a_t a_{t-k} + \beta a_{t-1} a_{t-2} a_{t-k} + \beta a_t a_{t-k-1} a_{t-k-2} + \beta^2 a_{t-1} a_{t-2} a_{t-k-1} a_{t-k-2}).$$

For all $k \neq 0$, each of the terms in the autocovariance function has zero expectation, so that, as far as its second-order properties are concerned,

η_t behaves just like an independent process. However, the MMSE forecast of a future observation, η_{t+1}, is not zero (the unconditional expectation), but is the conditional expectation

$$\hat{\eta}_{t+1} = E[\eta_{t+1} \mid \eta_t, \eta_{t-1}, \ldots] = \beta a_t a_{t-1}.$$

If η_t had been obtained as the 'residual' from a more general model, all the conventional tests for white noise based on the behaviour of the autocovariance function would confirm that the residuals were, in fact, white noise and hence that there was no model structure left to fit. However, the nonlinear structure of the η_t process could be exploited to improve the forecasts of the original series. This, of course, has been ruled out a priori in our previous models because it was assumed that the innovation process was always a sequence of independent random variables.

One problem with the process (16.1) is that, in order to express $\hat{\eta}_{t+1}$ in terms of past η_t's, we would have to invert (16.1) so as to write each a_t as a function of $\eta_t, \eta_{t-1}, \ldots$ Granger (1978) argues that, in general, this model is never invertible and, consequently, maximum likelihood estimates of the parameters cannot be obtained. Using arguments such as invertibility and nonexplosiveness, Granger (1978) rules out many classes of nonlinear models as being capable of modelling economic time series. A number of nonlinear time series models have been proposed, however, and found to be useful in a wide range of scientific applications and this chapter provides an introductory discussion of these.

16.2 Some special nonlinear models

An important class of nonlinear model is the *bilinear*, which takes the general form

$$x_t - \phi_1 x_{t-1} - \ldots - \phi_p x_{t-p} = a_t - \theta_1 a_{t-1} - \ldots - \theta_q a_{t-q}$$

$$+ \sum_{i=1}^{m} \sum_{j=1}^{k} \delta_{ij} x_{t-i} a_{t-j}, \tag{16.2}$$

where, as usual, a_t is a strict white noise process (i.e. a sequence of independent random variables). The second term on the right hand side of (16.2) is a bilinear form in a_{t-j} and x_{t-i}, and this accounts for the nonlinear character of the model: if all the δ_{ij} are zero then (16.2) reduces to the familiar ARMA model. These models have the property that although they involve only a finite number of parameters, they can approximate with arbitrary accuracy any 'reasonable' nonlinear relationship (Brockett, 1976).

Little analysis has been carried out on the general bilinear form (16.2), but Granger and Andersen (1978) have analysed the properties of several simple bilinear forms, characterised as

$$x_t = \delta x_{t-i} a_{t-j} + a_t. \tag{16.3}$$

If $i > j$ the model is called superdiagonal, if $i = j$ it is called diagonal, and if $i < j$, it is called subdiagonal. If the variance of a_t is σ^2, and if we define $\lambda = \delta\sigma$, then for superdiagonal models, x_t has zero mean and variance $\sigma^2/(1-\lambda^2)$, so that $|\lambda| < 1$ is a necessary condition for stability. Conventional identification techniques using SACFs, etc. would identify x_t as white noise, although if x_t^2 was analysed, we should, in theory at least, identify it as ARMA (i,j). Hence, analysis of the square of the series enables us to distinguish between white noise and a bilinear series, although analysing the original series does not allow this (cf. the use of x_t^2 in Chapter 15 for identifying ARCH processes).

Diagonal models will also be stationary if $|\lambda| < 1$. If $i = j = 1$ then x_t will be identified as MA (1), while x_t^2 will be identified as ARMA (1, 1). This result enables the diagonal model

$$x_t = \delta x_{t-1} a_{t-1} + a_t$$

to be distinguished from a (linear) MA (1) model, since Granger and Newbold (1976) prove that if $x_t \sim$ MA (1), then $x_t^2 \sim$ MA (1) also.

Subdiagonal models are essentially similar to superdiagonal models in that they appear to be white noise but generally have their square again following an ARMA (i,j) process. Further theoretical results on bilinear models may be found in Priestley (1980), Subba Rao (1981) and Li (1984) and forecasting applications are discussed in Subba Rao and Gabr (1980) and Poskitt and Tremayne (1986). Recently Stensholt and Tjostheim (1987) have analysed the class of multiple bilinear models.

A second class of nonlinear models that has created some interest is the *threshold AR* (Tong and Lim (1980), Tong (1983)). The basic idea here is to start with a linear model for x_t, and then allow the parameters to vary according to the values of a finite number of past values of x_t, or possibly of a finite number of past values of some other process. A first order threshold AR model (TAR (1)) would therefore be

$$x_t = \begin{cases} \phi^{(1)} x_{t-1} + a_t^{(1)}, & \text{if} \quad x_{t-1} < d \\ \phi^{(2)} x_{t-1} + a_t^{(2)}, & \text{if} \quad x_{t-1} \geqslant d, \end{cases}$$

and this can be extended to a 'k-threshold' form:

$$x_t = \phi^{(i)} x_{t-1} + a_t^{(i)}, \quad \text{if} \quad x_{t-1} \in R_i, \quad i = 1, \ldots, k, \tag{16.4}$$

where $R_1, \ldots, R_k$ are given subsets of the real line R^1. Looked at in this way, the k-threshold model may be regarded as a 'piecewise-linear' approximation to the general nonlinear first order model

$$x_t = \lambda(x_{t-1}) + a_t.$$

Higher order threshold autoregressions are similarly defined. The TAR(p) model, for example, is

$$\begin{aligned} &x_t - \phi_1^{(i)} x_{t-1} - \ldots - \phi_p^{(i)} x_{t-p} = a_t^{(i)}, \\ &\text{if} \quad (x_{t-1}, \ldots, x_{t-p}) \in R^{(i)}, \quad i = 1, \ldots, k, \end{aligned} \tag{16.5}$$

where $R^{(i)}$ is a given region of the p-dimensional Euclidean space R^p. Correspondingly, this model may be viewed as a piecewise-linear approximation to

$$x_t = f(x_{t-1}, x_{t-2}, \ldots, x_{t-p}) + a_t.$$

Usually, the thresholds are assumed to be given a priori, but K. S. Chan and Tong (1986) consider methods for estimating them jointly with the model.

Since TAR models have the property that, under suitable conditions, they can give rise to 'limit cycle' behaviour, they are well suited to the modelling of cyclical data.

A related class of models is the *exponential autoregressive* (Haggan and Ozaki (1981)). This begins with a (linear) AR(p) model and allows the coefficients to be exponential functions of x_{t-1}^2:

$$x_t - \phi_1 x_{t-1} - \ldots - \phi_p x_{t-p} = a_t, \tag{16.6}$$

where

$$\phi_i = (\alpha_i + \beta_i \exp(-\gamma x_{t-1}^2)).$$

These models behave similarly to the threshold AR models, except that the coefficients change smoothly between the segments. We note, however, that an alternative class of models having the same name has been proposed by Lawrence and Lewis (1985), and for an analysis of some further classes of nonlinear autoregressive models, see Lawrence and Lewis (1987).

16.3 State dependent models

All the above models are special cases of a general class of time series model termed by Priestley (1980) as *state dependent*. Using 'Volterra series' expansions, Priestley shows that a general relationship between an output series x_t and a white noise input a_t can be represented as

$$x_t = f(x_{t-1}, \ldots, x_{t-k}, a_{t-1}, \ldots, a_{t-l}) + a_t. \tag{16.7}$$

If f is assumed analytic, the right hand side of (16.7) can be expanded in a Taylor's series about an arbitrary but fixed time point, allowing the relationship to be written as the State Dependent Model (SDM) of order (k, l):

$$x_t - \sum_{i=1}^{k} \phi_i(\boldsymbol{x}_{t-1}) x_{t-i} = \mu(\boldsymbol{x}_{t-1}) + a_t - \sum_{i=1}^{l} \theta_i(\boldsymbol{x}_{t-1}) a_{t-i}, \tag{16.8}$$

where $\boldsymbol{x}_t$ denotes the state vector

$$\boldsymbol{x}_t = (a_{t-l+1}, \ldots, a_t, x_{t-k+1}, \ldots, x_t).$$

As Priestley (1980, page 54) remarks, this model 'has a natural and appealing interpretation as a locally linear ARMA model in which the evolution of the process at time $(t-1)$ is governed by a set of AR coefficients $\{\phi_i\}$, a set of MA coefficients $\{\theta_i\}$, and a local mean μ, all of which depend on the "state" of the process at time $(t-1)$'. Indeed, if μ, $\{\phi_i\}$ and $\{\theta_i\}$ are all taken as constants, i.e. independent of $\boldsymbol{x}_{t-1}$, then (16.8) reduces to the usual ARMA(k, l) model. The other special models discussed above are obtained from the SDM in the following ways:

(i) *Bilinear models*
Here we take μ and $\{\phi_i\}$ as constants and set

$$\theta_i(\boldsymbol{x}_{t-1}) = \theta_i + \sum_{j=1}^{k} \delta_{ij} x_{t-j}, \quad i = 1, \ldots, l.$$

The SDM model (16.8) then reduces to the bilinear model (16.2), with the value of l chosen as max (m, k). For the bilinear model, the $\theta_i(\boldsymbol{x}_{t-1})$ are linear functions of $\{x_{t-1}, \ldots, x_{t-k}\}$.

(ii) *Threshold AR models*
In this case we take $\theta_i = 0$, for all i, and

$$\mu_t = \mu^{(j)}, \phi_i(\boldsymbol{x}_{t-1}) = \phi_i^{(j)}, \quad \text{if} \quad \boldsymbol{x}_{t-1} \in R^{(j)}.$$

The model (16.8) then reduces to the TAR(p) model (16.5), with an additional parameter representing the local mean.

(iii) *Exponential AR models*
For these models we take $\theta_i = 0$, for all i, and

$$\phi_i(\boldsymbol{x}_{t-1}) = \alpha_i + \beta_i \exp(-\gamma x_{t-1}^2), \quad i = 1, \ldots, k.$$

The model (16.8) then reduces to (16.6), with the $\{\phi_i\}$ being exponential functions of a single past observation, x_{t-1}.

Priestley (1980) shows that an SDM has an equivalent state space representation (see Chapter 11), and this is useful for both the identification and estimation of such models. Haggan et al. (1984) provide an extensive study of the application of state-dependent models to a wide variety of nonlinear time series, although no economic data are used. Indeed, the only economic applications of any of these nonlinear models that we are aware of are the use of a 'subset' bilinear model to forecast West German unemployment by Gabr and Subba Rao (1981) and a forecasting example of a bilinear model in Maravall (1983). It must remain to be seen whether such models are found to be useful for the modelling of economic time series.

16.4 Testing for nonlinearity

The identification and estimation of nonlinear models of the SDM class is still very much in its infancy, as the paper by Haggan et al. (1984) clearly shows. It is therefore important to provide tests of whether a time series is linear or not. A number of tests have been proposed that are based on frequency domain estimation; W. S. Chan and Tong (1986) provide a useful survey. However, Keenan (1985) has proposed a simple time domain test that appears to be worth considering.

This test explicitly considers the Volterra series expansion of x_t used by Priestley (1980) to obtain equation (16.7). This allows x_t to be written in the very general form

$$x_t = \mu + \sum_{i=-\infty}^{\infty} \psi_i a_{t-i} + \sum_{i,j=-\infty}^{\infty} \psi_{ij} a_{t-i} a_{t-j} + \sum_{i,j,k=-\infty}^{\infty} \psi_{ijk} a_{t-i} a_{t-j} a_{t-k} + \ldots \tag{16.9}$$

Obviously, x_t is nonlinear if any of the higher order coefficients $\{\psi_{ij}\}$, $\{\psi_{ijk}\}, \ldots$ are nonzero; if not we obtain the usual linear filter representation of x_t. A test of nonlinearity is, therefore, equivalent to a test of no multiplicative terms in (16.9).

Since the Volterra series expansion in (16.9) is to a linear process what a polynomial is to a linear function, Keenan suggests using an analogue of Tukey's (1949) one degree of freedom for nonadditivity test. The mechanics of the test are as follows:

(i) Suppose we have the realisation $(x_1, \ldots, x_n)$ for large n. Regress x_t on $\{1, x_{t-1}, \ldots, x_{t-m}\}$ and calculate the fitted values $\{\hat{x}_t\}$ and the residuals $\{\hat{e}_t\}$, for $t = m+1, \ldots, n$, and calculate the residual sum of squares $\sum \hat{e}_t^2$.

(ii) Regress $\hat{x}_t^2$ on $\{1, x_{t-1}, \ldots, x_{t-m}\}$ and calculate the residuals $\{\hat{\xi}_t\}$ for $t = m+1, \ldots, n$.

(iii) Regress $\hat{e}_t$ on $\hat{\xi}_t$ and obtain $\hat{\eta}$ and $\hat{F}$ via

$$\hat{\eta} = \hat{\eta}_0 \left(\sum_{t=m+1}^{n} \hat{\xi}_t^2 \right)^{\frac{1}{2}},$$

where $\hat{\eta}_0$ is the regression coefficient, and

$$\hat{F} = \frac{\hat{\eta}^2(n-2m-2)}{\sum \hat{e}_t^2 - \hat{\eta}^2}.$$

Under the null hypothesis of linearity, $\hat{F}$ will be distributed as $F(1, n-2m-2)$.

In many cases, the power properties of this test are quite satisfactory, but for certain alternatives, the extensions to the test that have been proposed by Tsay (1986b) seem preferable.

Example 16.1: Testing for nonlinearity in daily interest rates

This test was performed on the daily interbank rate series using $m = 8$ lags. The value of $\hat{F}$ was only 0.38 and hence we cannot reject the hypothesis that a linear model is appropriate for this series, although we note that in Chapter 15 we found strong evidence of ARCH effects.

References

Abraham, B. (1980), 'Intervention Analysis and Multiple Time Series', *Biometrika*, 67, 73–80.

Abraham, B. (1984), 'Temporal Aggregation and Time Series', *International Statistical Review*, 52, 285–91.

Abraham, B. (1985), 'Seasonal Time Series and Transfer Function Modelling', *Journal of Business and Economic Statistics*, 3, 356–61.

Abraham, B. (1987), 'Application of Intervention Analysis to a Road Fatality Series in Ontario', *Journal of Forecasting*, 6, 211–20.

Abraham, B. and Ledolter, J. (1983), *Statistical Methods for Forecasting*, New York: Wiley.

Abraham, B. and Ledolter, J. (1984), 'A Note on Inverse Autocorrelations', *Biometrika*, 71, 609–14.

Akaike, H. (1974), 'A New Look at the Statistical Model Identification', *IEEE Transactions on Automatic Control*, AC–19, 716–23.

Akaike, H. (1976), 'Canonical Correlations Analysis of Time Series and the Use of an Information Criterion', in R. Mehra and D. G. Lainiotis (Editors), *Advances and Case Studies in System Identification*, 27–96, New York: Academic Press.

Ali, M. M. and Giaccotto, C. (1982), 'The Identical Distribution Hypothesis for Stock Market Prices – Location and Scale-Shift Alternatives', *Journal of the American Statistical Association*, 77, 19–28.

Ameen, J. R. M. and Harrison, P. J. (1984), 'Discount Weighted Estimation', *Journal of Forecasting*, 3, 285–96.

Anderson, O. D. (1976), *Time Series Analysis and Forecasting*, London: Butterworths.

Anderson, R. G., Johannes, J. M. and Rasche, R. H. (1983), 'A New Look at the Relationship Between Time-Series and Structural Econometric Models', *Journal of Econometrics*, 23, 235–51.

Anderson, T. W. (1971), *The Statistical Analysis of Time Series*, New York: Wiley.

Anderson, T. W. (1980), 'Maximum Likelihood Estimation for Vector Autoregressive Moving Average Models', in D. R. Brillinger and G. C. Tiao (Editors), *Directions in Time Series*, 49–59, Institute of Mathematical Statistics.

Ansley, C. F. (1979), 'An Algorithm for the Exact Likelihood of a Mixed Autoregressive Moving Average Process', *Biometrika*, 66, 547–55.

Ansley, C. F. and Newbold, P. (1980), 'Finite Sample Properties of Estimates for Autoregressive Moving Average Models', *Journal of Econometrics*, 13, 159–83.

Artis, M. J. and Lewis, M. K. (1984), 'How Unstable is the Demand for Money in the United Kingdom?', *Economica*, 51, 473–6.

Baillie, R. T. (1979), 'Asymptotic Prediction Mean Squared Error for Vector Autoregressive Models', *Biometrika*, 66, 675–8.

Banerjee, A., Dolado, J., Hendry, D. F. and Smith, G. (1986), 'Exploring Equilibrium Relationships in Econometrics Through Static Models: Some Monte Carlo Evidence', *Oxford Bulletin of Economics and Statistics*, 48, 253–77.

Bartlett, M. S. (1946), 'On the Theoretical Specification of Sampling Properties of Autocorrelated Time Series', *Journal of the Royal Statistical Society*, Series B, 8, 27–41.

Beaumont, C., Mahmoud, E. and McGee, V. E. (1985), 'Microcomputer Forecasting Software: A Survey', *Journal of Forecasting*, 4, 305–12.

Beguin, J.-M., Gourieroux, C. and Montfort, A. (1980), 'Identification of a Mixed Autoregressive-Moving Average Process: The Corner Method', in O. D. Anderson (Editor), *Time Series*, 423–36, Amsterdam: North Holland.

Bell, W. R. (1984), 'Signal Extraction for Nonstationary Time Series', *Annals of Statistics*, 12, 646–64.

Bell, W. R. and Hillmer, S. C. (1983), 'Modelling Time Series with Calendar Variation', *Journal of the American Statistical Association*, 78, 526–34.

Bell, W. R. and Hillmer, S. C. (1984), 'Issues Involved with the Seasonal Adjustment of Economic Time Series', *Journal of Business and Economic Statistics*, 2, 291–320.

Berndt, E. K., Hall, B. H., Hall, R. E. and Hausman, J. A. (1974), 'Estimation and Inference in Nonlinear Structural Models', *Annals of Economic and Social Measurement*, 4, 653–65.

Beveridge, S. and Nelson, C. R. (1981), 'A New Approach to Decomposition of Economic Time Series into Permanent and Transitory Components with Particular Attention to Measurement of the "Business Cycle"', *Journal of Monetary Economics*, 7, 151–74.

Bhansali, R. J. (1980), 'Autoregressive and Window Estimates of the Inverse Correlation Function', *Biometrika*, 67, 551–66.

Bhattacharyaa, M. N. and Layton, A. P. (1979), 'Effectiveness of Seat Belt Legislation on the Queensland Road Toll – an Australian Case Study in Intervention Analysis', *Journal of the American Statistical Association*, 74, 596–603.

Bilongo, R. and Carbone, R. (1985), 'Adaptive Model-Based Seasonal Adjustment of Time Series: An Empirical Comparison with X11–ARIMA and SIGEX', *International Journal of Forecasting*, 1, 165–78.

Bollerslev, T. (1986), 'Generalized Autoregressive Conditional Heteroskedasticity', *Journal of Econometrics*, 31, 307–27.

Bollerslev, T. (1987), 'A Conditionally Heteroskedastic Time Series Model for

Speculative Prices and Rates of Return', *Review of Economics and Statistics*, 69, 542–6.

Bollerslev, T. (1988), 'On the Correlation Structure for the Generalized Autoregressive Conditional Heteroskedastic Process', *Journal of Time Series Analysis*, 9, 121–32.

Box, G. E. P. and Cox, D. R. (1964), 'An Analysis of Transformations', *Journal of the Royal Statistical Society*, Series B, 26, 211–43.

Box, G. E. P., Hillmer, S. C. and Tiao, G. C. (1978), 'Analysis and Modelling of Seasonal Time Series', in A. Zellner (Editor), *Seasonal Analysis of Economic Time Series*, 309–33, Washington, DC: US Department of Commerce, Bureau of the Census.

Box, G. E. P. and Jenkins, G. M. (1976), *Time Series Analysis: Forecasting and Control*, Revised Edition, San Francisco: Holden-Day.

Box, G. E. P. and Pierce, D. A. (1970), 'Distribution of Residual Autocorrelations in Autoregressive Moving Average Time Series Models', *Journal of the American Statistical Association*, 65, 1509–26.

Box, G. E. P., Pierce, D. A. and Newbold, P. (1987), 'Estimating Trend and Growth Rates in Seasonal Time Series', *Journal of the American Statistical Association*, 82, 276–82.

Box, G. E. P. and Tiao, G. C. (1975), 'Intervention Analysis with Application to Economic and Environmental Problems', *Journal of the American Statistical Association*, 70, 70–9.

Brewer, K. R. W. (1973), 'Some Consequences of Temporal Aggregation and Systematic Sampling for ARMA and ARMAX Models', *Journal of Econometrics*, 1, 133–54.

Brockett, R. W. (1976), 'Volterra Series and Geometric Control Theory', *Automatica*, 12, 167–76.

Brown, R. G. (1963), *Smoothing, Forecasting and Prediction of Discrete Time Series*, Englewood-Cliffs, NJ: Prentice-Hall.

Burman, J. P. (1979), 'Seasonal Adjustment – A Survey', in *TIMS Studies in the Management Sciences, Volume 12: Forecasting*, 45–57, Amsterdam: North Holland.

Burman, J. P. (1980), 'Seasonal Adjustment by Signal Extraction', *Journal of the Royal Statistical Society*, Series A, 143, 321–37.

Burridge, P. and Wallis, K. F. (1984), 'Unobserved-Components Models for Seasonal Adjustment Filters', *Journal of Business and Economic Statistics*, 2, 350–9.

Capie, F. H., Mills, T. C. and Wood, G. E. (1986), 'Debt Management and Interest Rates: The British Stock Conversion of 1932', *Applied Economics*, 18, 1111–26.

Capie, F. H. and Webber, A. (1985), *A Monetary History of the United Kingdom, Volume 1: Data, Sources, Methods*, London: Allen and Unwin.

Chamberlain, G. (1982), 'The General Equivalence of Granger and Sims Causality', *Econometrica*, 50, 569–81.

Chambers, J. M., Cleveland, W. S., Kleiner, B. and Tukey, P. A. (1983), *Graphical Methods for Data Analysis*, Boston: Wadsworth/Duxbury.

Chan, K. H., Hayya, J. C. and Ord, J. K. (1977), 'A Note on Trend Removal Methods: The Case of Polynomial Versus Variate Differencing', *Econometrica*, 45, 737–44.

Chan, K. S. and Tong, H. (1986), 'On Estimating Thresholds in Autoregressive Models', *Journal of Time Series Analysis*, 7, 179–90.

Chan, W. S. and Tong, H. (1986), 'On Tests for Non-Linearity in Time Series Analysis', *Journal of Forecasting*, 5, 217–28.

Chan, W. Y. T. and Wallis, K. F. (1978), 'Multiple Time Series Modelling: Another Look at the Mink–Muskrat Interaction', *Applied Statistics*, 27, 168–75.

Chang, I., Tiao, G. C. and Chen, C. (1988), 'Estimation of Time Series Parameters in the Presence of Outliers', *Technometrics*, 30, 193–204.

Chatfield, C. (1978), 'The Holt–Winters Forecasting Procedure', *Applied Statistics*, 27, 264–79.

Chatfield, C. (1979), 'Inverse Autocorrelations', *Journal of the Royal Statistical Society*, Series A, 142, 363–77.

Chatfield, C. (1985), 'The Initial Examination of Data (with Discussion)', *Journal of the Royal Statistical Society*, Series A, 148, 214–53.

Chatfield, C. and Schimek, M. G. (1987), 'An Example of Model Formulation Using IDA', *The Statistician*, 36, 357–63.

Clark, P. K. (1973), 'A Subordinated Stochastic Process Model with Finite Variance for Speculative Prices', *Econometrica*, 41, 135–55.

Clark, P. K. (1987), 'The Cyclical Component of US Economic Activity', *Quarterly Journal of Economics*, 102, 797–814.

Cleveland, W. P. and Tiao, G. C. (1976), 'Decomposition of Seasonal Time Series: A Model for the Census X-11 Program', *Journal of the American Statistical Association*, 77, 63–70.

Cleveland, W. S. (1985), *The Elements of Graphing Data*, Monterey, California: Wadsworth.

Cleveland, W. S. (1987), 'Research in Statistical Graphics', *Journal of the American Statistical Association*, 82, 419–23.

Cleveland, W. S. and Devlin, S. J. (1982), 'Calendar Effects in Monthly Time Series: Modelling and Adjustment', *Journal of the American Statistical Association*, 77, 520–8.

Cleveland, W. S., Dunn, D. M. and Terpenning, I. J. (1978), 'SABL – A Resistant Seasonal Adjustment Procedure with Graphical Methods for Interpretation and Diagnosis', in A. Zellner (Editor), *Seasonal Analysis of Economic Time Series*, 201–31, Washington, DC: US Department of Commerce, Bureau of the Census.

Cleveland, W. S. and McGill, R. (1987), 'Graphical Perception: The Visual Decoding of Quantitative Information on Graphical Displays of Data (with Discussion)', *Journal of the Royal Statistical Society*, Series A, 150, 192–229.

Cochrane, W. J. (1952), 'The χ^2 Test of Goodness of Fit', *Annals of Mathematical Statistics*, 23, 315–45.

Cogger, K. O. (1974), 'The Optimality of General-Order Exponential Smoothing', *Operations Research*, 22, 858–67.

Cooley, T. F. and LeRoy, S. F. (1985), 'Atheoretical Macroeconometrics: A Critique', *Journal of Monetary Economics*, 16, 283–308.

Cooper, R. L. (1972), 'The Predictive Performance of Quarterly Econometric Models of the United States', in B. G. Hickman (Editor), *Econometric Models of Cyclical Behaviour*, 813–974, New York: Columbia University Press.

Cox, D. R. and Hinkley, D. V. (1974), *Theoretical Statistics*, London: Chapman and Hall.

Crafts, N. F. R., Leybourne, S. J. and Mills, T. C. (1989a), 'Trends and Cycles in British Industrial Production, 1700–1913', *Journal of the Royal Statistical Society*, Series A, 152, 43–60.

Crafts, N. F. R., Leybourne, S. J. and Mills, T. C. (1989b), 'Economic Growth in Nineteenth Century Britain: Comparisons with Europe in the Context of Gerschenkron's Hypotheses', *Warwick Economic Research Paper* 308.

Cramer, H. (1961), 'On Some Classes of Non-Stationary Processes', *Proceedings of the 4th Berkeley Symposium on Mathematical Statistics and Probability*, 57–78, University of California Press.

Cuddington, J. T. and Winters, L. A. (1987), 'The Beveridge–Nelson Decomposition of Economic Time Series: A Quick Computational Method', *Journal of Monetary Economics*, 19, 125–7.

Dagum, E. B. (1978), 'Modelling, Forecasting and Seasonally-Adjusting Economic Time Series with the X–11–ARIMA Method', *The Statistician*, 27, 203–16.

Dagum, E. B. and Morry, M. (1984), 'Basic Issues on the Seasonal Adjustment of the Canadian Consumer Price Index', *Journal of Business and Economic Statistics*, 2, 250–9.

Davidson, J. E. H. (1981), 'Problems with the Estimation of Moving Average Processes', *Journal of Econometrics*, 16, 295–310.

Davidson, J. E. H., Hendry, D. F., Srba, F. and Yeo, S. (1978), 'Econometric Modelling of the Aggregate Time-Series Relationship Between Consumers' Expenditure and Income in the United Kingdom', *Economic Journal*, 88, 661–92.

Davies, N. and Newbold, P. (1979), 'Some Power Studies of a Portmanteau Test of Time Series Model Specification', *Biometrika*, 66, 153–5.

Davies, N., Triggs, C. M. and Newbold, P. (1977), 'Significance Levels of the Box–Pierce Portmanteau Statistic in Finite Samples', *Biometrika*, 64, 517–22.

Davis, W. W. (1979), 'Robust Methods for Detection of Shifts of the Innovation Variance of a Time Series', *Technometrics*, 21, 313–20.

Deaton, A. S. (1977), 'Involuntary Saving through Unanticipated Inflation', *American Economic Review*, 67, 899–910.

Dickey, D. A., Bell, W. R. and Miller, R. B. (1986), 'Unit Roots in Time Series Models: Tests and Implications', *American Statistician*, 40, 12–26.

Dickey, D. A. and Fuller, W. A. (1979), 'Distribution of the Estimators for Autoregressive Time Series with a Unit Root', *Journal of the American Statistical Association*, 74, 427–31.

Diestler, M. (1985), 'General Structure and Parameterization of ARMA and

State-Space Systems and its Relation to Statistical Problems', in E. J. Hannan, P. R. Krishnaiah and M. M. Rao (Editors), *Handbook of Statistics, Volume 5: Time Series in the Time Domain*, 257–78, Amsterdam: North Holland.

Doan, T. A., Litterman, R. B. and Sims, C. A. (1984), 'Forecasting and Conditional Projection using Realistic Prior Distributions', *Econometric Reviews*, 3, 1–100.

Durbin, J. (1960), 'The Fitting of Time Series Models', *Review of the International Statistical Institute*, 28, 233–44.

Dwyer, G. P. (1985), 'Money, Income and Prices in the United Kingdom: 1870–1913', *Economic Inquiry*, 23, 415–35.

Edlund, P.-O. (1984), 'Identification of the Multi-input Box–Jenkins Transfer Function Model', *Journal of Forecasting*, 3, 197–208.

Engle, E. M. R. A. (1984), 'A Unified Approach to the Study of Sums, Products, Time-Aggregation and Other Functions of ARMA Processes', *Journal of Time Series Analysis*, 5, 159–71.

Engle, R. F. (1982), 'Autoregressive Conditional Heteroskedasticity with Estimates of the Variance of UK Inflation', *Econometrica*, 50, 987–1008.

Engle, R. F. (1983), 'Estimates of the Variance of US Inflation Based on the ARCH Model', *Journal of Money, Credit and Banking*, 15, 286–301.

Engle, R. F. and Bollerslev, T. (1986), 'Modelling the Persistence of Conditional Variances', *Econometric Reviews*, 5, 1–50.

Engle, R. F. and Granger, C. W. J. (1987), 'Co-Integration and Error Correction: Representation, Estimation and Testing', *Econometrica*, 55, 251–76.

Engle, R. F., Hendry, D. F. and Richard, J.-F. (1983), 'Exogeneity', *Econometrica*, 51, 277–304.

Engle, R. F., Hendry, D. F. and Trumble, D. (1985), 'Small-sample Properties of ARCH Estimators and Tests', *Canadian Journal of Economics*, 18, 66–93.

Engle, R. F. and Kraft, D. (1983), 'Multiperiod Forecast Error Variances of Inflation Estimated from ARCH Models', in A. Zellner (Editor), *Applied Time Series Analysis of Economic Data*, 293–302, Washington, DC: US Department of Commerce, Bureau of the Census.

Engle, R. F., Lilien, D. M. and Robbins, R. P. (1987), 'Estimating Time Varying Risk Premia in the Term Structure: The Arch-M Model', *Econometrica*, 55, 391–408.

Enns, P. G., Machak, J. A., Spivey, W. A. and Wrobleski, W. J. (1982), 'Forecasting Applications of an Adaptive Multiple Exponential Smoothing Model', *Management Science*, 28, 1035–44.

Fama, E. F. (1965), 'The Behaviour of Stock Market Prices', *Journal of Business*, 38, 34–105.

Fama, E. F. (1975), 'Short Term Interest Rates as Predictors of Inflation', *American Economic Review*, 65, 269–82.

Florens, J. P. and Mouchart, M. (1982), 'A Note on Noncausality', *Econometrica*, 50, 583–91.

Fox, A. J. (1972), 'Outliers in Time Series', *Journal of the Royal Statistical Society*, Series B, 34, 350–63.

Freeman, M. F. and Tukey, J. W. (1950), 'Transformations Related to the Angular and the Square Root', *Annals of Mathematical Statistics*, 21, 607–11.

Friedman, M. (1957), *A Theory of the Consumption Function*, New York: NBER.

Friedman, M. and Schwartz, A. J. (1982), *Monetary Trends in the United States and the United Kingdom*, Chicago: University of Chicago Press.

Fuller, W. A. (1976), *Introduction to Statistical Time Series*, New York: Wiley.

Fuller, W. A. (1985), 'Nonstationary Autoregressive Time Series', in E. J. Hannan, P. R. Krishnaiah and M. M. Rao (Editors), *Handbook of Statistics, Volume 5: Time Series in the Time Domain*, 1–24, Amsterdam: North-Holland.

Gabr, M. M. and Subba Rao, T. (1981), 'The Estimation and Prediction of Subset Bilinear Time Series Models with Applications', *Journal of Time Series Analysis*, 2, 155–71.

Gardner, E. S. Jr. (1985), 'Exponential Smoothing: The State of the Art', *Journal of Forecasting*, 4, 1–28.

Gebski, V. and McNeil, D. (1984), 'A Refined Method of Robust Smoothing', *Journal of the American Statistical Association*, 79, 616–23.

Geweke, J. (1978), 'Testing the Exogeneity Specification in the Complete Dynamic Simultaneous Equations Model', *Journal of Econometrics*, 7, 163–85.

Geweke, J. (1982), 'Measurement of Linear Dependence and Feedback Between Time Series', *Journal of the American Statistical Association*, 79, 304–24.

Geweke, J. (1984a), 'Inference and Causality in Economic Time Series Models', in Z. Griliches and M. D. Intriligator (Editors), *Handbook of Econometrics, Volume II*, 1101–44, Amsterdam: North Holland.

Geweke, J. (1984b), 'Measures of Conditional Linear Dependence and Feedback Between Time Series', *Journal of the American Statistical Association*, 79, 907–15.

Geweke, J., Meese, R. and Dent, W. (1983), 'Comparing Alternative Tests of Causality in Temporal Systems: Analytic Results and Experimental Evidence', *Journal of Econometrics*, 21, 161–94.

Geweke, J. and Porter-Hudak, S. (1983), 'The Estimation and Application of Long Memory Time Series Models', *Journal of Time Series Analysis*, 4, 221–38.

Gilbert, C. L. (1986), 'Professor Hendry's Methodology', *Oxford Bulletin of Economics and Statistics*, 48, 283–307.

Godfrey, L. G. (1979), 'Testing the Adequacy of a Time Series Model', *Biometrika*, 66, 67–72.

Godfrey, L. G. (1981), 'On the Invariance of the Lagrange Multiplier Test with Respect to Certain Changes in the Alternative Hypothesis', *Econometrica*, 47, 1443–56.

Godfrey, L. G. and Tremayne, A. R. (1988), 'Misspecification Tests for Time Series and Their Application in Econometrics', *Econometric Reviews*, 7, 1–42.

Granger, C. W. J. (1969a), 'Investigating Causal Relations by Econometric Models and Cross-Spectral Methods', *Econometrica*, 37, 424–38.

Granger, C. W. J. (1969b), 'Prediction with a Generalized Cost of Error Function', *Operations Research*, 20, 199–207.

Granger, C. W. J. (1978), 'New Classes of Time Series Models', *The Statistician*, 27, 237–54.

Granger, C. W. J. (1980a), 'Long Memory Relationships and the Aggregation of Dynamic Models', *Journal of Econometrics*, 14, 227–38.

Granger, C. W. J. (1980b), 'Testing for Causality: A Personal Viewpoint', *Journal of Economic Dynamics and Control*, 2, 329–52.

Granger, C. W. J. (1981), 'Some Properties of Time Series Data and Their Use in Econometric Model Specification', *Journal of Econometrics*, 16, 121–30.

Granger, C. W. J. (1982), 'Acronyms in Time Series Analysis (ATSA)', *Journal of Time Series Analysis*, 3, 103–7.

Granger, C. W. J. (1986a), 'Developments in the Study of Cointegrated Economic Variables', *Oxford Bulletin of Economics and Statistics*, 48, 213–28.

Granger, C. W. J. (1986b), 'Comment', *Journal of the American Statistical Association*, 81, 967–8.

Granger, C. W. J. and Andersen, A. P. (1978), *An Introduction to Bilinear Time Series Models*, Gottingen: Vandenhoeck and Ruprecht.

Granger, C. W. J. and Hughes, A. D. (1971), 'A New Look at Some Old Data: The Beveridge Wheat Price Series', *Journal of the Royal Statistical Society*, Series A, 134, 413–28.

Granger, C. W. J. and Joyeux, R. (1980), 'An Introduction to Long Memory Time Series Models and Fractional Differencing', *Journal of Time Series Analysis*, 1, 15–29.

Granger, C. W. J. and Morris, M. J. (1976), 'Time Series Modelling and Interpretation', *Journal of the Royal Statistical Society*, Series A, 139, 246–57.

Granger, C. W. J. and Newbold, P. (1974), 'Spurious Regressions in Econometrics', *Journal of Econometrics*, 2, 111–20.

Granger, C. W. J. and Newbold, P. (1975), 'Economic Forecasting: The Atheist's Viewpoint', in G. A. Renton (Editor), *Modelling the Economy*, 131–47, London: Heinemann.

Granger, C. W. J. and Newbold, P. (1976), 'Forecasting Transformed Series', *Journal of the Royal Statistical Society*, Series B, 38, 189–203.

Granger, C. W. J. and Newbold, P. (1977), *Forecasting Economic Time Series*, New York: Academic Press.

Granger, C. W. J. and Weiss, A. A. (1983), 'Time Series Analysis of Error-Correcting Models', in S. Karlin, T. Amemiya and L. A. Goodman (Editors), *Studies in Econometrics, Time Series, and Multivariate Statistics*, 255–78, New York: Academic Press.

Gray, H. L., Kelley, G. D. and McIntire, D. D. (1978), 'A New Approach to ARMA Modelling', *Communications in Statistics*, B7, 1–77.

Griliches, Z. (1967), 'Distributed Lags: A Survey', *Econometrica*, 35, 16–49.

Guilkey, D. K. and Salemi, M. K. (1982), 'Small Sample Properties of Three Tests for Granger–Causal Ordering in a Bivariate Stochastic System', *Review of Economics and Statistics*, 64, 68–80.

Haavelmo, T. (1943), 'The Statistical Implications of a System of Simultaneous Equations', *Econometrica*, 11, 1–12.

Haavelmo, T. (1944), 'The Probability Approach in Econometrics', *Econometrica*, 12, Supplement.

Haggan, V., Heravi, S. M. and Priestley, M. B. (1984), 'A Study of the Application of State-dependent Models in Non-linear Time Series Analysis', *Journal of Time Series Analysis*, 5, 69–102.

Haggan, V. and Ozaki, T. (1981), 'Modelling Non-linear Vibrations using an Amplitude-Dependent Autoregressive Time Series Model', *Biometrika*, 68, 189–96.

Hamilton, D. C. and Watts, D. G. (1978), 'Interpreting Partial Autocorrelation Functions of Seasonal Time Series Models', *Biometrika*, 65, 135–40.

Hannan, E. J. (1970), *Multiple Time Series*, New York: Wiley.

Hannan, E. J. (1980), 'The Estimation of the Order of an ARMA Process', *Annals of Statistics*, 8, 1071–81.

Hannsens, D. M. and Liu, L.-M. (1983), 'Lag Specification in Rational Distributed Lag Structural Models', *Journal of Business and Economic Statistics*, 1, 316–25.

Harrison, P. J. (1965), 'Short-term Sales Forecasting', *Applied Statistics*, 14, 102–39.

Harrison, P. J. (1967), 'Exponential Smoothing and Short-Term Sales Forecasting', *Management Science*, 13, 821–42.

Harvey, A. C. (1981), *Time Series Models*, Oxford: Philip Allan.

Harvey, A. C. (1984), 'A Unified View of Statistical Forecasting Procedures', *Journal of Forecasting*, 3, 245–75.

Harvey, A. C. (1985), 'Trends and Cycles in Macroeconomic Time Series', *Journal of Business and Economic Statistics*, 3, 216–27.

Harvey, A. C. (1986), 'Analysis and Generalization of a Multivariate Exponential Smoothing Model', *Management Science*, 32, 374–80.

Harvey, A. C. (1987), 'Applications of the Kalman Filter in Econometrics', in T. F. Bewley (Editor), *Advances in Econometrics – Fifth World Congress, Volume 1*, 285–313, Cambridge: Cambridge University Press.

Harvey, A. C. and Durbin, J. (1986), 'The Effects of Seat Belt Legislation on British Road Casualties: A Case Study in Structural Time Series Modelling (with Discussion)', *Journal of the Royal Statistical Society*, Series A, 149, 187–227.

Hasza, D. P. and Fuller, W. A. (1979), 'Estimation for Autoregressive Processes with Unit Roots', *Annals of Statistics*, 7, 1106–20.

Haugh, L. D. and Box, G. E. P. (1977), 'Identification of Dynamic Regression (Distributed Lag) Models Connecting Two Time Series', *Journal of the American Statistical Association*, 72, 121–30.

Hausman, J. A. and Watson, M. W. (1985), 'Errors in Variables and Seasonal Adjustment Procedures', *Journal of the American Statistical Association*, 80, 531–40.

Hendry, D. F. (1986), 'Econometric Modelling with Cointegrated Variables: An Overview', *Oxford Bulletin of Economics and Statistics*, 48, 201–12.

Hendry, D. F. (1987), 'Econometric Methodology: A Personal Perspective', in T. F. Bewley (Editor), *Advances in Econometrics – Fifth World Congress, Volume II*, 29–48, Cambridge: Cambridge University Press.

Hendry, D. F. and Mizon, G. E. (1978), 'Serial Correlation as a Convenient Simplification, not a Nuisance: A Comment on a Study of the Demand for Money by the Bank of England', *Economic Journal*, 88, 549–63.

Hendry, D. F., Pagan, A. R. and Sargan, J. D. (1984), 'Dynamic Specification', in Z. Griliches and M. D. Intriligator (Editors), *Handbook of Econometrics, Volume II*, 1023–1100, Amsterdam: North-Holland.

Hendry, D. F. and Richard, J.-F. (1982), 'On the Formulation of Empirical Models in Dynamic Econometrics', *Journal of Econometrics*, 20, 3–33.

Hendry, D. F. and Richard, J.-F. (1983), 'The Econometric Analysis of Economic Time Series (with Discussion)', *International Statistical Review*, 51, 111–63.

Hendry, D. F. and von Ungern-Sternberg, T. (1981), 'Liquidity and Inflation Effects on Consumer's Expenditure', in A. S. Deaton (Editor), *Essays in the Theory and Measurement of Consumer's Behaviour*, 237–61, Cambridge: Cambridge University Press.

Heyse, J. F. and Wei, W. W. S. (1985), 'Modelling the Advertising–Sales Relationship Through the Use of Multiple Time Series Techniques', *Journal of Forecasting*, 4, 165–82.

Hill, G. and Fildes, R. (1984), 'The Accuracy of Extrapolation Methods: An Automatic Box–Jenkins Package SIFT', *Journal of Forecasting*, 3, 319–24.

Hillmer, S. C., Bell, W. R. and Tiao, G. C. (1983), 'Modelling Considerations in the Seasonal Adjustment of Economic Time Series', in A. Zellner (Editor), *Applied Time Series Analysis of Economic Data*, 74–100, Washington, DC: US Department of Commerce, Bureau of the Census.

Hillmer, S. C., Larcker, D. F. and Schroeder, D. A. (1983), 'Forecasting Accounting Data: A Multiple Time Series Analysis', *Journal of Forecasting*, 2, 389–404.

Hillmer, S. C. and Tiao, G. C. (1982), 'An ARIMA–Model Based Approach to Seasonal Adjustment', *Journal of the American Statistical Association*, 77, 63–70.

Holland, P. W. (1986), 'Statistics and Causal Inference', *Journal of the American Statistical Association*, 81, 945–60.

Holt, C. C., Modigliani, F., Muth, J. F. and Simon, H. (1960), *Planning Production, Inventories and Work Force*, Englewood Cliffs, NJ: Prentice-Hall.

Hood, W. C. and Koopmans, T. C. (1953) (Editors), *Studies in Econometric Method*, Volume 14, Cowles Commission Monograph, New York: Wiley.

Hosking, J. R. M. (1980), 'The Multivariate Portmanteau Statistic', *Journal of the American Statistical Association*, 75, 602–8.

Hosking, J. R. M. (1981a), 'Fractional Differencing', *Biometrika*, 68, 165–76.

Hosking, J. R. M. (1981b), 'Lagrange Multiplier Tests of Multivariate Time Series Models', *Journal of the Royal Statistical Society*, Series B, 43, 219–30.

Hosking, J. R. M. (1982), 'Some Models of Persistence in Time Series', in O. D. Anderson (Editor), *Time Series Analysis: Theory and Practice 1*, 641–54, Amsterdam: North-Holland.

Hosoya, Y. (1977), 'On the Granger Condition for Non-causality', *Econometrica*, 45, 1735–6.

Hsiao, C. (1979), 'Autoregressive Modelling of Canadian Money and Income Data', *Journal of the American Statistical Association*, 74, 553–60.

Hsu, D. A. (1977), 'Tests for Variance Shift at an Unknown Time Point', *Applied Statistics*, 26, 279–84.

Hsu, D. A. (1979), 'Detecting Shifts of Parameter in Gamma Sequences with Applications to Stock Price and Air Traffic Flow Analysis', *Journal of the American Statistical Association*, 74, 31–40.

Hsu, D. A. (1982), 'A Bayesian Robust Detection of Shift in the Risk Structure of Stock Market Returns', *Journal of the American Statistical Association*, 77, 29–39.

Hylleberg, S. (1986), *Seasonality in Regression*, New York: Academic Press.

Jeffreys, H. (1961), *Theory of Probability*, 3rd Edition, Oxford: Clarendon Press.

Jenkins, G. M. (1979), *Practical Experiences with Modelling and Forecasting Time Series*, Jersey: GJP.

Jenkins, G. M. and Alavi, A. S. (1981), 'Some Aspects of Modelling and Forecasting Multivariate Time Series', *Journal of Time Series Analysis*, 2, 1–47.

Jenkins, G. M. and McLeod, G. (1982), *Case Studies in Time Series Analysis*, Lancaster: GJP.

Johannes, J. M. and Rasche, R. (1979), 'Predicting the Money Multiplier', *Journal of Monetary Economics*, 5, 301–25.

Jones, R. H. (1966), 'Exponential Smoothing for Multivariate Time Series', *Journal of the Royal Statistical Society*, Series B, 28, 241–51.

Judge, G. G., Griffiths, W. E., Carter Hill, R., Lutkepohl, H. and Lee, T.-C. (1985), *The Theory and Practice of Econometrics*, 2nd Edition, New York: Wiley.

Kalman, R. E. (1960), 'A New Approach to Linear Filtering and Prediction Problems', *Journal of Basic Engineering*, 82, 35–45.

Kang, H. (1981), 'Necessary and Sufficient Conditions for Causality Testing in Multivariate ARMA Models', *Journal of Time Series Analysis*, 2, 95–101.

Kashyap, R. L. and Eom, K.-B. (1988), 'Estimation in Long Memory Time Series Models', *Journal of Time Series Analysis*, 9, 35–42.

Keenan, D. M. (1985), 'A Tukey Nonadditivity-Type Test for Time Series Nonlinearity', *Biometrika*, 72, 39–44.

Kenny, P. and Durbin, J. (1982), 'Local Trend Estimation and Seasonal Adjustment of Economic and Social Time Series (with Discussion)', *Journal of the Royal Statistical Society*, Series A, 145, 1–41.

Klein, L. R. and Burmeister, E. (1976) (Editors), *Econometric Model Performance*, Philadelphia: University of Pennsylvania Press.

Koopmans, T. C. (1937), *Linear Regression Analysis of Economic Time Series*, Haarlem: Netherlands Economic Institute.

Koopmans, T. C. (1950) (Editor), *Statistical Inference in Dynamic Economic Models*, Volume 10, Cowles Commission Monograph, New York: Wiley.

Lawrence, A. J. and Lewis, P. A. W. (1985), 'Modelling and Residual Analysis of

Nonlinear Autoregressive Time Series in Exponential Variables (with Discussion)', *Journal of the Royal Statistical Society*, Series B, 47, 165–202.

Lawrence, A. J. and Lewis, P. A. W. (1987), 'Higher-order Residual Analysis for Nonlinear Time Series with Autoregressive Correlation Structures', *International Statistical Review*, 55, 21–36.

Leamer, E. E. (1983), 'Let's Take the Con out of Econometrics', *American Economic Review*, 73, 31–44.

Leamer, E. E. (1985), 'Vector Autoregressions for Causal Inference?', in K. Brunner and A. H. Meltzer (Editors), *Understanding Monetary Regimes*, Carnegie–Rochester Conference Series on Public Policy, 22, 255–304.

Leamer, E. E. (1987), 'Econometric Metaphors', in T. F. Bewley (Editor), *Advances in Econometrics – Fifth World Congress, Volume II*, 1–28, Cambridge: Cambridge University Press.

Ledolter, J. and Abraham, B. (1984), 'Some Comments on the Initialization of Exponential Smoothing', *Journal of Forecasting*, 3, 79–84.

Levenbach, H. and Cleary, J. P. (1981), *The Beginning Forecaster: The Forecasting Process Through Data Analysis*, Belmont, California: Wadsworth.

Lewandowski, R. (1982), 'Sales Forecasting by FORSYS', *Journal of Forecasting*, 1, 205–14.

Li, W. K. (1984), 'On the Autocorrelation Structure and Identification of Some Bilinear Time Series', *Journal of Time Series Analysis*, 5, 173–81.

Lii, K.-S. (1985), 'Transfer Function Model Order and Parameter Estimation', *Journal of Time Series Analysis*, 6, 153–69.

Liu, L.-M. (1986), 'Identification of Time Series Models in the Presence of Calendar Variation', *International Journal of Forecasting*, 2, 357–72.

Liu, L.-M. and Hanssens, D. M. (1982), 'Identification of Multiple-Input Transfer Function Models', *Communications in Statistics*, A, 11, 297–314.

Liu, L.-M. and Hudak, G. B. (1983), 'An Integrated Time Series Analysis Computer Program: The SCA Statistical System', in O. D. Anderson (Editor), *Time Series Analysis: Theory and Practice*, 4, 291–310, Amsterdam: North-Holland.

Ljung, G. M. and Box, G. E. P. (1978), 'On a Measure of Lack of Fit in Time Series Models', *Biometrika*, 65, 297–303.

Ljung, G. M. and Box, G. E. P. (1979), 'The Likelihood Function of Stationary Autoregressive-Moving Average Models', *Biometrika*, 66, 265–70.

Lucas, R. E. (1976), 'Econometric Policy Evaluation: A Critique', in K. Brunner and A. H. Meltzer (Editors), *The Phillips Curve and Labour Markets*, Carnegie–Rochester Conference Series on Public Policy, 1, 19–46.

Lutkepohl, H. (1982a), 'Non-Causality due to Omitted Variables', *Journal of Econometrics*, 19, 367–78.

Lutkepohl, H. (1982b), 'The Impact of Omitted Variables on the Structure of Multiple Time Series: Quenouille's Data Revisited', in O. D. Anderson (Editor), *Time Series Analysis: Theory and Practice*, 2, 143–59, Amsterdam: North Holland.

Lutkepohl, H. (1982c), 'Discounted Polynomials for Multiple Time Series Model Building', *Biometrika*, 69, 107–15.

Lutkepohl, H. (1982d), 'Differencing Multiple Time Series: Another Look at Canadian Money and Income Data', *Journal of Time Series Analysis*, 3, 235–43.

Lutkepohl, H. (1985), 'Comparison of Criteria for Estimating the Order of a Vector Autoregressive Process', *Journal of Time Series Analysis*, 6, 35–52.

McGill, R., Tukey, J. W. and Larsen, W. A. (1978), 'Variations of Box Plots', *American Statistician*, 32, 12–16.

McKenzie, E. (1974), 'A Comparison of Standard Forecasting Systems with the Box–Jenkins Approach', *The Statistician*, 23, 107–16.

McKenzie, E. (1976), 'An Analysis of General Exponential Smoothing', *Operations Research*, 24, 131–40.

McKenzie, E. (1984), 'General Exponential Smoothing and the Equivalent ARMA Process', *Journal of Forecasting*, 3, 333–44.

McKenzie, E. (1985), 'Comments on "Exponential Smoothing: The State of the Art" by E. S. Gardner, Jr.', *Journal of Forecasting*, 4, 32–6.

McKenzie, E. (1988), 'A Note on Using the Integrated Form of ARIMA Forecasts', *International Journal of Forecasting*, 4, 117–24.

McKenzie, S. K. (1984), 'Concurrent Seasonal Adjustment with Census X–11', *Journal of Business and Economic Statistics*, 2, 235–49.

McLeod, A. I. (1978), 'On the Distribution of Residual Autocorrelations in Box–Jenkins Models', *Journal of the Royal Statistical Society*, Series B, 40, 296–302.

McLeod, A. I. and Hipel, K. W. (1978), 'Preservation of the Rescaled Adjusted Range, I: A Reassessment of the Hurst Phenomenon', *Water Resources Research*, 14, 491–518.

McLeod, A. J. and Li, W. K. (1983), 'Diagnostic Checking ARMA Time Series Models Using Squared-Residual Correlations', *Journal of Time Series Analysis*, 4, 269–73.

McLeod, G. (1983), *Box–Jenkins in Practice*, Jersey: GJP.

Makridakis, S., Andersen, A., Carbone, R., Fildes, R., Hibon, M., Lewandowski, R., Newton, J., Parzen, E. and Winkler, R. (1982), 'The Accuracy of Extrapolation (Time Series) Methods: Results of a Forecasting Competition', *Journal of Forecasting*, 1, 111–53.

Makridakis, S. and Hibon, M. (1979), 'Accuracy of Forecasting: An Empirical Investigation (with Discussion)', *Journal of the Royal Statistical Society*, Series A, 142, 97–145.

Makridakis, S., Wheelwright, S. C. and McGee, V. E. (1983), *Forecasting: Methods and Applications*, 2nd Edition, New York: Wiley.

Mandelbrot, B. B. (1963), 'New Methods in Statistical Economics', *Journal of Political Economy*, 71, 421–40.

Mandelbrot, B. B. (1969), 'Long-Run Linearity, Locally Gaussian Process, *H*-Spectra, and Infinite Variances', *International Economic Review*, 10, 82–111.

Mandelbrot, B. B. (1972), 'Statistical Methodology for Nonperiodic Cycles: From the Covariance to *R/S* Analysis', *Annals of Economic and Social Measurement*, 1/3, 259–90.

Mandelbrot, B. B. and Taqqu, M. S. (1979), 'Robust *R/S* Analysis of Long-Run Serial Correlation', *Bulletin of the International Statistical Institute*, 48, 69–99.

Mandelbrot, B. B. and Van Ness, J. W. (1968), 'Fractional Brownian Motions, Fractional Noises, and Applications', *SIAM Review*, 10, 422–37.

Mandelbrot, B. B. and Wallis, J. R. (1969), 'Some Long-Run Properties of Geophysical Records', *Water Resources Research*, 5, 321–40.

Maravall, A. (1983), 'An Application of Nonlinear Time Series Forecasting', *Journal of Business and Economic Statistics*, 1, 66–74.

Maravall, A. (1984), 'Comment', *Journal of Business and Economic Statistics*, 2, 337–9.

Maravall, A. (1985), 'On Structural Time Series Models and the Characterization of Components', *Journal of Business and Economic Statistics*, 3, 350–5.

Maravall, A. (1986a), 'An Application of Model-Based Estimation of Unobserved Components', *International Journal of Forecasting*, 2, 305–18.

Maravall, A. (1986b), 'Revisions in ARIMA Signal Extraction', *Journal of the American Statistical Association*, 81, 736–40.

Maravall, A. and Pierce, D. A. (1987), 'A Prototypical Seasonal Adjustment Model', *Journal of Time Series Analysis*, 8, 177–93.

Martin, R. D. and Yohai, V. J. (1985), 'Robustness in Time Series and Estimating ARMA Models', in E. J. Hannan, P. R. Krishnaiah and M. M. Rao (Editors), *Handbook of Statistics, Volume 5: Time Series in the Time Domain*, 119–56, Amsterdam: North-Holland.

Mendenhall, W., Schaeffer, R. L. and Wackerley, D. D. (1986), *Mathematical Statistics with Applications*, 3rd Edition, Boston: Duxbury.

Menzefricke, U. (1981), 'A Bayesian Analysis of a Change in the Precision of a Sequence of Independent Random Normal Variables at an Unknown Time Point', *Applied Statistics*, 30, 141–6.

Miller, S. M. (1988), 'The Beveridge–Nelson Decomposition of Economic Time Series: Another Economical Computational Method', *Journal of Monetary Economics*, 21, 141–2.

Mills, T. C. (1981), 'Modelling the Formation of Australian Inflation Expectations', *Australian Economic Papers*, 20, 150–60.

Mills, T. C. (1982), 'The Use of Unobserved Components and Signal Extraction Techniques in Modelling Economic Time Series', *Bulletin of Economic Research*, 34, 92–108.

Mills, T. C. (1983), 'Composite Monetary Indicators for the United Kingdom: Construction and Empirical Analysis', *Bank of England Technical Paper*, 3.

Mills, T. C. (1986), 'A Forecasting Model for the UK Narrow Money Supply', *University of Leeds Discussion Paper*, Series A, 86/1.

Mills, T. C. and Stephenson, M. J. (1985), 'An Empirical Analysis of the UK Treasury Bill Market', *Applied Economics*, 17, 689–703.

Mills, T. C. and Stephenson, M. J. (1987), 'A Time Series Forecasting System for the UK Money Supply', *Economic Modelling*, 4, 355–69.

Mills, T. C. and Wood, G. E. (1982), 'Econometric Evaluation of Alternative Money Stock Series, 1880–1913', *Journal of Money, Credit and Banking*, 14, 265–77.

Mills, T. C. and Wood, G. E. (1987), 'Money Growth, Interest Rates and Prices

Under the Gold Standard: A Re-examination', *University of Leeds Discussion Paper*, Series A, 87/8.

Mizon, G. E. (1977), 'Model Selection Procedures', in M. J. Artis and A. R. Nobay (Editors), *Studies in Modern Economic Analysis*, 97–120, Oxford: Blackwell.

Mizon, G. E. and Hendry, D. F. (1980), 'An Empirical Application and Monte Carlo Analysis of Tests of Dynamic Specification', *Review of Economic Studies*, 47, 21–46.

Montgomery, D. C. and Johnson, L. A. (1976), *Forecasting and Time Series Analysis*, New York: McGraw-Hill.

Moore, G. H., Box, G. E. P., Kaitz, H. B., Stephenson, J. A. and Zellner, A. (1981), *Seasonal Adjustment of the Monetary Aggregates: Report of the Committee of Experts on Seasonal Adjustment Techniques*, Washington, DC: Board of Governors of the Federal Reserve System.

Muth, J. F. (1960), 'Optimal Properties of Exponentially Weighted Forecasts', *Journal of the American Statistical Association*, 55, 299–305.

Naylor, T. H., Seaks, T. G. and Wichern, D. W. (1972), 'Box–Jenkins Methods: An Alternative to Econometric Models', *International Statistical Review*, 40, 123–37.

Nelson, C. R. (1973), *Applied Time Series Analysis for Managerial Forecasting*, San Francisco: Holden-Day.

Nelson, C. R. (1976), 'The Interpretation of R^2 in Autoregressive-Moving Average Time Series Models', *American Statistician*, 30, 175–80.

Nelson, C. R. and Kang, H. (1981), 'Spurious Periodicity in Inappropriately Detrended Time Series', *Econometrica*, 49, 741–51.

Nelson, C. R. and Kang, H. (1984), 'Pitfalls in the Use of Time as an Explanatory Variable in Regression', *Journal of Business and Economic Statistics*, 2, 73–82.

Nelson, C. R. and Plosser, C. I. (1982), 'Trends and Random Walks in Macroeconomic Time Series: Some Evidence and Implications', *Journal of Monetary Economics*, 10, 139–62.

Nelson, C. R. and Schwert, G. W. (1977), 'Short-term Interest Rates as Predictors of Inflation: On Testing the Hypothesis that the Real Rate of Interest is Constant', *American Economic Review*, 67, 478–86.

Nelson, C. R. and Schwert, G. W. (1982), 'Tests for Predictive Relationships between Time Series Variables: A Monte Carlo Investigation', *Journal of the American Statistical Association*, 77, 11–18.

Nelson, H. L. and Granger, C. W. J. (1979), 'Experience with Using the Box–Cox Transformation when Forecasting Economic Time Series', *Journal of Econometrics*, 10, 57–69.

Nerlove, M. (1972), 'Lags in Economic Behaviour', *Econometrica*, 39, 359–82.

Nerlove, M., Grether, D. M. and Carvalho, J. L. (1979), *Analysis of Economic Time Series: A Synthesis*, New York: Academic Press.

Nerlove, M. and Wage, S. (1964), 'On the Optimality of Adaptive Forecasting', *Management Science*, 10, 207–24.

Newbold, P. (1974), 'The Exact Likelihood Function for a Mixed Autoregressive Moving Average Process', *Biometrika*, 61, 423–6.

Newbold, P. (1978), 'Feedback Induced by Measurement Errors', *International Economic Review*, 19, 787–91.

Newbold, P. (1980), 'The Equivalence of Two Tests of Model Adequacy', *Biometrika*, 67, 463–5.

Newbold, P. (1988), 'Predictors Projecting Linear Trend Plus Seasonal Dummies', *The Statistician*, 37, 111–27.

Nickell, S. (1985), 'Error Correction, Partial Adjustment and all That: An Expository Note', *Oxford Bulletin of Economics and Statistics*, 47, 119–29.

O'Donovan, T. M. (1983), *Short Term Forecasting: An Introduction to the Box–Jenkins Approach*, New York: Wiley.

Osborn, D. R. (1982), 'On the Criteria Functions Used for the Estimation of Moving Average Processes', *Journal of the American Statistical Association*, 77, 388–92.

Pagan, A. R. (1987), 'Three Econometric Methodologies: A Critical Appraisal', *Journal of Economic Surveys*, 1, 3–24.

Pagan, A. R., Hall, A. D. and Trivedi, P. K. (1983), 'Assessing the Variability of Inflation', *Review of Economic Studies*, 50, 585–96.

Pankratz, A. (1983), *Forecasting with Univariate Box–Jenkins Models: Concepts and Cases*, New York: Wiley.

Pankratz, A. and Dudley, U. (1987), 'Forecasts of Power-Transformed Series', *Journal of Forecasting*, 6, 239–48.

Parzen, E. (1982), 'ARARMA Models for Time Series Analysis and Forecasting', *Journal of Forecasting*, 1, 67–82.

Peña, D. (1984), 'The Autocorrelation Function of Seasonal ARMA Models', *Journal of Time Series Analysis*, 5, 269–72.

Penm, J. H. W. and Terrell, R. D. (1982), 'On the Recursive Fitting of Subset Autoregressions', *Journal of Time Series Analysis*, 3, 43–59.

Phillips, P. C. B. (1986), 'Understanding Spurious Regressions in Econometrics', *Journal of Econometrics*, 33, 311–40.

Phillips, P. C. B. (1987a), 'Time Series Regression with a Unit Root', *Econometrica*, 55, 227–301.

Phillips, P. C. B. (1987b), 'Towards a Unified Asymptotic Theory for Autoregression', *Biometrika*, 74, 535–47.

Phillips, P. C. B. and Durlauf, S. N. (1986), 'Multiple Time Series Regression with Integrated Processes', *Review of Economic Studies*, 53, 473–95.

Pierce, D. A. (1972), 'Least Squares Estimation in Dynamic-Disturbance Time Series Models', *Biometrika*, 59, 73–8.

Pierce, D. A. (1975a), 'On Trend and Autocorrelation', *Communications in Statistics*, 4, 163–75.

Pierce, D. A. (1975b), 'Forecasting in Dynamic Models with Stochastic Regressors', *Journal of Econometrics*, 3, 349–74.

Pierce, D. A. (1977), 'Relationships – and the Lack Thereof – Between Economic Time Series, with Special Reference to Money and Interest Rates', *Journal of the American Statistical Association*, 72, 11–22.

Pierce, D. A. (1978), 'Seasonal Adjustment when both Deterministic and Stochastic Seasonality are Present', in A. Zellner (Editor), *Seasonal Analysis of Economic Time Series*, 365–97, Washington, DC: US Department of Commerce, Bureau of the Census.

Pierce, D. A. (1979a), 'Signal Extraction Error in Nonstationary Time Series', *Annals of Statistics*, 7, 1303–20.

Pierce, D. A. (1979b), 'R^2 Measures for Time Series', *Journal of the American Statistical Association*, 74, 901–10.

Pierce, D. A. (1980), 'A Survey of Recent Developments in Seasonal Adjustment', *American Statistician*, 34, 125–34.

Pierce, D. A. (1982), 'Comment', *Journal of the American Statistical Association*, 77, 315–16.

Pierce, D. A., Grupe, M. R. and Cleveland, W. P. (1984), 'Seasonal Adjustment of the Weekly Monetary Aggregates: A Model-Based Approach', *Journal of Business and Economic Statistics*, 2, 260–70.

Pierce, D. A. and Haugh, L. D. (1977), 'Causality in Temporal Systems: Characterizations and a Survey', *Journal of Econometrics*, 5, 265–93.

Plosser, C. I. and Schwert, G. W. (1977), 'Estimation of a Non-Invertible Moving Average Process: The Case of Overdifferencing', *Journal of Econometrics*, 6, 199–224.

Plosser, C. I. and Schwert, G. W. (1978), 'Money, Income, and Sunspots: Measuring Economic Relationships and the Effects of Differencing', *Journal of Monetary Economics*, 4, 637–60.

Polasek, W. (1980), 'ACF-Patterns in Seasonal MA-Processes', in O. D. Anderson (Editor), *Time Series*, 259–76, Amsterdam: North-Holland.

Poskitt, D. S. and Tremayne, A. R. (1980), 'Testing the Specification of a Fitted Autoregressive-Moving Average Model', *Biometrika*, 67, 359–63.

Poskitt, D. S. and Tremayne, A. R. (1981), 'An Approach to Testing Linear Time Series Models', *Annals of Statistics*, 9, 974–86.

Poskitt, D. S. and Tremayne, A. R. (1982), 'Diagnostic Tests for Multiple Time Series Models', *Annals of Statistics*, 10, 114–20.

Poskitt, D. S. and Tremayne, A. R. (1983), 'On the Posterior Odds of Time Series Models', *Biometrika*, 70, 157–62.

Poskitt, D. S. and Tremayne, A. R. (1986), 'The Selection and Use of Linear and Bilinear Time Series Models', *International Journal of Forecasting*, 2, 101–14.

Poskitt, D. S. and Tremayne, A. R. (1987), 'Determining a Portfolio of Linear Time Series Models', *Biometrika*, 74, 125–37.

Poulos, L., Kvanli, A. and Pavur, R. (1987), 'A Comparison of the Accuracy of the Box–Jenkins Method with that of Automated Forecasting Methods', *International Journal of Forecasting*, 3, 261–8.

Priestley, M. B. (1980), 'State-dependent Models: A General Approach to Nonlinear Time Series Analysis', *Journal of Time Series Analysis*, 1, 47–71.

Prothero, D. L. and Wallis, K. F. (1976), 'Modelling Macroeconomic Time Series (with Discussion)', *Journal of the Royal Statistical Society*, Series A, 139, 468–500.

Reinsel, G. (1980), 'Asymptotic Properties of Prediction Errors for the Multivariate Autoregressive Model Using Estimated Parameters', *Journal of the Royal Statistical Society*, Series B, 42, 328–33.

Reinsel, G. (1983), 'Some Results on Multivariate Autoregressive Index Models', *Biometrika*, 70, 145–56.

Riise, T. and Tjostheim, D. (1985), 'Theory and Practice of Multivariate ARMA Forecasting', *Journal of Forecasting*, 3, 309–17.

Rissanen, J. (1978), 'Modelling by Shortest Data Description', *Automatica*, 14, 465–71.

Roberts, S. A. (1982), 'A General Class of Holt–Winters Type Forecasting Models', *Management Science*, 28, 808–20.

Roberts, S. A. and Harrison, P. J. (1984), 'Parsimonious Modelling and Forecasting of Seasonal Time Series', *European Journal of Operational Research*, 16, 365–77.

Ryan, B. F., Joiner, B. L. and Ryan, T. A., Jr. (1985), *Minitab Handbook*, 2nd Edition, Boston: Duxbury.

Said, S. E. and Dickey, D. A. (1984), 'Testing for Unit Roots in Autoregressive Moving-Average Models with Unknown Order', *Biometrika*, 71, 599–607.

Said, S. E. and Dickey, D. A. (1985), 'Hypothesis Testing in ARIMA$(p, 1, q)$ Models', *Journal of the American Statistical Association*, 80, 369–74.

Sargan, J. D. (1964), 'Wages and Prices in the United Kingdom: a Study in Econometric Methodology', in P. E. Hart, G. Mills and J. K. Whittaker (Editors), *Econometric Analysis for National Economic Planning*, 25–64, London: Butterworths.

Sargan, J. D. (1980a), 'Some Tests of Dynamic Specification for a Single Equation', *Econometrica*, 48, 879–97.

Sargan, J. D. (1980b), 'The Consumer Price Equation in the Post War British Economy: An Exercise in Equation Specification Testing', *Review of Economic Studies*, 47, 113–35.

Sargan, J. D. and Bhargava, A. S. (1983), 'Testing Residuals from Least Squares Regression for being Generated by the Gaussian Random Walk', *Econometrica*, 51, 153–74.

Sargent, T. J. (1979), *Macroeconomic Theory*, New York: Academic Press.

Sargent, T. J. (1981), 'Interpreting Economic Time Series', *Journal of Political Economy*, 89, 213–48.

Sargent, T. J. and Sims, C. A. (1977), 'Business Cycle Modelling Without Pretending to Have Too Much A Priori Economic Theory', in C. A. Sims (Editor), *New Methods of Business Cycle Research: Proceedings from a Conference*, 45–109, Minnesota: Federal Reserve Bank of Minneapolis.

SAS (1985a), *SAS User's Guide*: *Basics*, Version 5 Edition, Cary NC: SAS Institute.

SAS (1985b), *SAS User's Guide*: *Statistics*, Version 5 Edition, Cary NC: SAS Institute.

SAS (1985c), *SAS/ETS User's Guide*, Version 5 Edition, Cary NC: SAS Institute.

SAS (1985d), *SAS/GRAPH User's Guide*, Version 5 Edition, Cary NC: SAS Institute.

SAS (1985e), *SAS/IML User's Guide*, Version 5 Edition, Cary NC: SAS Institute.

Schmid, C. F. (1983), *Statistical Graphics: Design Principles and Practices*, New York: Wiley.

Schwarz, G. (1978), 'Estimating the Dimension of a Model', *Annals of Statistics*, 6, 461–4.

Schwert, G. W. (1987), 'Effects of Model Specification on Tests for Unit Roots in Macroeconomic Data', *Journal of Monetary Economics*, 20, 73–103.

Shibata, R. (1985), 'Various Model Selection Techniques in Time Series Analysis', in E. J. Hannan, P. R. Krishnaiah and M. M. Rao (Editors), *Handbook of Statistics, Volume 5: Time Series in the Time Domain*, 179–87, Amsterdam: North-Holland.

Shiller, R. J. (1979), 'The Volatility of Long-Term Interest Rates and Expectations Models of the Term Structure', *Journal of Political Economy*, 87, 1190–219.

Shiskin, J., Young, A. H. and Musgrave, J. C. (1967), 'The X-11 Variant of the Census Method II Seasonal Adjustment Program', Technical Paper No. 15, US Department of Commerce, Bureau of Economic Analysis.

Sims, C. A. (1972), 'Money, Income and Causality', *American Economic Review*, 62, 540–52.

Sims, C. A. (1974), 'Seasonality in Regression', *Journal of the American Statistical Association*, 69, 618–26.

Sims, C. A. (1977), 'Exogeneity and Causal Ordering in Macroeconomic Models', in C. A. Sims (Editor), *New Methods in Business Cycle Research: Proceedings from a Conference*, 23–43, Minneapolis: Federal Reserve Bank of Minneapolis.

Sims, C. A. (1980), 'Macroeconomics and Reality', *Econometrica*, 48, 1–48.

Sims, C. A. (1981), 'An Autoregressive Index Model for the US 1948–1975', in J. Kmenta and J. B. Ramsey (Editors), *Large-Scale Macroeconometric Models*, 283–327, Amsterdam: North-Holland.

Sims, C. A. (1982), 'Policy Analysis with Econometric Models', *Brookings Papers on Economic Activity*, 1, 107–64.

Sims, C. A. (1987), 'Making Economics Credible', in T. F. Bewley (Editor), *Advances in Econometrics – Fifth World Congress, Volume II*, 49–60, Cambridge: Cambridge University Press.

Slutsky, E. (1937), 'The Summation of Random Causes as the Source of Cyclic Processes', *Econometrica*, 5, 105–46.

Spanos, A. (1986), *Statistical Foundations of Econometric Modelling*, Cambridge: Cambridge University Press.

SPSS^X (1983), *SPSS^X User's Guide*, Chicago: SPSS Inc/McGraw-Hill.

Stensholt, B. K. and Tjostheim, D. (1987), 'Multiple Bilinear Time Series Models', *Journal of Time Series Analysis*, 8, 221–34.

Stigler, G. J. and Kindahl, J. K. (1970), *The Behaviour of Industrial Prices*, New York: NBER.

Stock, J. H. (1987), 'Asymptotic Properties of Least Squares Estimators of Cointegrating Vectors', *Econometrica*, 55, 1035–56.

Stram, D. O. and Wei, W. W. S. (1986), 'Temporal Aggregation in the ARIMA Process', *Journal of Time Series Analysis*, 7, 279–92.

Subba Rao, T. (1981), 'On the Theory of Bilinear Models', *Journal of the Royal Statistical Society*, Series B, 43, 244–55.

Subba Rao, T. and Gabr, M. M. (1980), 'A Test for Linearity of Stationary Time Series', *Journal of Time Series Analysis*, 1, 145–58.

Sweet, A. L. (1981), 'Adaptive Smoothing for Forecasting Seasonal Series', *AIIE Transactions*, 13, 243–8.

Tauchen, G. E. and Pitts, M. (1983), 'The Price Variability–Volume Relationship on Speculative Markets', *Econometrica*, 51, 485–505.

Taylor, S. J. (1986), *Modelling Financial Time Series*, New York: Wiley.

Theil, H. (1966), *Applied Economic Forecasting*, Amsterdam: North-Holland.

Theil, H. and Wage, S. (1964), 'Some Observations on Adaptive Forecasting', *Management Science*, 10, 198–206.

Thompson, H. E. and Tiao, G. C. (1971), 'Analysis of Telephone Data: A Case Study of Forecasting Seasonal Time Series', *Bell Journal of Economics and Management Science*, 2, 515–41.

Tiao, G. C. (1972), 'Asymptotic Behaviour of Temporal Aggregates of Time Series', *Biometrika*, 59, 525–31.

Tiao, G. C. (1985), 'Autoregressive Moving Average Models, Intervention Problems and Outlier Detection in Time Series', in E. J. Hannan, P. R. Krishnaiah and M. M. Rao, *Handbook of Statistics, Volume 5: Time Series in the Time Domain*, 85–118, Amsterdam: North-Holland.

Tiao, G. C. and Box, G. E. P. (1981), 'Modelling Multiple Time Series with Applications', *Journal of the American Statistical Association*, 76, 802–16.

Tiao, G. C., Box, G. E. P., Grupe, M. R., Hudak, G. B., Bell, W. R. and Chang, I. (1979), 'The Wisconsin Multiple Time Series (WMTS-1) Program: A Preliminary Guide', Department of Statistics, Madison, Wisconsin.

Tiao, G. C. and Hillmer, S. C. (1978), 'Some Consideration of Decomposition of a Time Series', *Biometrika*, 65, 497–502.

Tiao, G. C. and Grupe, M. R. (1980), 'Hidden Periodic Autoregressive-Moving Average Models in Time Series Data', *Biometrika*, 67, 365–73.

Tiao, G. C. and Tsay, R. S. (1983), 'Multiple Time Series Modelling and Extended Sample Cross-Correlations', *Journal of Business and Economic Statistics*, 1, 43–56.

Tiao, G. C. and Wei, W. W. S. (1976), 'Effects of Temporal Aggregation on the Dynamic Relationship of Two Time Series Variables', *Biometrika*, 63, 513–23.

Tong, H. (1983), *Threshold Models in Non-Linear Time Series Analysis*, Berlin: Springer-Verlag.

Tong, H. and Lim, K. S. (1980), 'Threshold Autoregression Limit Cycles, and Cyclical Data', *Journal of the Royal Statistical Society*, Series B, 42, 245–92.

Tsay, R. S. (1985), 'Model Identification in Dynamic Regression (Distributed Lag) Models', *Journal of Business and Economic Statistics*, 3, 228–37.

Tsay, R. S. (1986a), 'Time Series Model Specification in the Presence of Outliers', *Journal of the American Statistical Association*, 81, 132–41.

Tsay, R. S. (1986b), 'Nonlinearity Tests for Time Series', *Biometrika*, 73, 461–6.

Tsay, R. S. (1987), 'Conditional Heteroskedastic Time Series Models', *Journal of the American Statistical Association*, 82, 590–604.

Tsay, R. S. (1988), 'Outliers, Level Shifts, and Variance Changes in Time Series', *Journal of Forecasting*, 7, 1–20.

Tsay, R. S. and Tiao, G. C. (1984), 'Consistent Estimates of Autoregressive Parameters and Extended Sample Autocorrelation Function for Stationary and

Nonstationary ARMA Models', *Journal of the American Statistical Association*, 79, 84–96.

Tufte, E. R. (1983), *The Visual Display of Quantitative Information*, Cheshire, Connecticut: Graphics Press.

Tukey, J. W. (1949), 'One Degree of Freedom for Non-Additivity', *Biometrics*, 5, 232–42.

Tukey, J. W. (1972), 'Some Graphic and Semigraphic Displays', in T. A. Bancroft (Editor), *Statistical Papers in Honor of George W. Snedecor*, 292–316, Iowa: Iowa State University Press.

Tukey, J. W. (1977), *Exploratory Data Analysis*, Reading, Mass: Addison-Wesley.

Tukey, J. W. and Wilk, M. B. (1970), 'Data Analysis and Statistics: Techniques and Approaches', in E. R. Tufte (Editor), *The Quantitative Analysis of Social Problems*, 370–90, Reading, Mass: Addison-Wesley.

Tukey, P. A. and Tukey, J. W. (1981), 'Graphical Display of Data Sets in 3 or more Dimensions', in V. Barnett (Editor), *Interpreting Multivariate Data*, 189–278 (part III, chapters 10–12), Chichester, UK: Wiley.

Tyssedal, J. S. and Tjostheim, D. (1982), 'Autoregressive Processes with a Time Dependent Variance', *Journal of Time Series Analysis*, 3, 209–17.

Vandaele, W. (1983), *Applied Time Series and Box–Jenkins Models*, New York: Academic Press.

Velleman, P. F. (1980), 'Definition and Comparison of Robust Nonlinear Data Smoothers', *Journal of the American Statistical Association*, 75, 609–15.

Velleman, P. F. and Hoaglin, D. C. (1981), *Applications, Basics, and Computing of Exploratory Data Analysis*, Boston, Mass: Duxbury.

Wallis, K. F. (1974), 'Seasonal Adjustment and Relations Between Variables', *Journal of the American Statistical Association*, 69, 18–32.

Wallis, K. F. (1977), 'Multiple Time Series Analysis and the Final Form of Econometric Models', *Econometrica*, 45, 1481–97.

Wallis, K. F. (1982), 'Seasonal Adjustment and Revision of Current Data: Linear Filters for the X-11 Method', *Journal of the Royal Statistical Society*, Series A, 145, 74–85.

Wallis, K. F. (1987), 'Time Series Analysis of Bounded Economic Variables', *Journal of Time Series Analysis*, 8, 115–23.

Watson, M. W. (1986), 'Univariate Detrending Methods with Stochastic Trends', *Journal of Monetary Economics*, 18, 44–75.

Wecker, W. E. (1981), 'Asymmetric Time Series', *Journal of the American Statistical Association*, 76, 16–21.

Wei, W. W. S. (1978), 'Some Consequences of Temporal Aggregation in Seasonal Time Series Models', in A. Zellner (Editor), *Seasonal Analysis of Economic Time Series*, 433–44, Washington, DC: US Department of Commerce, Bureau of the Census.

Weiss, A. A. (1984), 'ARMA Models with ARCH Errors', *Journal of Time Series Analysis*, 5, 129–43.

Whittle, P. (1983), *Prediction and Regulation by Linear Least-Square Methods*, 2nd Edition, Revised, Oxford: Blackwell.

Wickens, M. R. and Breusch, T. S. (1988), 'Dynamic Specification, the Long-Run and the Estimation of Transformed Regression Models', *Economic Journal*, 98 (Conference 1988), 189–205.

Wiener, N. (1956), 'The Theory of Prediction', in E. F. Beckenback (Editor), *Modern Mathematics for Engineers*, 165–90, New York: McGraw-Hill.

Wilk, M. B. and Gnanadesikan, R. (1968), 'Probability Plotting Methods for the Analysis of Data', *Biometrika*, 55, 1–17.

Winters, P. R. (1960), 'Forecasting Sales by Exponentially Weighted Moving Averages', *Management Science*, 6, 324–42.

Wold, H. (1938), *A Study in the Analysis of Stationary Time Series*, Stockholm: Almqvist and Wiksell.

Wold, H. (1960), 'A Generalization of Causal Chain Models', *Econometrica*, 28, 443–63.

Woodward, W. A. and Gray, H. L. (1981), 'On the Relationship between the *S* Array and the Box–Jenkins Method of ARMA Model Identification', *Journal of the American Statistical Association*, 76, 579–87.

Working, H. (1960), 'Note on the Correlation of First Differences of Averages in a Random Chain', *Econometrica*, 28, 916–18.

Young, A. H. (1968), 'Linear Approximations of the Census and BLS Seasonal Adjustment Methods', *Journal of the American Statistical Association*, 63, 445–71.

Yule, G. U. (1927), 'On a Method of Investigating Periodicities in Disturbed Series, with Special Reference to Wolfer's Sunspot Numbers', *Philosophical Transactions of the Royal Society*, Series A, 226, 267–98.

Zellner, A. (1962), 'An Efficient Method of Estimating Seemingly Unrelated Regressions and Tests of Aggregation Bias', *Journal of the American Statistical Association*, 57, 348–68.

Zellner, A. (1979), 'Causality and Econometrics', in K. Brunner and A. H. Meltzer (Editors), *Three Aspects of Policy and Policymaking: Knowledge, Data and Institutions*, Carnegie–Rochester Conference Series on Public Policy, 10, 9–54.

Zellner, A. and Palm, F. (1974), 'Time Series Analysis and Simultaneous Equation Econometric Models', *Journal of Econometrics*, 2, 17–54.

Subject index

Author index